Analytic Geometry and Calculus

Analytic Geometry and Calculus
With Technical Applications

JERRY D. STRANGE

Technical Institute
University of Dayton

BERNARD J. RICE

Dept. of Mathematics
University of Dayton

JOHN WILEY & SONS

New York • Chichester • Brisbane • Toronto

Library of Congress Catalogue Card Number: 78–105382

SBN 471 83190 5

Printed in the United States of America

10   9

Preface

This book is intended for those interested in learning enough calculus for most elementary applications. The material is presented with the assumption that the student has a working knowledge of college-level algebra and trigonometry.

The primary objective of a technical calculus book is to relate mathematical concepts to practical engineering problems. We have attempted to attain this goal by using physical problems, which are well within the scope of the beginning student, as a springboard to mathematical concepts. Our approach is exemplified in Chapters 2 and 6 in connection with the concepts of the derivative and the definite integral, respectively. Both of these chapters are designed to lead the student to the desired mathematical concept by identifying the common thread in several apparently unrelated physical problems. For example, in Chapter 2 the similarity in the solutions to the slope of a tangent line, the velocity of a particle, and the current in a circuit is used to motivate the definition of the derivative.

Proper ordering is particularly important for books of this level. If difficult manipulative problems are introduced too early, the student is hampered in learning the new mathematical ideas. Therefore, only the calculus of polynomials is discussed in the introductory chapters. Algebraic and transcendental functions are not considered until after a good background in the elementary concepts has been established.

Analytic geometry is presented as it is needed to facilitate an understanding of calculus applications. Aside from Chapters 1 and 12, the two subjects are interwoven. The conic sections of Chapter 12 can be taught at any time after finishing Chapter 1, but are delayed in the arrangement of topics to allow earlier coverage of the transcendental functions. The book is designed to satisfy the needs of a two-semester course; however, those desiring a one-semester coverage of calculus will find the first eight chapters ideal for this purpose.

Because repetition is an important part of the learning process, we have included an abundant supply of graded problems in each exercise. Many problems are stated in the terminology of the applications, so that the student will begin to feel comfortable with the association of mathematics to some applications. Most exercise sets contain a balance between manipulative and applied problems. The answers to the odd-numbered problems have been provided.

No specific bibliography is presented, but we acknowledge the influence of many calculus books, both old and new.

We are particularly indebted to Hylda Strange for typing not only the original manuscript, but also a rather extensive revision.

University of Dayton   JERRY D. STRANGE
March 1970   BERNARD J. RICE

Contents

Analytic Geometry and Calculus

1
Basic Concepts in Analytic Geometry

1.1 INTRODUCTION

Up to this point in mathematics you have studied algebra and plane geometry as separate courses, possibly without recognizing any relationship between the two subjects. Historically, algebra and geometry were studied separately until the French mathematician René Descartes (1596–1650) discovered that they were related. He found that by associating points in the plane with pairs of numbers he could describe many geometric figures algebraically and also represent algebraic expressions geometrically. The study of the relationship between algebra and geometry has come to be known as *analytic geometry*. In analytic geometry we shall be concerned with two questions.

1. Given an algebraic expression, how may it be represented geometrically?
2. Given a geometric figure, how may it be described algebraically?

1.2 THE RECTANGULAR COORDINATE SYSTEM

The basic feature of analytic geometry is the use of a coordinate system to locate points in the plane. There are many coordinate systems available for this purpose but the one that is most commonly used is the *rectangular coordinate system.** In order to construct such a system, we first draw a pair of mutually perpendicular intersecting lines as shown in Figure 1.1. Normally the horizontal line is called the *x-axis*, the vertical line is called the *y-axis*, and their intersection is called the *origin*. When considered together, the two axes are referred to as the *coordinate* axes. As you can see, the coordinate axes divide the plane into four zones or *quadrants*. The upper right quadrant is called the *first* quadrant and the others are numbered consecutively from this one in a counterclockwise direction as in Figure 1.1.

We can now locate the points in a plane by using the origin as a reference point and laying off a suitable scale on each of the coordinate

---

*This system is also called the *Cartesian* coordinate system, in honor of Descartes, who invented it.

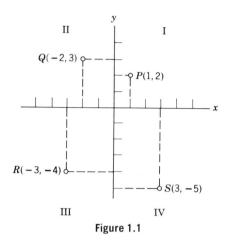

Figure 1.1

axes. The displacement of a point in the plane to the right or left of the
y-axis is called the *x-coordinate*, or *abscissa* of the point, and is denoted
by *x*. Values of *x* measured to the right of the *y*-axis are considered to be
positive, and to the left, *negative*. The displacement of a point in the plane
above or below the *x*-axis is called the *y-coordinate*, or *ordinate* of the
point, and is denoted by *y*. Values of *y* above the *x*-axis are considered to
be *positive*, and below the *x*-axis, *negative*. Considered together, the
abscissa and ordinate of a point are called the *coordinates* of the point. It
is conventional to write the coordinates of a point in parenthesis, with the
abscissa written first and separated from the ordinate by a comma, that
is, (*x, y*). Because of this ordering, the coordinates of a point are referred
to as an *ordered pair* of numbers.

By employing the rectangular coordinate system, we can establish a
one-to-one correspondence between the points in a plane and an ordered
pair of numbers (*x, y*); that is, each point in the plane can be described by
an ordered pair of numbers (*x, y*), and each ordered pair of numbers
(*x, y*) can be represented by a point in the plane.

Example 1.   The locations of the points $P(1,2)$, $Q(-2,3)$, $R(-3,-4)$,
and $S(3,-5)$ are shown in Figure 1.1.

1.3   THE DISTANCE BETWEEN TWO POINTS

Consider two points P_1 and P_2 on the *x*-axis as shown in Figure 1.2.
The distance between these two points can be found by counting the
number of units between them. In analytic geometry, distances measured
from left to right are considered to be positive, and from right to left,
negative. Thus, the distance from P_1 to P_2, which is denoted by P_1P_2, is

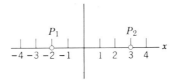

Figure 1.2

counted as 5 units, while the distance from P_2 to P_1, which is denoted by P_2P_1, is counted as -5 units. The distance P_1P_2 can also be found algebraically by subtracting P_1 from P_2; that is, $P_1P_2 = 3 - (-2) = 5$. It should also be observed that if we subtract P_2 from P_1, the distance P_2P_1 is given by $P_2P_1 = -2 - 3 = -5$, the minus sign indicating the direction of measurement. In order to avoid negative distances, we shall always subtract the leftmost point in the plane from the rightmost point. In general, the horizontal distance between two points in the plane can be found by the following rule.

Rule 1. The horizontal distance between two points in the plane is the abscissa of the rightmost point in the plane minus the abscissa of the leftmost point.

Similarly, the vertical distance between two points in the plane can be found by Rule 2.

Rule 2. The vertical distance between two points in the plane is the ordinate of the uppermost point in the plane minus the ordinate of the lowermost point.

Now let us consider two points $P_1(x_1, y_1)$ and $P_2(x_2, y_2)$ which determine a slant line as shown in Figure 1.3. Draw a line through P_1 parallel to the x-axis and a line through P_2 parallel to the y-axis. These two lines

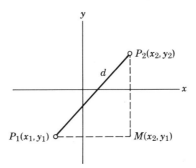

Figure 1.3

intersect at the point $M(x_2, y_1)$. Hence, by the Pythagorean theorem, the distance P_1P_2 is given by

$$(P_1P_2)^2 = (P_1M)^2 + (MP_2)^2 \tag{1}$$

We see from Figure 1.3 that P_1M is the horizontal distance between P_1 and P_2. Therefore, by Rule 1, the distance P_1M is given by

$$P_1M = x_2 - x_1$$

Likewise, by Rule 2, the vertical distance MP_2 is given by

$$MP_2 = y_2 - y_1$$

Making these substitutions into Equation 1, and denoting P_1P_2 by d, we have

$$d^2 = (x_2 - x_1)^2 + (y_2 - y_1)^2$$
$$d = \sqrt{(x_2 - x_1)^2 + (y_2 - y_1)^2} \tag{2}$$

Equation 2 is called the *distance equation* and is used to find the distance between two points in the plane directly from the coordinates of the points. The order in which the two points are labeled is immaterial to the outcome of Equation 2, since $(x_2 - x_1)^2 = (x_1 - x_2)^2$ and $(y_2 - y_1)^2 = (y_1 - y_2)^2$.

Example 2.   Find the distance between $(-3, -6)$ and $(5, -2)$ (Figure 1.4).

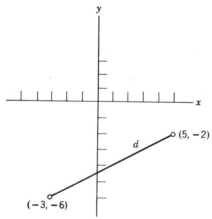

Figure 1.4

Solution.   Here we let $(x_1, y_1) = (-3, -6)$ and $(x_2, y_2) = (5, -2)$. Substituting these values into the distance formula, we have

$$d = \sqrt{(x_1 - x_2)^2 + (y_1 - y_2)^2}$$
$$= \sqrt{(-3 - 5)^2 + (-6 - (-2))^2}$$
$$d = \sqrt{64 + 16} = \sqrt{80} = 4\sqrt{5}$$

Notice that in substituting the values into the distance formula, the numerical sign of each number is included.

Example 3.   Find the distance between $(2, 5)$ and $(2, -1)$ (Figure 1.5).

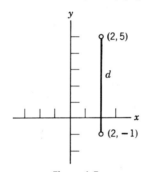

Figure 1.5

Solution.   In this case the two given points lie on a vertical line since they have the same abscissa. The distance between the two points there-fore, can be found by Rule 2; that is,

$$d = 5 - (-1) = 5 + 1 = 6 \text{ units}$$

The distance can also be found by the distance equation. Letting $(x_1, y_1) = (2, 5)$ and $(x_2, y_2) = (2, -1)$, we have

$$d = \sqrt{(2 - 2)^2 + (-1 - 5)^2} = \sqrt{36} = 6 \text{ units}$$

Thus, the distance equation can be used to find the distance between any two points in the plane.

EXERCISES

Plot each of the following ordered pairs.
1. $(3, 2)$ **4.** $(-6, -5)$
2. $(4, 6)$ **5.** $(\frac{1}{4}, -\frac{1}{2})$
3. $(-2, \frac{1}{2})$ **6.** $(-2.5, 1.7)$

7. In what two quadrants do the points have positive abscissas?
8. In what two quadrants do the points have negative ordinates?
9. In what quadrant are the abscissa and ordinate both negative?
10. In what quadrants is the ratio y/x negative?
11. What is the ordinate of a point on the x-axis?

Plot each of the following pairs of points and find the distance between the points.

12. $(1, 2), (5, 4)$

13. $(0, 4), (-1, 3)$

14. $(-1, 5), (-1, -6)$

15. $(\frac{1}{2}, \frac{1}{2}), (\frac{1}{2}, -\frac{3}{4})$

16. $(-5, 3), (2, -1)$

17. $(0.5, 1.6), (6.2, 7.5)$

18. $(-3, 4), (0, 4)$

19. $(2, -6), (-\sqrt{3}, -3)$

1.4 FUNCTIONS

There are two kinds of symbols used to represent numbers or numerical quantities: *constants* and *variables*. As we learned in algebra, constants have fixed values throughout a discussion, while variables may take on certain admissible values. We frequently use the letters toward the end of the alphabet to denote variables and letters at the beginning of the alphabet to denote constants. However, in physical applications it is common practice to denote a variable quantity by the first letter in its name. Thus, velocity is denoted by v, acceleration by a, capacitance by c, etc.

In science and technology we often talk about one variable quantity as a *function* of some other variable quantity. For example, in physics we say that atmospheric pressure is a function of altitude or in electronics that the plate current in a diode is a function of the plate voltage. The functional relationship between two variables is usually given by an equation, but it may also be given by other means such as a table of values. The notion that two variables may be related to one another is basic to our understanding of the physical world. It is, therefore, essential that we know precisely what we mean when we use the word "function." The following definition of a function is used in this book.

Definition
A variable y is said to be a *function* of the variable x if for each value of x there corresponds a unique value of y.

Notice that this definition does not require that the value of y change when the value of x changes, but only that y have a unique value corresponding to each value of x. In this light, we may consider constants to be functions of some variable. Thus, $y = 5$ is a valid function since y has the unique value 5 for any value of x that we choose. On the other hand, $y = \pm\sqrt{x}$ does not satisfy the definition because there are two values of y corresponding to each value of x, namely, the positive and negative square roots of x. In order to make equations like this comply with our definition, we will use the symbol $\sqrt{\ }$ to denote only positive square roots.

The variable to which values are assigned is commonly called the *independent* variable and the corresponding value of the function is called the *dependent* variable. The set of all values taken on by the independent variable is called the *domain* of the function and the corresponding set of values taken on by the dependent variable is called the *range* of the function. Unless otherwise noted, the domain and the range of a function will be restricted to real numbers.

1.5. FUNCTIONAL NOTATION

In some mathematical discussions we are interested in indicating that y is a function of x without actually specifying the particular formula. This is particularly true in discussions of the general mathematical properties of functions. The notation commonly used to indicate that a functional relationship exists between x and y is

$$y = f(x)$$

This is read "y is the f function of x" or simply "y equals f of x." The letter f is the name of the functional relationship between x and y; it is not a variable. While there is a tendency to use the letter f as the name of the function, it should be emphasized that any other letter or symbol will serve as well. As a matter of fact, it is frequently necessary to use different functional names to avoid confusion. For instance, to indicate that the velocities of two different atomic particles are both functions of time, we might denote the velocity of particle p_1 by $v_1 = f(t)$ and the velocity of particle p_2 by $v_2 = g(t)$. In this way we make it clear that the relationship between velocity and time is different for each particle. Had we used f or g for both functional names, this fact would not be obvious. The worth of functional notation lies in its flexibility.

The function designated by $f(x)$ may or may not have a definite mathematical expression. If a specific expression is available, functional notation offers a convenient way of indicating substitutions into the expression.

Example 4.   Find the value of the function $f(x) = x^2 + 3x$ at $x = 2$.

Solution.   We denote the value of $f(x)$ at $x = 2$ by $f(2)$. To find $f(2)$, we substitute 2 for x in $x^2 + 3x$; that is,

$$f(2) = (2)^2 + 3(2) = 10$$

The dependent variable may be a function of two or more independent variables. The equation for the area of a triangle is $A = \frac{1}{2}bh$, where b is the length of the base and h is the altitude. Thus, the area of a triangle is a function of both the base and the altitude of the triangle. This could be noted in general by using functional notation and writing $A = f(b, h)$. If we want to indicate that the acceleration a of a body is a function of force f and mass m, we could write $a = g(f, m)$. The following example shows how a function of two independent variables can be evaluated numerically.

Example 5.   Evaluate $f(x, y) = x^2 + 3xy^2$ at $x = 2, y = -3$.

Solution.   Making the appropriate substitutions, we have

$$f(2, -3) = (2)^2 + 3(2)(-3)^2 = 58$$

EXERCISES

1. Given $f(x) = x^2 - 5x$, find $f(2)$ and $f(-1)$.
2. Given $g(x) = x^2 - 4$, find $g(2)$ and $g(-4)$.
3. Given $f(t) = (t-2)(t+3)$, find $f(-4)$ and $f(0)$.
4. Given $\phi(y) = y(y+1)$, find $\phi(3)$ and $\phi(-3)$.
5. Given $h(x) = a(x^2 - 3)$, find $h(5)$ and $h(1)$.
6. Given $g(z) = (1 - z^2)/(1 + z^2)$, find $g(-1)$ and $g(4)$.
7. Given $F(s, t) = s^2 - t^2$, find $F(3, 1)$ and $F(2, 2)$.
8. Given $f(s, t) = 2st - 3$, find $f(2, -1)$ and $f(3, 2)$.
9. Given $\theta(\alpha, \beta) = \alpha^2 - \alpha\beta + \beta^2$, find $\theta(3, -2)$ and $\theta(-5, -2)$.
10. Given $G(u, v) = u/(u - v)$, find $G(4, 2)$ and $G(-1, 2)$.

Symbolize the following expressions in functional notation.

11. The area A of a circle as a function of the radius r.
12. The circumference C of a circle as a function of its diameter d.
13. The perimeter P of a square as a function of its side length S.
14. The volume V of a cylinder as a function of its altitude h and the radius of the base r.
15. The current I as a function of the voltage V and the resistance R.
16. The velocity v of a particle as a function of its acceleration a and the time elapsed t.

1.6 THE SLOPE OF A STRAIGHT LINE

The rectangular coordinate system can be used to establish a correspondence between algebraic equations and certain geometric figures. If we plot all of those points in the plane whose coordinates satisfy a given equation, a definite path will be traced in the plane. This path or curve is referred to as the *graph* of the equation. One of the most important graphs that we encounter in technological applications is the straight line. In succeeding articles we will show that a correspondence exists between straight lines and first degree equations, but for now let us concentrate on identifying some of the important characteristics of straight lines in the rectangular plane.

The location of a straight line in the Cartesian plane is described by (1) the point at which the straight line crosses the *y*-axis, and (2) the inclination of the straight line with respect to the coordinate axes. What do we mean by the inclination of a line? Intuitively, the inclination of a straight line is the extent of deviation of the line from a vertical or horizontal position. It is, therefore, natural to describe the inclination of a straight line by the angle that the straight line makes with another fixed line. In analytic geometry, the *angle of inclination* of a straight line is defined as the smallest positive angle between the line and the positive *x*-axis. Hence the angle of inclination, which is denoted by α in Figure 1.6, is always less than 180°. If a line is parallel to the *x*-axis, its angle of inclination is defined to be zero.

Another quantity that can be used to describe the inclination of a straight line is its *slope*. The concept of slope is used frequently in everyday life and is, therefore, already familiar to you on an intuitive basis. For instance, we talk about the slope of a hill or the slope of a roof top and understand that the slope describes the steepness of the inclined plane. The slope of an inclined plane is defined as the ratio of the vertical *rise* of the incline to the corresponding horizontal *run*; that is,

$$\text{Slope} = \frac{\text{vertical rise}}{\text{horizontal run}}$$

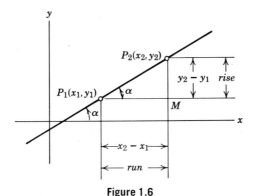

Figure 1.6

In analytic geometry, we use the same definition of slope as we used in the preceding paragraph but we express it in terms of the coordinates of two points on a line. Thus, if the coordinates of two points on a line are known, *the slope of the line is defined as the difference in the ordinates of the two points divided by the difference in the corresponding abscissas.* Applying this definition to the line segment P_1P_2 in Figure 1.6, the slope m of the line is expressed by the equation

$$m = \frac{y_2 - y_1}{x_2 - x_1} \tag{3}$$

Since the slope expresses the ratio of a change in y to the corresponding change in x, we also refer to this ratio as the *rate of change* of y with respect to x. The concept of rate of change is utilized extensively in the study of the calculus.

An interesting and important relationship exists between the angle of inclination of a line and its slope. Referring to Figure 1.6, we see that α is also an angle of the right triangle P_1MP_2. The side opposite angle α is $y_2 - y_1$, and the side adjacent angle α is $x_2 - x_1$. From trigonometry, the ratio of the side opposite an angle to the side adjacent to the angle is the tangent of the angle. Thus

$$\tan \alpha = \frac{y_2 - y_1}{x_2 - x_1} \tag{4}$$

But the ratio on the right side of (4) is also the slope of the line. Therefore, the slope of a line is equal to the tangent of the angle of inclination, and we write

$$m = \tan \alpha \tag{5}$$

From (5) we see that lines that have angles of inclination greater than $90°$ and less than $180°$ have negative slopes because the tangent of an obtuse angle is negative. Also, the slope of a line parallel to the y-axis is undefined because $\tan 90°$ is undefined.

Example 6.   Find the slope of the straight line passing through the points $(-5, 1)$ and $(2, -3)$ (Figure 1.7).

Solution.   Letting $(x_1, y_1) = (-5, 1)$ and $(x_2, y_2) = (2, -3)$ and using Equation 3, the desired slope is

$$m = \frac{y_2 - y_1}{x_2 - x_1} = \frac{-3 - 1}{2 - (-5)} = \frac{-4}{7} = -\frac{4}{7}$$

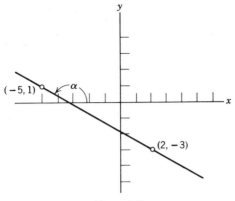

Figure 1.7

If we interchange the labels of the given points and let $(x_1, y_1) = (2, -3)$ and $(x_2, y_2) = (-5, 1)$, the result is

$$m = \frac{y_2 - y_1}{x_2 - x_1} = \frac{1 - (-3)}{-5 - 2} = \frac{4}{-7} = -\frac{4}{7}$$

Hence, the order in which we label the given points is immaterial.

We interpret a slope of $-\frac{4}{7}$ to mean that, for every 7 units we move to the right in the direction of the x-axis, the straight line moves down 4 units. Or, for every 7 units that we move to the left, the line moves up 4 units.

Example 7.   Find the angle of inclination of the straight line shown in Figure 1.7.

Solution.   The slope of the line is $m = -\frac{4}{7}$. Therefore, by Equation 5, we have

$$\tan \alpha = m = -\frac{4}{7} = -0.5714$$

The angle α will be an obtuse angle since $\tan \alpha$ is negative. Using a table of trigonometric functions, the angle whose tangent is 0.5714 is 29°45′. The desired angle is then

$$\alpha = 180° - 29°45' = 150°15'$$

Example 8.   Draw the straight line passing through the point (2, 1) with a slope $m = 2$ (Figure 1.8).

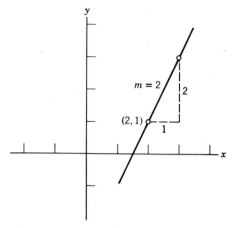

Figure 1.8

Solution.   The slope $m = 2$ can be written as $m = \frac{2}{1}$. To lay off the indicated slope, we start at the point $(2, 1)$ and move 1 unit to the right and 2 units upward. This brings us to the point $(3, 3)$. The desired straight line is then the line that passes through the points $(2, 1)$ and $(3, 3)$.

In working with straight lines, the following facts prove to be helpful.

1. The slope of a straight line parallel to the x-axis is zero. (A fact that will have important applications in Chapter 3.)

2. The slope of a straight line parallel to the y-axis is undefined.

3. The slope of a straight line is positive if, as you follow the curve from left to right, you move upward on the coordinate axes.

4. The slope of a straight line is negative if, as you follow the curve from left to right, you move downward on the coordinate axes.

5. Parallel lines have equal slopes.

6. The slopes of perpendicular lines are negative reciprocals of each other; that is, if a line l_1 has a slope m_1 and line l_2 has a slope m_2, and if l_1 and l_2 are perpendicular to each other, then

$$m_1 = -\frac{1}{m_2}$$

This last statement may not be immediately obvious to the student so we will present the following proof.

Let us construct a straight line l_1 with a slope m_1 and a second straight line l_2 with a slope m_2 such that the two lines are perpendicular to each other. Referring to Figure 1.9, we see that l_1 makes an angle of α_1 with the

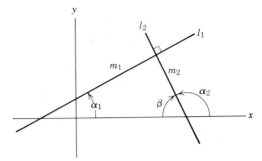

Figure 1.9

x-axis and l_2 makes an angle of α_2 with the x-axis. The following reasoning leads to the desired result.

$$\beta = 90° - \alpha_1$$

$$m_1 = \tan \alpha_1 \tag{6}$$

$$m_2 = \tan \alpha_2 = -\tan (90° - \alpha_1) \tag{7}$$

Multiplying Equations 6 and 7

$$m_1 m_2 = \tan \alpha_1 [-\tan (90° - \alpha_1)]$$

But, since α_1 and $90° - \alpha_1$ are complementary angles,

$$m_1 m_2 = \tan \alpha_1 [-\cot \alpha_1] = -1$$

Finally, dividing both sides by m_2, we have

$$m_1 = -\frac{1}{m_2} \tag{8}$$

Example 9.   Find the slope of a line that is drawn perpendicular to a line passing through the points $(1,1)$ and $(-1,-2)$.

Solution.   Letting $(x_1,y_1) = (1,1)$ and $(x_2,y_2) = (-1,-2)$, the slope of the line through these two points is given by

$$m = \frac{y_2 - y_1}{x_2 - x_1} = \frac{-2-1}{-1-1} = \frac{-3}{-2} = \frac{3}{2}$$

The slope m' of a line drawn perpendicular to this line then has a slope of

$$m' = -\frac{1}{m} = -\frac{1}{3/2} = -\frac{2}{3}$$

EXERCISES

Plot each of the following pairs of points and compute the slope and angle of inclination of the line passing through them.

1. $(1,2), (5,4)$ 5. $(-5,2), (3,-7)$
2. $(-1,-1), (3,-6)$ 6. $(7,3), (0,5)$
3. $(-2,-3), (-5,-7)$ 7. $(3,-2), (7,6)$
4. $(-2,3), (5,3)$ 8. $(\frac{1}{2},\frac{1}{3}), (\frac{1}{4},-\frac{2}{5})$

9. Draw the straight line passing through $(2,1)$ with a slope of $\frac{1}{3}$.
10. Draw the straight line passing through $(-3,4)$ with a slope of 3.
11. Draw the straight line passing through $(2,-5)$ with a slope of $-\frac{2}{5}$.

Determine the slope of the straight line having the given angle of inclination.

12. $\alpha = 35°$ 14. $\alpha = 125°$
13. $\alpha = 45°$ 15. $\alpha = 150°$

16. Find the slope of a line which is drawn perpendicular to a line which has a slope $m = -\frac{3}{2}$.
17. Find the slope of a line perpendicular to the line drawn through $(2,3)$ and $(7,2)$.
18. Find the slope of a 25-ft ladder that is leaning against a building if the foot of the ladder is 12 ft from the building.
19. The acceleration of a particle is defined as the change in velocity divided by the corresponding change in time. If the initial velocity is 25 ft/sec and the velocity is 40 ft/sec at the end of 10 sec, what is the acceleration? What units does acceleration have?
20. The current (amperes) in a resistor is defined as the charge transferred (coulombs) divided by the corresponding time (seconds). Find the current in a 5 ohm resistor if the charge varies 15 coulombs in 3 sec.

1.7   THE EQUATION OF A STRAIGHT LINE

In the previous articles we have presented some basic definitions and characteristics of straight lines in the Cartesian plane. We continue our study of straight lines by discussing the correspondence between straight lines and first degree equations. The following examples are intended to illustrate the fact that every straight line can be represented by a first degree equation (Example 10) and, conversely, every first degree equation can be represented by a straight line (Example 11).

Example 10.   Show that the equation of the straight line passing through $(2,1)$ with a slope of $\frac{1}{3}$ is of the first degree (Figure 1.10).

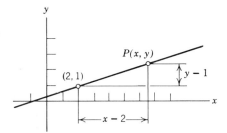

Figure 1.10

Solution.   Let $P(x, y)$ represent any point on the straight line other than $(2, 1)$. It should be emphasized that $P(x, y)$ is a general point and may be chosen anywhere on the straight line. Then, by definition, the slope of the given line can be written as $(y-1)/(x-2)$. This quantity must be equal to the given slope, so that

$$\frac{y-1}{x-2} = \frac{1}{3}$$

Simplifying, we have

$$3y - 3 = x - 2$$
$$3y - x = 1$$

which is a first degree equation. This equation is then the expression of the relationship that exists between the abscissa and the ordinate of any point on the given straight line.

Example 11.   Show that the graph of the first degree equation $4x - 2y = 8$ is a straight line.

Solution.   To demonstrate this fact, the given equation is solved for selected values of the independent variable and the resulting ordered pairs plotted in the plane. As you can see in Figure 1.11, a straight line can be drawn through the plotted points. This straight line is then the geometric representation of the given first degree equation. Since first degree equations have straight line graphs, they are often called *linear* equations.

Now that we have illustrated the correspondence between straight lines in the Cartesian plane and first degree equations, let us develop two general forms of a linear equation.

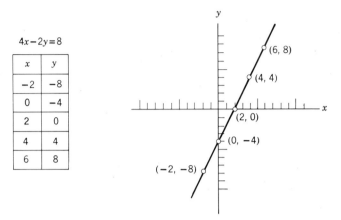

$4x - 2y = 8$

x	y
-2	-8
0	-4
2	0
4	4
6	8

Figure 1.11

1.8 POINT-SLOPE FORM

Consider a straight line passing through the given point $P(x_1, y_1)$ with a slope of m as shown in Figure 1.12. Choose another point $P(x, y)$ on this straight line. The location of the point $P(x, y)$ on the given straight line is arbitrary. Then by the same reasoning that we used in Example 10, we have

$$\frac{y - y_1}{x - x_1} = m$$

Therefore

$$y - y_1 = m(x - x_1) \tag{9}$$

Thus, Equation 9 represents the given line and is called the *point-slope form* of a linear equation. The point-slope form can be used to find the equation of a given straight line if the slope of the line and one point on the line are known.

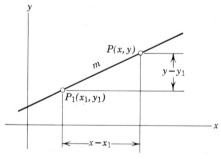

Figure 1.12

Example 12.   Find the equation of the straight line passing through $(-1, 2)$ with a slope of 3.

Solution.   Since $(-1, 2)$ is the given point, we substitute $x_1 = -1$, $y_1 = 2$ and $m = 3$ into the point-slope form. Thus

$$y - 2 = 3[x - (-1)]$$
$$y - 2 = 3x + 3$$
$$y - 3x = 5$$

is the equation of the given line.

Physical situations frequently arise in which two variable quantities are directly proportional to one another. For instance, the current in a resistor is directly proportional to the applied voltage. Any physical situation in which one variable is directly proportional to the other can be represented by a linear equation. To illustrate a situation of this kind, let us consider the following example.

Example 13.   The amount that a spring stretches is directly proportional to the applied force. Find the equation relating the displacement of the spring to the applied force if the spring is displaced 3 in. by a force of 6 lb, and 7 in. by a force of 14 lb.

Solution.   We can plot the given conditions if we treat the displacement d as the dependent variable and the force f as the independent variable (Figure 1.13). The slope of the straight line between these two points is then given by

$$m = \frac{d_2 - d_1}{f_2 - f_1} = \frac{7 - 3}{14 - 6} = \frac{4}{8} = \frac{1}{2}$$

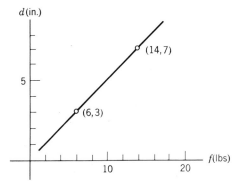

Figure 1.13

The point-slope form of this straight line can be written in the form

$$d - d_1 = m(f - f_1)$$

Now, letting $(d_1, f_1) = (3, 6)$ and $m = \frac{1}{2}$, we obtain

$$d - 3 = \tfrac{1}{2}(f - 6)$$

or

$$d = \tfrac{1}{2}f$$

as the required equation.

1.9 SLOPE-INTERCEPT FORM

The second form of a straight line that we shall study is called the *slope-intercept form*. As we shall see, this is also a special case of the point-slope form.

Consider the straight line in Figure 1.14. Let the slope of this line be m, and assume that the line crosses the y-axis at the point $(0, b)$. This assumption is valid as long as the straight line is not parallel to the y-axis. Substituting the slope m and the y-intercept $(0, b)$ into the point-slope form of a straight line, we have

$$y - b = m(x - 0)$$
$$y = mx + b \tag{10}$$

Equation 10 is referred to as the *slope-intercept* form of a straight line since both of these quantities can be determined directly from the equation. When a first degree equation is rearranged into form 10, the slope of its graph is equal to the coefficient of the x-variable and the y-intercept of its graph is equal to the constant term. The following examples demonstrate the use of Equation 10 in plotting the graph of a linear equation.

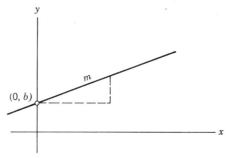

Figure 1.14

Example 14. Draw the graph of the linear equation $3x + 2y = -5$.

Solution. Rearranging into slope-intercept form, we have

$$2y = -3x - 5$$

$$y = -\frac{3}{2}x - \frac{5}{2} = \left(\overset{\overset{\displaystyle m}{\downarrow}}{-\frac{3}{2}}\right)x + \left(\overset{\overset{\displaystyle b}{\downarrow}}{-\frac{5}{2}}\right)$$

Hence, the graph crosses the y-axis at $(0, -\frac{5}{2})$ with a slope $m = -\frac{3}{2}$. By plotting the indicated y-intercept and slope, we may determine the graph of the given equation. Notice that the y-intercept is plotted first, and then the slope laid off from this point. The desired graph is then obtained by drawing a straight line through $(0, -\frac{5}{2})$ and $(2, -\frac{11}{2})$, as illustrated in Figure 1.15.

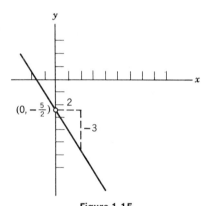

Figure 1.15

Example 15. The breaking load of a new alloy is found to vary with temperature according to the equation $3p + 2t = 600$, where p is the breaking load and t is the temperature. Show the relationship between p and t graphically.

Solution. Considering the load p as the dependent variable, we can write

$$p = -\frac{2}{3}t + 200$$

Hence, the graph of the given equation intercepts the p-axis at $(0, 200)$ with a slope $m = -\frac{2}{3}$. This is depicted in Figure 1.16. There is no need to

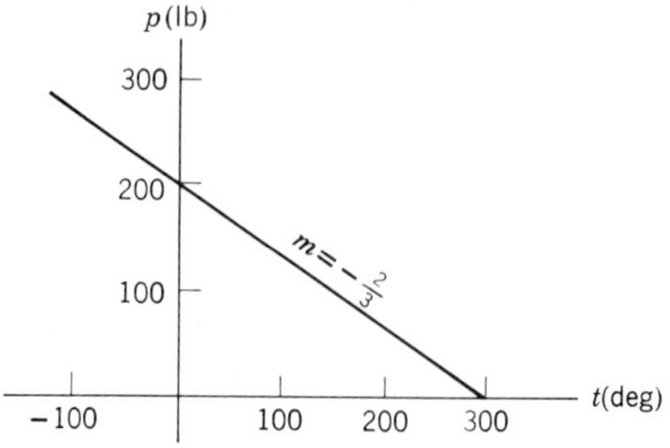

Figure 1.16

extend the graph below the *t*-axis because a negative breaking load is meaningless.

EXERCISES

Draw the line passing through the given point with the given slope and then determine its equation.

1. $(2,5)$, $m = \frac{1}{2}$
2. $(-1,-3)$, $m = 3$

3. $(5,-2)$, $m = -7$
4. $(3,4)$, $m = -\frac{2}{5}$

Draw the line through the given points and then find its equation.

5. $(1,3), (6,2)$
6. $(2,5), (-3,7)$
7. $(-1,-1), (1,2)$

8. $(0,0), (3,-2)$
9. $(0,2), (-5,0)$
10. $(\frac{1}{2},\frac{1}{3}), (\frac{1}{2},-\frac{1}{3})$

Find the equation of the line passing through the given point with the given angle of inclination.

11. $(1,-2)$, $\alpha = 30°$
12. $(-3,-4)$, $\alpha = 75°$

13. $(-3,0)$, $\alpha = 110°$
14. $(3,5)$, $\alpha = 135°$

Find the slope and *y*-intercept of the following equations and then draw the line.

15. $2x - 3y = 5$
16. $x + y = 2$
17. $4y - 2x + 8 = 0$

18. $3x + 4y = 0$
19. $5x + 2y = -3$
20. $-5x - y - 2 = 0$

Find the equation of the line perpendicular to the given line at the indicated points.

21. $3x + 2y = 7$ at $(1,2)$
22. $x - y = 2$ at $(5,3)$

23. $x + 3y = 11$ at $(2,3)$
24. $2x - 5y = 2$ at $(-4,-2)$

25. The current I in a resistor is directly proportional to the applied voltage V. Find the equation relating the current to the voltage if the current is $\frac{1}{2}$ amp when the voltage is 6 volts and $\frac{2}{3}$ amp when the voltage is 8 volts.

26. In an experiment to determine the coefficient of friction, it is found that a 10-lb block has a frictional force of 3 lb and a 25-lb block has a frictional

force of 7.5 lb. Find the equation relating frictional force to weight if the frictional force is a linear function of weight.

27. The resistance R of a certain resistor varies linearly with the temperature T of the resistor. Find the equation relating resistance and temperature if $R = 100$ ohms when $T = 0°C$ and $R = 104$ ohms when $T = 200°C$. What is the resistance when the temperature is $90°C$?

28. The velocity of a rocket sled is given by the equation $v = 20 - 5t$ where v is the velocity in m/sec and t is the time in seconds. Draw the graph of this function. From the graph find the time at which the velocity is zero.

29. The pressure at a point below the surface of a body of water is given by $P = 0.04D + 14.7$, where P is the pressure in psi, and D is the depth in inches. Draw the graph of this function.

30. The error E of a radar is found to vary with slant range R according to the equation $5E - R = 3$. Show the relationship between radar error and slant range graphically.

1.10 GRAPHS OF NONLINEAR FUNCTIONS

A function whose graph is not a straight line is called a *nonlinear* function. The graph of a nonlinear function $y = f(x)$ can be constructed by plotting all of the points in the plane whose coordinates satisfy the indicated equation. However, this approach is impractical, since there is no limit to the number of such points that can be plotted. In practice, the graph of a nonlinear function is constructed by plotting a few selected solution points and then connecting these points with a smooth curve.* To illustrate this technique, consider the function $y = x^2$. By assigning values to x in the given function, we can obtain the corresponding values of y, as shown below.

x	-3	-2	-1	0	1	2	3
y	9	4	1	0	1	4	9

We now plot the set of ordered pairs in the Cartesian plane as shown in Figure 1.17a. The representation of the graph of $y = x^2$ can then be obtained by connecting the points with a smooth curve as in Figure 1.17b. The graph obtained in this way is, of course, only an approximation of the true graph of the function. The accuracy of the graph will depend on the number of points plotted and the care with which the smooth curve connecting the points is drawn.

Usually only a segment of the graph in the vicinity of the origin is plotted, even though the graph may extend much farther. For instance, in

*A smooth curve is a curve without any sharp corners.

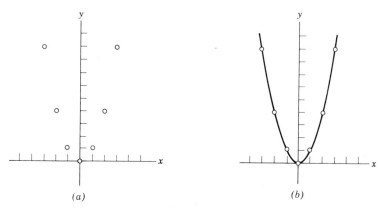

Figure 1.17

Figure 1.17 only the portion of the curve from $x = -3$ to $x = 3$ is plotted, although the domain of the function is the entire real line.

A great deal of information can be obtained about the properties of a function by plotting its graph. The following example shows the use of the graphical approach to analyzing the properties of a functional relationship between two physical variables.

Example 16.   Consider a car moving at a constant speed of 50 mph along a highway. By following a point on the tread of one of the tires through one complete revolution, we find that the linear speed of this point relative to the road varies with time. Find the maximum linear speed of the point if Figure 1.18 is the graphical representation of the functional relationship between linear speed and time.

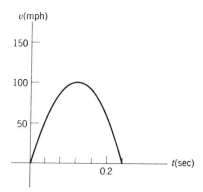

Figure 1.18

Solution.   We conclude that the maximum linear speed of a point on the tread of the tire is 100 mph since this is the upper extremity of the graph. The graph of a function usually demonstrates the properties of the function more clearly than any other form of description.

1.11   THE DOMAIN OF A FUNCTION

In defining a function $y = f(x)$, it is understood that the domain of the function contains all values of x for which the formula $f(x)$ yields real values. Thus the function $y = x^2$ is defined for all real values of x while the function $y = \sqrt{x-2}$ is defined only for $x \geq 2$. Since the domain of a function comprises all values of x for which the function is defined, it also describes the physical extent of the graph of the function in the plane. Hence, in many cases we can facilitate the plotting process by finding the domain of the function.

Example 17.   Draw the graph of $y = \sqrt{4-x}$.

Solution.   We see from the equation that y is real for all values of $x \leq 4$ and imaginary for $x > 4$. Hence, the graph of the given function lies entirely to the left of $x = 4$. Computing y for some convenient values of $x \leq 4$, we have the graph shown in Figure 1.19.

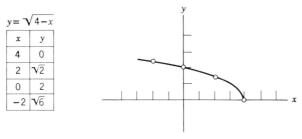

$y = \sqrt{4-x}$

x	y
4	0
2	$\sqrt{2}$
0	2
-2	$\sqrt{6}$

Figure 1.19

Example 18.   A numerically controlled milling machine produces a die whose cross-sectional area is described by the function $y = \sqrt{16-4x^2}$. Let y represent the height of the piece above the bed of the milling machine and x the horizontal distance from the center line of the piece to the outer extreme. Draw the shape of the die.

Solution.   The domain of this function is $-2 \leq x \leq 2$ since y is imaginary for $x < -2$ and $x > 2$. The values in the table are computed from the given function and then plotted in Figure 1.20.

In many physical situations the domain of the function is restricted by the nature of the problem. For instance, the motion of a ball that is thrown

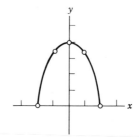

$y = \sqrt{16 - 4x^2}$

x	y
-2	0
-1	$\sqrt{12}$
0	4
1	$\sqrt{12}$
2	0

Figure 1.20

upward with a velocity of 32 ft/sec is described by the function $h = 32t - 16t^2$, where t is the elapsed time and h is the vertical height. An analysis of this problem reveals that the ball will strike the ground 2 seconds after it is thrown upward. The domain of the indicated function is, therefore, $0 \leq t \leq 2$ since the equation has no meaning before the ball was thrown or after it hits the ground.

Example 19.   Draw the graph of $h = 32t - 16t^2$.

Solution.   Computing some convenient values of h in the interval $0 \leq t \leq 2$, we have the following table and graph. A glance at the graph in Figure 1.21 reveals that the maximum height to which the ball rises is 16 ft.

$h = 32t - 16t^2$

t	h
0	0
0.5	12
1.0	16
1.5	12
2.0	0

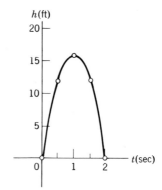

Figure 1.21

EXERCISES

Graph each of the following functions:

1. $y = x^3$
2. $f(x) = -x^2$
3. $z = t^2 + 4$
4. $i = r - r^2$

5. $y(x) = \sqrt{x}$

6. $\phi = \dfrac{w^2}{2}$

7. $p = z^2 - z - 6$

8. $v = 10 + 2t$

9. $y = \sqrt{16 - 4x^2}$

10. $y = \sqrt{25 - x^2}$

11. $\alpha = \theta^{1/3}$

12. $z = \sqrt[3]{t^2}$

13. $y(x) = \sqrt{x - 2}$

14. $y = \sqrt[3]{2 - x}$

15. $v = s^3 - 4s^2$

16. $a = \frac{1}{4}b^4$

17. $y = -x^3$

18. $\beta = -\alpha^{1/2}$

19. The work w done in moving an object varied with the distance s according to $w = \sqrt[3]{2s}$. Show this relationship graphically.

20. The path of a certain projectile is described by the function $h = 100x - 2x^2$, where h is the vertical height in feet and x is the horizontal displacement in feet. Draw the path of the projectile.

21. The power in a resistor is given by $P = I^2R$ where I is the current in amps and R is the resistance of the resistor in ohms. Plot the power in a 25 ohm resistor as a function of current.

22. An archway is to be built in the form of a semiellipse. Draw the shape of the archway if it is described analytically by the function $y = \sqrt{64 - 4x^2}$, where y is the vertical height of the arch and x is the horizontal distance from the centerline of the arch. Both x and y are measured in feet.

23. Torricelli's law states that the velocity of a stream of liquid issuing from an orifice is given by $v = \sqrt{2gy}$, where g is the acceleration of gravity and y is the depth, in feet, of the orifice below the surface of the liquid. Show the relationship between v and y if $g = 32$ ft/sec^2.

24. The functional relationship between the safe speed for a car to round a curve and the degree of the curve is described in the following table. Draw the graph of safe speed as a function of the degree of the curve.

Degrees of Curve	5	10	15	20	25	30	35
Safe Speed (mph)	70	68	66	63	58	50	34

25. A thermocouple generates a voltage when its two ends are kept at different temperatures. In laboratory experiments, the cold end of the thermocouple is usually kept at 0°C, and the temperature of the other end varied. The table represents the results of an experiment in which the output voltage of the thermocouple was recorded for various thermocouple temperatures. Draw the graph of voltage versus temperature.

$T(°C)$	0	50	100	150	200	250
$V(\text{mv})$	0	2.1	4.1	5.9	7.0	7.8

1.12   DIVISION BY ZERO   *DON'T DO IT*

The quotient of two numbers a and b is denoted by a/b and is defined as a unique number c such that $a = b \cdot c$. The following argument shows that this definition is not satisfied when the divisor is zero. If a is assumed to be any number other than zero, then $b = 0$ requires that c be a number such that $a = 0 \cdot c$. Since there is no value of c that when multiplied by zero will yield a nonzero value of a, we conclude that division by zero is meaningless. If a and b are both zero, we have $0 = 0 \cdot c$. Consequently, c can be any number whatever and satisfy the definition. Therefore, in this case, division by zero is meaningless because the ratio does not yield a unique number c.

1.13   INFINITY   *CONCEPT LIMIT*

In mathematics a variable that ultimately becomes greater than any fixed positive number is said to *increase without bound*, or to *become positively infinite*. Similarly, a variable that ultimately becomes less than any fixed negative number is said to *decrease without bound*, or *become negatively infinite*. We use ∞ as the symbol for infinity and we write

$$x \to \infty \qquad \text{and} \qquad x \to -\infty$$

to indicate that x becomes positively infinite and negatively infinite, respectively.

The concept of infinity is useful in many situations where large numbers are involved. However, in using the symbol ∞, we must realize that it is not a number but a symbol representing the idea that a variable increases without bound. Because ∞ is not a number in the ordinary sense of the word, it cannot be combined arithmetically with the real numbers; that is, we cannot add 5 to ∞ or divide ∞ by 2.

1.14   ASYMPTOTES   *- STRAIGHT LINES - GRAPHIC LIMITS*

To introduce the concept of an asymptote of a curve, let us consider what happens to the graph of $y = 1/x$ as the value of x increases. We see from the following table that y tends toward zero as the value of x increases. However, no matter how large x becomes, y can never equal zero. The graph of this function will, therefore, approach closer and closer to the line $y = 0$ as we extend the graph to the right. See Figure 1.22. The line $y = 0$ is referred to as an asymptote of the graph. In general, a line is called an *asymptote* of a graph if the distance between the line and a point $P(x, y)$ on the graph approaches zero as the point P recedes

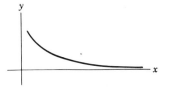

$y = \frac{1}{x}$					
x	1	10	100	1000	10000
y	1	0.1	0.01	0.001	0.0001

Figure 1.22

along the graph. If the graph of a function possesses asymptotes, it is to our advantage to determine their location because the process of drawing the graph is made much simpler by this knowledge.

In this chapter we shall restrict our study of asymptotes to those that are either vertical (parallel to the y-axis) or horizontal (parallel to the x-axis). The following example illustrates both types of asymptotes and shows how they are used in the plotting scheme.

Example 20.   Find the horizontal and vertical asymptotes of the graph of $y = 1/x$.

Solution

Horizontal Asymptote. If we allow x to increase without bound in this function, the value of the function approaches zero. But no matter how large x becomes, $1/x$ will not equal zero. Similarly, if we allow x to decrease without bound, the value of the function approaches zero. The line $y = 0$, is, therefore, a horizontal asymptote of the graph of the given function.

Vertical Asymptotes. The function $1/x$ is not defined at $x = 0$; however, it is defined for all other values of x. The value of the function becomes

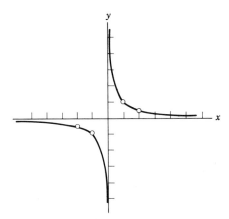

Figure 1.23

positively infinite as x approaches zero through the positive numbers since $1/x$ can be made as large as we wish by making x sufficiently small. Similarly, the value of the function becomes negatively infinite as x approaches zero through the negative numbers. The line $x = 0$ is, therefore, a vertical asymptote of the desired graph.

The graph of the function can now be sketched by plotting a few convenient points and drawing the curve through these points in such a way that it approaches both its vertical and its horizontal asymptote. The resulting graph is shown in Figure 1.23.

1.15   RATIONAL FUNCTIONS

A single term of the form cx^n, where c is a constant and n is a positive integer or zero, is called a *monomial* in x. The sum of a finite number of monomial terms is called a *polynomial* in x. For example, $3x^2 + 2$ and $x^3 + 2x - 4$ are polynomials; $x + \sqrt{x}$ is not a polynomial.

In this article we are interested in graphs of functions of the form

$$y = \frac{N(x)}{D(x)} \tag{11}$$

where $N(x)$ and $D(x)$ are polynomials *without* any common factors. Functions of this form are called *rational* functions of x. Some examples of rational functions are

$$\frac{3x^5 + 5}{15x^3 - 7x + 1}, \qquad \frac{1}{x^2 - 1}, \qquad \frac{3x^2 - x + 13}{x^2 + 12x - 2}$$

The graphs of many rational functions have asymptotes. These asymptotes are usually easy to find and are very helpful in sketching the graph. The following rule can be used to find any vertical asymptotes of the graph of a rational function.

Rule 3.   The vertical asymptotes of the graph of a rational function of the form

$$y = \frac{N(x)}{D(x)}$$

are indicated by the values of x for which $D(x) = 0$. Thus, if $D(k) = 0$, the line $x = k$ is a vertical asymptote. If no real values of x exist such that $D(x) = 0$, the graph has no vertical asymptotes.

Applying this rule to the rational function

$$y = \frac{2x^2 + 5}{x^2 - 9} \qquad (12)$$

we see that the denominator $x^2 - 9$ is equal to zero when $x = 3$ and $x = -3$. Therefore, the graph of this function has two vertical asymptotes: the lines $x = 3$ and $x = -3$.

The graph of a rational function of form (11) may have more than one vertical asymptote, but it may not have more than one horizontal asymptote. (We are assuming here the conventional orientation of the x and y axes.) A convenient method for finding the horizontal asymptote of the graph of a rational function, if it exists, is given in Rule 4.

Rule 4. To find the horizontal asymptote of the graph of a rational function in x, first divide $N(x)$ and $D(x)$ by the highest power of x which occurs in the function. If the resulting function approaches a finite number c as x becomes infinite, then the line $y = c$ is a horizontal asymptote. If the function does not approach a finite number as x becomes infinite, the graph has no horizontal asymptote.

To illustrate the use of Rule 4, let us once again consider the function given in (12). Dividing the numerator and denominator of this function by x^2, we obtain

$$y = \frac{2 + 5/x^2}{1 - 9/x^2}$$

Now as x becomes infinite, both $5/x^2$ and $9/x^2$ approach zero. Hence, the value of y approaches 2, and we conclude that $y = 2$ is a horizontal asymptote. As an example of a rational function that does not have a horizontal asymptote, consider

$$y = \frac{x^3 + 4}{3x + 2}$$

Dividing the numerator and denominator by x^3, we get

$$y = \frac{1 + 4/x^3}{3/x^2 + 1/x^3}$$

As x becomes infinite, both terms in the denominator approach zero and, therefore, y cannot approach a finite number.

Example 21.   Sketch the graph of $y = \dfrac{3x}{x-2}$.

Solution

Vertical Asymptote. We see that the denominator of the fraction is zero for $x = 2$. Since the denominator and numerator have no common factors, the line $x = 2$ is a vertical asymptote.

Horizontal Asymptote. To find the horizontal asymptote, we divide the numerator and denominator of the given function by x to get

$$y = \frac{3}{1 - 2/x}$$

Now as x increases without bound, the fraction $2/x$ approaches zero and the value of y approaches 3. Therefore, the line $y = 3$ is a horizontal asymptote. Finally, let us determine some additional points in the vicinity of the vertical asymptote as shown in the table. The graph of the function can now readily be plotted (Figure 1.24).

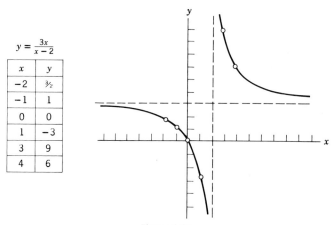

$y = \dfrac{3x}{x-2}$

x	y
-2	$\frac{3}{2}$
-1	1
0	0
1	-3
3	9
4	6

Figure 1.24

Example 22.   Sketch the graph of $y = \dfrac{4x}{x^2-4}$.

Solution

Vertical Asymptote. The graph of this function has two vertical asymptotes. They are the lines $x = 2$ and $x = -2$ since the denominator is zero for both of these values of x.

Horizontal Asymptote. Dividing numerator and denominator by x^2, we have

$$y = \frac{4/x}{1 - 4/x^2}$$

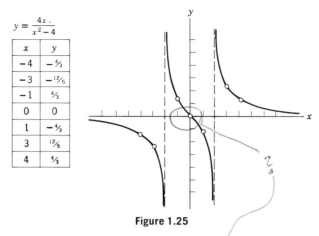

$$y = \frac{4x}{x^2 - 4}$$

x	y
-4	$-\frac{4}{3}$
-3	$-\frac{12}{5}$
-1	$\frac{4}{3}$
0	0
1	$-\frac{4}{3}$
3	$\frac{12}{5}$
4	$\frac{4}{3}$

Figure 1.25

As x increases without bound the numerator approaches zero and the denominator approaches 1. Therefore, y approaches zero which means that the x-axis is the horizontal asymptote. The graph may then be determined by computing a few additional points (Figure 1.25).

From the preceding examples we see that we can get a good approximation of the graph of a rational function by finding its asymptotes and plotting a few additional points.

Example 23.   The heat generated by a certain chemical process varies with time according to

$$H = \frac{t}{t^2 + 4}$$

where H is the heat in calories and t is the time in seconds. Draw the graph of this function and find the maximum heat output.

Solution.   This rational function has no vertical asymptotes since there is no real value of t for which $t^2 + 4 = 0$. However, it does have a horizontal asymptote. Dividing numerator and denominator by t^2, we have

$$H = \frac{1/t}{1 + 4/t^2}$$

In this form, we see that H approaches zero as t increases without bound and, therefore, the t-axis is a horizontal asymptote. Since a negative time has no physical significance, we consider only positive values of t. Plotting a few additional points and drawing the curve so that it approaches zero as t increases, we have the graph in Figure 1.26. We find from the graph that the heat output increases rapidly at first reaching a peak of 0.28 cal at about 2.5 sec. From this point, the heat output decreases gradually toward zero.

$$H = \frac{t}{t^2 + 4}$$

t	H
0	0
2	0.25
4	0.20
6	0.15
8	0.12
10	0.09

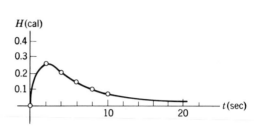

Figure 1.26

EXERCISES

Graph the following rational functions.

1. $y = -\dfrac{1}{x}$ X is not zero.

9. $y = \dfrac{x}{x^2 - 9}$ Va ±3

2. $y = \dfrac{x}{2 - x}$ ha = 2   Va = 1

10. $y = \dfrac{-2x}{x^2 - 4}$

3. $y = \dfrac{x - 3}{x - 4}$ Ha. 4   ha DE

11. $s = \dfrac{t + 2}{t^2 + 2t}$

4. $z = \dfrac{t^2}{t - t}$ ha = 1   va = 0

12. $w = \dfrac{3}{x^2}$

5. $p = \dfrac{w + 1}{w - 1}$ Va   ha NE

13. $y = \dfrac{3}{w^2 - 9w}$ 3/w²   9/w

6. $v = \dfrac{1}{t^2 + 1}$ ha - none   va = 0

14. $r = \dfrac{5z}{3z - z^2}$

7. $s = \dfrac{5/w^2}{4 + w^2}$ Va ±2   x axis

15. $y = \dfrac{1}{p^2 - 3p - 4}$ w = 2   w = -2

8. $m = \dfrac{u + 1}{u^2 - 1}$ ha = 1

16. $n = \dfrac{1}{s^2 - s - 2}$

17. The force between two unit masses varies with the distance between them according to the formula $F = 1/d^2$. The force is in dynes when the distance is measured in centimeters. Show this relationship graphically.

18. Due to leakage, the pressure in a hydraulic system varies with time according to $P = 10/(t^2 + 1)$ psi, where t is the time in seconds. Show this pressure variation graphically.

19. The coefficient of friction μ of a plastic block sliding on an aluminum table is found to vary according to $\mu = (v^2 + 3)/(4v^2 + 5)$ where v is the velocity of the block in cm/sec. Draw the graph of this function. What is the coefficient of friction when the block is static? What is the coefficient of friction when the block is moving very rapidly?

2
Limits

2.1 LIMITS

In the previous chapter we discussed the behavior of a function over its entire domain of definition by referring to its graph. In graphing the functional value at, say $x = a$, a careful examination was not made of what happens to these values close to $x = a$. We just assumed that the function would behave nicely and that its graph could be obtained by drawing a smooth curve through a few known points. The technique of describing the behavior of a function near a given point is one of the really important ideas in mathematics. Indeed, it is this idea which separates calculus from algebra (and also technicians from craftsmen). To give you some idea of what we are driving at, consider the following example.

Example 1. Determine as well as you can the behavior of x^2 for values of x near to 2 (but not at 2 – we know that answer).

Solution. Numerically, this means that we should evaluate x^2 for values of x, say equal to 1.9, 1.99, 1.999, etc. The table below shows the result of these computations.

x	1.9	1.99	1.999	1.9999
x^2	3.61	3.9601	3.9960	3.999+

Now, of course, we should compute x^2 for $x = 2.1$, 2.01, and indeed for many more values. However, practically speaking, we probably have enough information to answer the given question as follows: when x is near 2, then x^2 is near 4. Notice that the answer is intentionally vague – the vagueness stemming from the use of the word "near." A different approach to the problem might be to make a graph of the functional values in order to get an idea of what happens when x is near 2 (Figure 2.1).

You may be wondering why we have given so much discussion to showing that x^2 is near 4, when x is near 2, when it is obvious that $x^2 = 4$ if $x = 2$. The reason is that the behavior of a function near a given point can be studied even though the function is not defined at that point. While the foregoing example represented no real challenge, the following one should make you take notice.

33

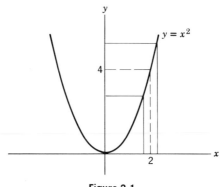

Figure 2.1

Example 2.   Determine the behavior of $(x^2-9)/(x-3)$ when x is near 3, but $x \neq 3$.

Solution.   It is important that you first realize that the given function is not defined at $x = 3$. When $x = 3$ is substituted into $(x^2-9)/(x-3)$, we get the meaningless symbol 0/0. However, while the function is not defined at $x = 3$, it is defined for all other values of x and, therefore, we can investigate its behavior near 3. To proceed in the same way that we did in the foregoing example, we consider the following table of values of x near 3.

			3.0			
x	3.10	3.01	3.001	2.999	2.990	2.900
$\dfrac{x^2-9}{x-3}$	6.10	6.01	6.001	5.999	5.990	5.900

The table seems to show that the nearer x is to 3, the nearer $(x^2-9)/(x-3)$ is to 6. Intuitively, we feel that $(x^2-9)/(x-3)$ can be made as near to 6 as we please by simply taking x to be sufficiently close to 3. The graphical representation of this function is shown in Figure 2.2.

The numerical procedure described in the foregoing examples is tedious and not very practical, but it typifies the analysis involved in determining the behavior of a function near a given point. This procedure, as you may have guessed, is of more than just passing interest. In fact, it is the basis of the ensuing discussion and motivates the following definition.

Definition
We say that "L is the limit of a function $f(x)$ as x approaches a" if, whenever x is close to a, then $f(x)$ is close to L, $x \neq a$. We write $\lim\limits_{x \to a} f(x) = L$.

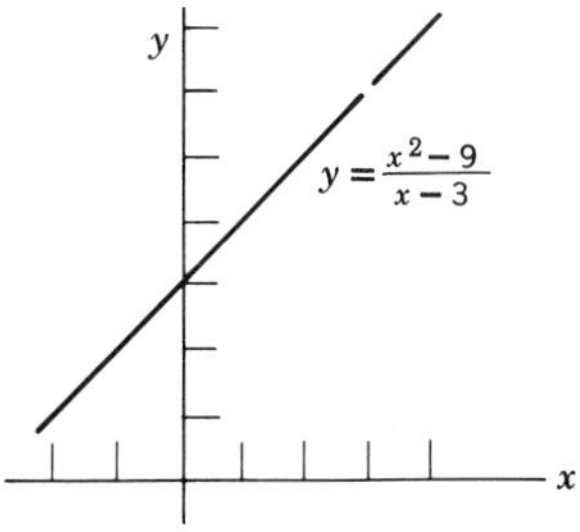

$$y = \frac{x^2 - 9}{x - 3}$$

Figure 2.2

Using this definition and notation, we may rewrite the results of Examples 1 and 2 as follows

$$\lim_{x \to 2} x^2 = 4 \quad \text{and} \quad \lim_{x \to 3} \frac{x^2 - 9}{x - 3} = 6$$

The definition given for the meaning of $\lim_{x \to a} f(x) = L$ is imprecise and informal. It is meant to appeal to your intuitive idea of "closeness." Whereas a precise definition is necessary to insure complete understanding and for dealing with delicate proofs, the definition given here will usually suffice. One point about the definition of which you should take particular note is the fact that the definition rules out considering the value of the function at $x = a$.

Before presenting any examples, we shall state some theorems that will assist us in computing limits. Most of the problems involving limits that we shall encounter can be solved quickly and easily with the aid of the following theorems that we state without proof.

THEOREM 2.1

If $\lim_{x \to a} f(x)$ and $\lim_{x \to a} g(x)$ exist, then the following statements are true.

* (a) $\lim_{x \to a} cf(x) = c \lim_{x \to a} f(x)$, where c is any constant.

   READ.
   UTILIZE.
   KNOW!

* (b) $\lim_{x \to a} [f(x) + g(x)] = \lim_{x \to a} f(x) + \lim_{x \to a} g(x)$

* (c) $\lim_{x \to a} f(x)g(x) = \lim_{x \to a} f(x) \lim_{x \to a} g(x)$

* (d) $\lim_{x \to a} \frac{f(x)}{g(x)} = \frac{\lim_{x \to a} f(x)}{\lim_{x \to a} g(x)}$ provided $\lim_{x \to a} g(x) \neq 0$

Students often tend to use this theorem incorrectly because they fail to read the hypothesis.

Example 3.   Find $\lim\limits_{x\to 0} (x+3)(x-2)$.

Solution.   Applying Theorem 2.1, we have

$$\lim_{x\to 0} (x+3)(x-2) = \lim_{x\to 0} (x+3) \lim_{x\to 0} (x-2)$$

We see from this that when x is near 0, $(x+3)$ is near 3, $(x-2)$ is near -2, and the product $(x+3)(x-2)$ is near -6.

Hence

$$\lim_{x\to 0} (x+3)(x-2) = \lim_{x\to 0} (x+3) \lim_{x\to 0} (x-2) = (3)(-2) = -6$$

Example 4.   Find $\lim\limits_{x\to 4} \dfrac{3x^2-x}{\sqrt{x}}$.

Solution.   Applying Theorem 2.1, we have

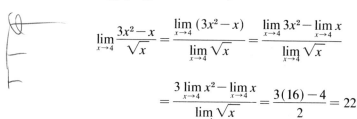

$$\lim_{x\to 4} \frac{3x^2-x}{\sqrt{x}} = \frac{\lim\limits_{x\to 4}(3x^2-x)}{\lim\limits_{x\to 4}\sqrt{x}} = \frac{\lim\limits_{x\to 4}3x^2 - \lim\limits_{x\to 4}x}{\lim\limits_{x\to 4}\sqrt{x}}$$

$$= \frac{3\lim\limits_{x\to 4}x^2 - \lim\limits_{x\to 4}x}{\lim\limits_{x\to 4}\sqrt{x}} = \frac{3(16)-4}{2} = 22$$

Example 5.   Show that $\lim\limits_{x\to 3} (x^2-9)/(x-3) = 6$.

Solution.   If we attempt to apply Theorem 2.1d to this problem, we obtain

$$\lim_{x\to 3} \frac{x^2-9}{x-3} = \frac{\lim\limits_{x\to 3}(x^2-9)}{\lim\limits_{x\to 3}(x-3)} = \frac{0}{0}$$

which in itself is a meaningless symbol. Our first inclination is to conclude that the limit does not exist. However, further analysis of the limit reveals this conclusion incorrect. To solve this particular problem, we use the fact that $(x^2-9)/(x-3)$ is defined for all values of x, except $x=3$. Therefore, as long as we exclude $x=3$ from our consideration, we can write

$$\frac{x^2-9}{x-3} = \frac{(x-3)(x+3)}{x-3} = x+3; \quad x \neq 3$$

The desired limit is then given by

$$\lim_{x \to 3} \frac{x^2 - 9}{x - 3} = \lim_{x \to 3} (x+3) = 6$$

Keep in mind that in finding such a limit we are not finding the value of $(x^2-9)/(x-3)$ when $x = 3$, but the value approached by $(x^2-9)/(x-3)$ as x approaches arbitrarily close to 3.

The definition of limit is stated for the case of x approaching a finite number a. The definition may also be used to describe the behavior of a function as x increases (or decreases) without bound.

Definition
We say that "L is the limit of a function $f(x)$ as x increases (or decreases) without bound," if $f(x)$ is near to L when x is very large. We denote this by $\lim_{x \to \infty} f(x) = L$.

Under no circumstances should you think of infinity as the representative of a large number. The symbol ∞ is defined within the context of x becoming large. We have not, and we shall not define ∞ as a number.

Theorem 2.1 was stated for the case of x approaching a. An analogous theorem is valid for the case where x becomes arbitrarily large, that is, as $x \to \infty$.

Example 6.   Find $\lim_{x \to \infty} 1/x$. (See Example 20 in the previous chapter.)

Solution.   The larger the value of x in the fraction $1/x$, the closer the value of the fraction to zero. Since we can make the value of $1/x$ approach arbitrarily close to zero, we conclude that $\lim_{x \to \infty} (1/x) = 0$.

Example 7.   Find

$$\lim_{x \to \infty} \frac{x^2 + 5x + 3}{4x - 3x^2}$$

Solution.   The analogue to Theorem 2.1d is not applicable since neither the limit of the numerator or denominator exists. We therefore attempt algebraic simplification again. The trick we try is similar to that in finding horizontal asymptotes, that is divide numerator and denominator by x^2 to obtain

$$\lim_{x \to \infty} \frac{x^2 + 5x + 3}{4x - 3x^2} = \lim_{x \to \infty} \frac{1 + \dfrac{5}{x} + \dfrac{3}{x^2}}{\dfrac{4}{x} - 3} = \frac{1+0+0}{0-3} = -\frac{1}{3}$$

The technique used in Example 7 is applicable to all rational functions. It is left for you to show that, if the degree of $f(x)$ and $g(x)$ are equal, then

$$\lim_{x \to \infty} \frac{f(x)}{g(x)} = \frac{\text{coefficient of highest order term of } f(x)}{\text{coefficient of highest order term of } g(x)}$$

EXERCISES

Find the indicated limits.

1. $\lim\limits_{x \to \infty} \dfrac{3x}{2x+2}$

2. $\lim\limits_{t \to 0} \dfrac{t-3}{t^2+t+4}$

3. $\lim\limits_{t \to 0} \dfrac{t^3 - 3t^2 + 4t}{t^2 + 2t}$

4. $\lim\limits_{x \to \infty} \dfrac{x^3 + 2x}{x^4 + 5x^2 + 3}$

5. $\lim\limits_{x \to \infty} \dfrac{7 + x^2}{3 + x + 3x^2}$

6. $\lim\limits_{i \to 0} \dfrac{4i+5}{3i^3 + 4i^2}$

7. $\lim\limits_{x \to 3} \dfrac{x^2 - 2x - 3}{x - 3}$

8. $\lim\limits_{x \to \infty} \dfrac{ax^2 + bx + c}{e + dx^2}$

9. $\lim\limits_{w \to 2} \dfrac{w^2 + w - 6}{w^2 - 4}$

10. $\lim\limits_{x \to 4} \dfrac{\sqrt{x} - 2}{x - 4}$

11. $\lim\limits_{x \to 1} \dfrac{\sqrt{x} - 1}{x - 1}$

12. $\lim\limits_{x \to 1} \dfrac{x^3 - 1}{x - 1}$

13. $\lim\limits_{x \to 1} \dfrac{\dfrac{1}{x} - 1}{x - 1}$

14. $\lim\limits_{x \to \frac{1}{2}} \dfrac{\dfrac{2}{x} - 4}{x - \frac{1}{2}}$

2.2 CONTINUITY

When sketching the graph of a function by very elementary methods, you are told to plot a few points and then connect the points by a nice *continuous, connected* set of points. The reason you were able to do this is because most functions in practice exhibit these properties of "continuity" or "smoothness." The behavior of a function at a point is usually described by one of the following graphical characteristics.

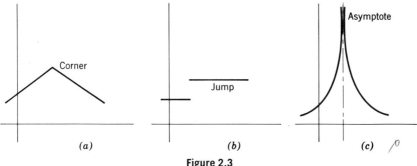

Figure 2.3

(a) The graph has a "corner" (see Figure 2.3*a*).
(b) The graph has a "jump" (see Figure 2.3*b*).
(c) The graph has a vertical asymptote (see Figure 2.3*c*).

None of the functions graphed in Figure 2.3 are "smooth," but the one with a "corner" does exhibit the property of being the graph of a "continuous" set of points. Intuitively, we say a function is continuous if we can sketch its graph without removing our pencil from the paper as we draw it. Points where a function is not continuous are said to be points of *discontinuity*. Points at which there are "jumps," or points at which there is a vertical asymptote, are examples of points of discontinuity. Now see if you can point out the discontinuities in the graphs of Figure 2.3.

An analytic formulation of the property of continuity is best stated using the idea of a limit.

Definition
A function $f(x)$ is said to be continuous at $x = a$ if the function is defined at $x = a$ and if

$$\lim_{x \to a} f(x) = f(a)$$

An alternate definition that is often used is as follows.

Alternate Definition
A function $f(x)$ is said to be continuous at $x = a$ if the function is defined at $x = a$ and if

$$\lim_{\Delta x \to 0} f(a + \Delta x) = f(a)$$

If you have either of these definitions clearly in mind, it should be obvious that a function is better "behaved" if it is continuous at a point than if it is discontinuous.

Example 8. Is $f(x) = x^3 - 3x$ continuous at $x = 2$?

Solution. The answer is *yes*, since $f(2) = 8 - 6 = 2$, and $\lim_{x \to 2} f(x) = 2$.

Example 9. Let $f(x) = x - 2; \; x \neq 0$. Examine for continuity at $x = 0$.

$$= 2; \qquad x = 0$$

Solution. In this case, $f(0) = 2$ by definition of the function. But $\lim_{x \to 0} f(x) = \lim_{x \to 0} (x - 2) = 0 - 2 = -2$. Therefore, this function is discontinuous at $x = 0$ (Figure 2.4).

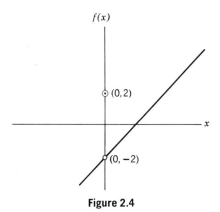

Figure 2.4

Example 10. Let $f(x) = (x^2 - 9)/(x - 3)$ (see Example 5). Examine for continuity at $x = 3$.

Solution. Example 5 showed us that the value of $\lim_{x \to 3} (x^2 - 9)/(x - 3) = 6$. But the function is undefined at $x = 3$. Therefore, this function is discontinuous at $x = 3$.

Example 11. Examine for points of discontinuity:

$$f(x) = \frac{1}{x^2}, \quad \text{if } x \neq 0$$

$$= 1, \quad \text{if } x = 0$$

Solution. The graph of the function is shown in Figure 2.5. From the graph it is easy to see that the function is "unbounded" in the neighborhood of zero, even though $f(0)$ exists. Therefore, the function is discontinuous at $x = 0$.

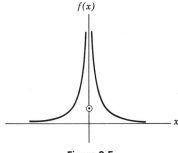

Figure 2.5

EXERCISES

Investigate the continuity of the indicated functions at the given point. In each case, sketch the graph of the function.

1. $y = x^2$ at $x = 2$

2. $f(x) = x^2 - 3x$ at $x = 0$

3. $y = \dfrac{1}{x-2}$ at $x = 2$

4. $y = \dfrac{1}{x-2}$ at $x = 1$

5. $y = x;\ x \neq 0$ at $x = 0$
   $= 1;\ x = 0$

6. $y = x^2;\ x \neq 1$ at $x = 1$
   $= 2;\quad x = 1$

7. $f(x) = \dfrac{x^2 - 4}{x-2};\ x \neq 2$ at $x = 2$
   $= 4;\qquad x = 2$

8. $y = \dfrac{1}{x};\ x \neq 0$ at $x = 0$
   $= 4;\ x = 0$

2.3   THE TANGENT PROBLEM

It is possible to define the *tangent line* to a circle at a point P on the circle as the line drawn perpendicular to the radius of the circle at P (Figure 2.6). Although this definition gives us an intuitive notion of what

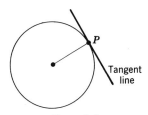

Figure 2.6

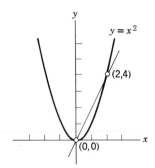

Figure 2.7

is meant by a tangent line, it is restricted to circles. Our purpose in this section is to motivate a definition for a tangent line that is applicable to more general curves. Historically, the problem of constructing a tangent to a curve at a given point was one of the problems that led to the formulation of differential calculus. While this problem may seem to be of only minor importance, many of the problems of science and technology are in some way related to the tangent problem.

The Secant Line to a Curve.   A straight line drawn between two points on a curve is called a *secant line* to the curve. For example, the secant line to the curve $y = x^2$ between $(0,0)$ and $(2,4)$ is shown in Figure 2.7. The determination of the slope of the secant line is easy, since we are given two points on the line. Thus the slope of the secant line in Figure 2.7 is

$$m_{\text{sec}} = \frac{y_2 - y_1}{x_2 - x_1} = \frac{4 - 0}{2 - 0} = 2$$

The Tangent Line to a Curve.   Let us now discuss the problem of finding the slope of a line tangent to a curve at a given point, say the curve $y = x^2$ at the point $(2,4)$. (See Figure 2.8.)

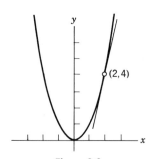

Figure 2.8

One approach to this problem is to solve it graphically by carefully drawing the line tangent to the curve at (2,4) and then estimating its slope, but this is both cumbersome and imprecise. Another approach consists of using a secant line to *estimate* the slope of the tangent line. Thus, in the indicated problem, we can estimate the slope of the tangent at (2,4) by using this point and some other point on the curve to compute the slope of a secant line. For example, one estimate of the slope of the tangent line at (2,4) is the slope of the secant line between (2,4) and (2.5, 6.25). Numerically, we have

$$m_{\text{sec}} = \frac{x^2 - x_0{}^2}{x - x_0} = \frac{6.25 - 4.00}{2.5 - 2.0} = 4.5$$

The indicated secant line is shown in Figure 2.9.

We can also estimate the slope of the tangent at (2,4) by choosing a point to the left of the given point, for example (1.5,2.25). Using (1.5,2.25) and (2,4) in the equation for the slope of a secant line, our estimate of the slope of the tangent at (2,4) is

$$m_{\text{sec}} = \frac{2.25 - 4.00}{1.5 - 2.0} = \frac{-1.75}{-0.5} = 3.5$$

These computations and Figure 2.9 show the slope of the tangent line at (2,4) somewhere between 3.5 and 4.5. (It might be 3.7, or 4.0, or 4.3, or any other number between 3.5 and 4.5.) We suspect that a better estimate would be obtained if we would choose our points closer to (2,4). The following table will help you visualize what happens to the slope of the

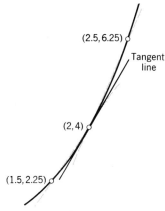

Figure 2.9

secant line through (2,4) as we choose points on the curve closer and closer to this point.

2.00

x	1.50	1.90	1.99	2.01	2.10	2.50
$m_{\text{sec}} = \dfrac{x^2 - 4}{x - 2}$	3.5	3.9	3.99	4.01	4.1	4.5

The table indicates that the slope of the tangent line at (2,4) is between 3.99 and 4.01. Any number in this interval will be a good estimate of the slope of the tangent at (2,4). Later we will find its exact value.

The above discussion is not only a description of an important means of estimating the slope of a tangent line to a curve at a given point, but it is also a prelude to a precise definition of the slope of a tangent line. If we have a precise definition of the slope of a tangent line to a curve at a given point, we will have no problem in constructing the tangent line. Indeed, it will even be easy to write the equation of the line. (Figures 2.10*a* and 2.10*b*.)

Suppose that the equation of the curve in Figure 2.10*a* is $y = g(x)$ and let it be required to find the slope of the tangent line to this curve at *P*. We construct a secant line through *P* and another point on the curve, say *Q*. The slope of the secant line through *P* and *Q* is then

$$m_{\text{sec}} = \frac{g(x) - g(x_0)}{x - x_0} \qquad (1)$$

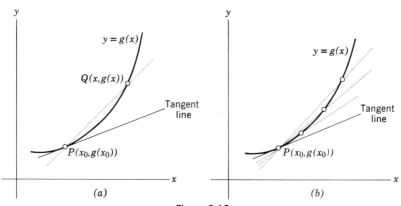

(a)   (b)

Figure 2.10

Now if we move Q along the curve toward P, the secant line rotates about P (see Figure 2.10b). As Q approaches P the slope of the secant line approaches that of the tangent line at P, that is

$$m_{\text{tan}} = \lim_{Q \to P} m_{\text{sec}} = \lim_{Q \to P} \frac{g(x) - g(x_0)}{x - x_0} \qquad (2)$$

Since $Q \to P$ implies $x \to x_0$, this result can also be written

$$m_{\text{tan}} = \lim_{x \to x_0} \frac{g(x) - g(x_0)}{x - x_0} \qquad (3)$$

alt method

From the foregoing results, it seems reasonable to *define* the slope of a tangent line as the limit of m_{sec}. The precise definition of the slope of a tangent line to a curve is given below.

Definition

The slope $m(x_0)$ of the tangent line to the graph of the function $y = g(x)$ at the point $P(x_0, y_0)$ is

$$m(x_0) = \lim_{x \to x_0} \frac{g(x) - g(x_0)}{x - x_0}$$

SEE NOTES

provided that this limit exists.

Example 12.   Find the slope of the tangent line to the graph of $y = x^2$ at $P(2,4)$. See Figure 2.8.

Solution.   The slope at $(2,4)$ can be found using the above definition with $x_0 = 2$. Thus

$$m(2) = \lim_{x \to 2} \frac{g(x) - g(2)}{x - 2}$$

Since $g(x) = x^2$ and $g(2) = 4$, we have

$$m(2) = \lim_{x \to 2} \frac{x^2 - 4}{x - 2}$$

$$= \lim_{x \to 2} \frac{(x - 2)(x + 2)}{x - 2}$$

$$= \lim_{x \to 2} (x + 2) = 4$$

Example 13.   Find the equation of the tangent line to the curve $y = x^2$ at the point $(2,4)$.

Solution.   From the preceeding problem, the slope of the tangent line to this curve at (2,4) is $m(2) = 4$. The equation of the tangent line may then be found by using the point-slope form of a straight line with $m = 4$ and $(x_1,y_1) = (2,4)$. This gives

$$y - 4 = 4(x - 2)$$

$$4x - y = 4$$

Example 14.   Find the slope of the tangent line to the curve in Example 12 at $x_0 = 1$ (Figure 2.11).

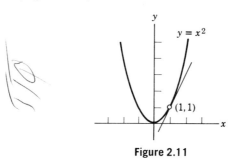

Figure 2.11

Solution.   Letting $x_0 = 1$ in the definition, we have

$$m(1) = \lim_{x \to 1} \frac{g(x) - g(1)}{x - 1}$$

Then $g(x) = x^2$ and $g(1) = 1$, therefore

$$m(1) = \lim_{x \to 1} \frac{x^2 - 1}{x - 1}$$

$$= \lim_{x \to 1} \frac{(x - 1)(x + 1)}{x - 1}$$

$$= \lim_{x \to 1} (x + 1) = 2$$

EXERCISES

1. Using a convenient scale draw the graph of $y = x^2$, and as carefully as you can, draw a tangent line at (1,1). Estimate the slope of the tangent line. Use the grid of the graph paper to obtain this estimate.

2. Draw the graph of $y = \frac{1}{2}x^2$, and as carefully as you can, draw a tangent line at $(3, \frac{9}{2})$. Estimate the slope of this line as you did in Exercise 1.

3. Estimate the slope of the tangent line in Exercise 1 by using $x_0 = 1$ and $x = 1.1$ in the formula for m_{sec}. What estimate do you get if you use $x = 0.9$ and $x_0 = 1$ in the formula for m_{sec}?

4. Using the results of Exercise 3, complete the statement. The slope of the tangent line to $y = x^2$ at $(1,1)$ is greater than _____ and less than _____. How does your estimate in Exercise 1 compare?

 5. Estimate the slope of the tangent line in Exercise 2 by using $x_0 = 3$ and $x = 3.01$ in the formula for m_{sec}. What estimate do you get if you use $x_0 = 3$ and $x = 2.99$?

6. How does your estimate in Exercise 2 compare with your estimates in Exercise 5?

Draw the graph of each of the following equations, and compute the slope $m(x_0)$ of the tangent line at the indicated point by using the definition of $m(x_0)$.

7. $y = 3x^2$ at $x_0 = -1$
8. $y = 3x^2$ at $x_0 = 2$
9. $y = x^2 + 5$ at $x_0 = 1$
10. $y = 9 - x^2$ at $x_0 = -4$
11. $y = x^2 - 2x$ at $x_0 = 0$
12. $g(x) = x - 4x^2$ at $x_0 = \frac{1}{4}$

13. $y = x^2 - 6x + 5$ at $x_0 = 3$
14. $y = x^2 + 2x + 1$ at $x_0 = 0$
15. $y = x^3$ at $x_0 = 2$
16. $y = \frac{1}{2}x^3$ at $x_0 = 1$
17. $y = 1/x$ at $x_0 = 2$
18. $y = 1/x^2$ at $x_0 = 1$

19. Write the equation of the tangent line in Exercise 7.
20. Write the equation of the tangent line in Exercise 9.
21. Find the equation of the line that is tangent to the curve $y = x^2$, and is parallel to the line $y = 2x$.

2.4 THE RATE PROBLEM

An idea seemingly unrelated to the tangent problem (but mathematically identical) is rate of change, a concept used frequently in everyday life as well as in science and technology. The terminology of rates is part of ordinary language describing such things as velocity (miles per hour), density (pounds per cubic inch), and cost of steel (dollars per ton). You may be surprised to learn that concepts such as electric current and power also fit into the general category of rates. Possibly the most familiar rate is that of velocity and, therefore, we approach the rate problem by first discussing the velocity of a particle moving in a straight line.

Instantaneous Velocity. The motion of a particle may be defined as the continuous change in the position of the particle. The ratio of the distance traveled by the particle to the elapsed time is then called the *average velocity* of the particle and is denoted by

$$v_{avg} = \frac{\Delta s}{\Delta t} \tag{4}$$

Figure 2.12

To illustrate the use of this definition, let us consider the particle shown in Figure 2.12, whose displacement varies with time according to $s = 3t^2$ ft. After 1 sec the particle has traveled 3 ft, after 2 sec it has traveled 12 ft, after 3 sec it has traveled 27 ft, etc. We can find the velocity between any two successive times by dividing the distance traveled by the elapsed time. Thus, the velocity of the particle from $t = 1$ to $t = 2$ is

$$v_{avg} = \frac{12 - 3}{2 - 1} = 9 \text{ ft/sec}$$

Now consider the problem of finding the velocity of the particle at some instant, say $t = 2$ sec. We call this the *instantaneous velocity* and denote it by $v(2)$. Equation 4 fails, in this case, because the elapsed time Δt is equal to zero. However, the equation is valid for any $\Delta t > 0$. With this in mind, we can use Equation 4 to estimate the instantaneous velocity at $t = 2$ sec. We have already seen that the average velocity of the particle from $t = 1$ to $t = 2$ sec is 9 ft/sec. Similarly, its velocity from $t = 2$ to $t = 3$ sec is

$$v_{avg} = \frac{27 - 12}{3 - 2} = 15 \text{ ft/sec}$$

The instantaneous velocity is therefore greater than 9 ft/sec and less than 15 ft/sec. We can improve our estimate of the instantaneous velocity by choosing smaller time intervals about $t = 2$ sec. From the following table, we see that the velocity at $t = 2$ is between 11.7 and 12.3 ft/sec.

				2 sec			
t	1	1.5	1.9	2.1	2.5	3.0
$v_{avg} = \dfrac{3t^2 - 12}{t - 2}$	9	10.5	11.7	12.3	13.5	15

The foregoing results suggest that we can make the average velocity come as close as we please to the instantaneous velocity at $t = 2$ by

LIKELY

choosing t sufficiently close to 2. We make this notion more precise in the following definition.

Definition

If the displacement of a particle varies with time according to the formula $s = s(t)$, then the instantaneous velocity of the particle at time t_0 is given by

$$v(t_0) = \lim_{t \to t_0} \frac{s(t) - s(t_0)}{t - t_0}$$

The symbol $v(t_0)$ is used to indicate the velocity at the time $t = t_0$.

The instantaneous velocity of the above particle at $t = 2$ can now be found. Thus,

$$v(2) = \lim_{t \to 2} \frac{3t^2 - 12}{t - 2} = \lim_{t \to 2} \frac{3(t^2 - 4)}{t - 2}$$

$$= \lim_{t \to 2} \frac{3(t - 2)(t + 2)}{t - 2} = \lim_{t \to 2} 3(t + 2)$$

$$= 12 \text{ ft/sec.}$$

Example 15. An object is thrown vertically upward with a velocity of 64 ft/sec. The height of the object at any time is shown in physics to be given by the formula $h = 64t - 16t^2$. What is the velocity of the object 1 sec after it is thrown upward?

Solution. Applying the definition of instantaneous velocity with $t_0 = 1$, we have

$$v(1) = \lim_{t \to 1} \frac{s(t) - s(1)}{t - 1}$$

Here, $s(t) = 64t - 16t^2$ and $s(1) = 64 - 16 = 48$. Hence

$$v(1) = \lim_{t \to 1} \frac{64t - 16t^2 - 48}{t - 1}$$

$$= \lim_{t \to 1} \frac{-16(t^2 - 4t + 3)}{t - 1}$$

$$= \lim_{t \to 1} \frac{-16(t - 3)(t - 1)}{t - 1}$$

$$= \lim_{t \to 1} -16(t - 3) = 32 \text{ ft/sec}$$

Electric Current.   The approach used to find the instantaneous velocity of a particle is typical of rate problems in general; that is, any problem involving an instantaneous rate can be solved in the same way we solved the velocity problem. As an example, the electric current in a circuit is the time rate of transferring electric charge past a point in the circuit. If i denotes current, q charge, and t time, the average current over some time interval $\Delta t = t - t_0$ is

$$i_{avg} = \frac{\Delta q}{\Delta t} = \frac{q - q_0}{t - t_0} \tag{5}$$

The unit of current is the *ampere*, when the charge is measured in *coulombs* and the time in *seconds*. We can approximate the current at any instant t_0 by computing the average current over a very small time interval about t_0. Concluding that we can make the average current as close as we please to the instantaneous current by taking t sufficiently close to t_0, we have the following definition of current.

Definition

If the electric charge transferred through a circuit varies with time according to the formula $q = h(t)$, then the instantaneous current in the circuit at time t_0 is given by

$$i(t_0) = \lim_{t \to t_0} \frac{q - q_0}{t - t_0} = \lim_{t \to t_0} \frac{h(t) - h(t_0)}{t - t_0}$$

Example 16.   The electric charge in a circuit varies with time according to the formula $q = 3t^2$ coul. What is the current in the circuit $\frac{1}{2}$ sec after the switch is closed?

Solution.   By the definition

$$i(\tfrac{1}{2}) = \lim_{t \to \frac{1}{2}} \frac{h(t) - h(\tfrac{1}{2})}{t - \tfrac{1}{2}}$$

We have $h(t) = 3t^2$, so that $h(\tfrac{1}{2}) = \tfrac{3}{4}$, and

$$i(\tfrac{1}{2}) = \lim_{t \to \frac{1}{2}} \frac{3t^2 - \tfrac{3}{4}}{t - \tfrac{1}{2}}$$

$$= \lim_{t \to \frac{1}{2}} \frac{3(t^2 - \tfrac{1}{4})}{t - \tfrac{1}{2}}$$

$$= \lim_{t \to \frac{1}{2}} \frac{3(t - \tfrac{1}{2})(t + \tfrac{1}{2})}{t - \tfrac{1}{2}}$$

$$= \lim_{t \to \frac{1}{2}} 3(t + \tfrac{1}{2}) = 3 \text{ amp}$$

EXERCISES

1. The displacement-time equation of an object is $s = 3t^2 - t$ feet. Estimate the velocity of the object when $t_0 = 2$ sec by computing v_{avg} between $t_0 = 2.0$ and $t = 2.5$.
2. Use v_{avg} between $t_0 = 2.0$ and $t = 1.5$ sec to estimate the velocity of the object in Exercise 1.
3. Using the results of Exercises 1 and 2, complete the statement $v(t_0)$ is greater than _____ and less than _____.
4. Make two more estimates of the velocity of the object at $t_0 = 2$ sec by using points 0.1 sec to either side of t_0. You may now say that $v(t_0)$ is greater than _____ and less than _____.

In Exercises 5–9, find the velocity $v(t_0)$ at the given instant if the distance-time equation is as follows.

5. $s = 6 + 9t; t_0 = 2$ sec   8. $s = 3t^2/2; t_0 = 3$
6. $s = t^2 + 2t; t_0 = 1$ 9. $s = 1/t^2; t_0 = 2$
7. $s = t^2 + 2t; t_0 = 2$

In Exercises 10–14, find the current $i(t_0)$ if the charge-time equation is given.

10. $q = 3t^2 - t; t_0 = 2$ sec   13. $q = 2t^3; t_0 = 1$
11. $q = 3t^2 - t + 1; t_0 = 1$ 14. $q = 5/t; t_0 = 0.5$
12. $q = t^2; t_0 = 0.1$

2.5   THE DERIVATIVE OF A FUNCTION

In the three previous sections we have defined m_{tan}, $v(t_0)$, and $i(t_0)$. Let us write all three definitions together, so that their similarity can be seen by inspection.

Slope of a curve.

$$m(x_0) = \lim_{x \to x_0} \frac{g(x) - g(x_0)}{x - x_0}$$

Velocity of a particle.

$$v(t_0) = \lim_{t \to t_0} \frac{s(t) - s(t_0)}{t - t_0}$$

Electric current.

$$i(t_0) = \lim_{t \to t_0} \frac{h(t) - h(t_0)}{t - t_0}$$

The three are obviously identical except for notational differences. We say that the three are abstractly identical or mathematically indistinguishable. Thus, instead of studying tangents to curves, velocities of particles, electric current, and many other problems separately, we study the one unifying thread of all these seemingly different problems.

Definition

The *difference quotient* of the function $f(x)$ at x_0 is the ratio of the change in the functional values to the change in x; that is,

$$\frac{\Delta y}{\Delta x} = \frac{f(x) - f(x_0)}{x - x_0} \tag{6}$$

Definition

If the limit of the difference quotient exists as x approaches x_0, then the value of the limit is called the *derivative* of $f(x)$ at x_0, and is denoted by $f'(x_0)$, that is,

$$f'(x_0) = \lim_{x \to x_0} \frac{f(x) - f(x_0)}{x - x_0} \tag{7}$$

An alternate definition that is often used is

$$f'(x_0) = \lim_{\Delta x \to 0} \frac{f(x_0 + \Delta x) - f(x_0)}{\Delta x} \tag{8}$$

where $\Delta x = x - x_0$. (Notice that $\Delta x \to 0$ as $x \to x_0$.)

From the discussion prior to the definition, you should see that the derivative could stand for any number of physical quantities. One of the purposes of this book is to put you at ease in calculating the particular limit which we have called the derivative.

The derivative of a function $f(x)$ is defined at all those points for which the limit of the difference quotient exists. Consequently, we have a new function defined; a function derived from $f(x)$ by evaluating the limit of the difference quotient at each point. The derived function is denoted by $f'(x)$; it is called the *derivative function* of $f(x)$ or, simply, the *derivative* of $f(x)$. Practically, the formula for $f'(x)$ is the formula for $f'(x_0)$ with x_0 replaced by x. The function $f(x)$ is said to be *differentiable* at those points where the limit of the difference quotient exists. You should be able to convince yourself that the functions $f'(x)$ and $f(x)$ are defined at the same points except for the points of nondifferentiability.

Example 17. Find the derivative of $y = x^2$.

Solution. Here, $f(x) = x^2$; therefore

$$f(x + \Delta x) = (x + \Delta x)^2 = x^2 + 2x(\Delta x) + (\Delta x)^2$$

at least 5-
10 pts.

The derivative is then

$$f'(x) = \lim_{\Delta x \to 0} \frac{x^2 + 2x(\Delta x) + (\Delta x)^2 - x^2}{\Delta x}$$

$$= \lim_{\Delta x \to 0} \frac{2x(\Delta x) + (\Delta x)^2}{\Delta x}$$

$$= \lim_{\Delta x \to 0} (2x + \Delta x) = 2x$$

Example 18. Find the slope of the curve $y = x^2$ at $x = -\frac{2}{3}$ (see Figure 2.13).

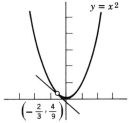

$$\left(-\frac{2}{3}, \frac{4}{9}\right)$$

Figure 2.13

Solution.   From the preceeding example, the derivative of the given function is $f'(x) = 2x$. The slope of the curve at $x = -\frac{2}{3}$ is then given by $f'(-\frac{2}{3}) = 2(-\frac{2}{3}) = -\frac{4}{3}$.

Example 19.   The displacement of a car varies with time according to the formula $s = t^2 - 2t$. Find a formula for the instantaneous velocity of the car.

Solution

$$v = \lim_{\Delta t \to 0} \frac{f(t + \Delta t) - f(t)}{\Delta t}$$

$$= \lim_{\Delta t \to 0} \frac{[(t + \Delta t)^2 - 2(t + \Delta t)] - [t^2 - 2t]}{\Delta t}$$

$$= \lim_{\Delta t \to 0} \frac{t^2 + 2t(\Delta t) + (\Delta t)^2 - 2t - 2(\Delta t) - t^2 + 2t}{\Delta t}$$

$$= \lim_{\Delta t \to 0} \frac{2t(\Delta t) + (\Delta t)^2 - 2(\Delta t)}{\Delta t}$$

$$= \lim_{\Delta t \to 0} (2t + \Delta t - 2) = 2t - 2$$

Example 20.   Find a formula for the current in a circuit if the charge transferred varies with time according to $q = 1/(t+1)$.

Solution

$$i = \lim_{\Delta t \to 0} \frac{f(t+\Delta t) - f(t)}{\Delta t}$$

Here $f(t) = \dfrac{1}{t+1}$, and $f(t+\Delta t) = \dfrac{1}{t+\Delta t+1}$. The formula for the current is then given by

$$i = \lim_{\Delta t \to 0} \frac{\dfrac{1}{t+\Delta t+1} - \dfrac{1}{t+1}}{\Delta t}$$

$$= \lim_{\Delta t \to 0} \frac{\dfrac{(t+1) - (t+\Delta t+1)}{(t+1)(t+\Delta t+1)}}{\Delta t}$$

$$= \lim_{\Delta t \to 0} \frac{-1}{(t+1)(t+\Delta t+1)} = \frac{-1}{(t+1)(t+1)} = \frac{-1}{(t+1)^2}$$

Example 21.   What is the current in the circuit in Example 20 when $t = \frac{1}{2}$ sec?

Solution.   The current at time t is given by $i = -1/(t+1)^2$. Letting $t = \frac{1}{2}$ in this formula, we get

$$i = \frac{-1}{(\frac{1}{2}+1)^2} = \frac{-1}{\frac{9}{4}} = -\frac{4}{9} = -0.444 \text{ amps}$$

What is the significance of the negative sign?

EXERCISES

Use Equation 8 to find the derivative function of each of the following functions.

1. $y = \frac{1}{3}x^2$
2. $f(x) = 5x^2$
3. $y = x^2 + 3x$
4. $y = 2x - x^2$
5. $y = 7 + 5x + 3x^2$
6. $y(x) = -1/x$
7. $y = 3/x$
8. $f(x) = 2/x^2$
9. $f(x) = x^3$
10. $y = x^2 - 2x + 3$
11. $y = x^3 - x^2$
12. $y = 1/(x+5)$

Use the results obtained in Exercises 1–12 to assist you in solving the remaining problems.

13. Find the slope of the curve $y = \frac{1}{3}x^2$ at: (a) $x = -1$; (b) $x = 0$; and (c) $x = 1$.
14. Find the slope of the curve $y = 7 + 5x + 3x^2$ at: (a) $x = \frac{1}{2}$; (b) $x = 1$; and (c) $x = 2$.
15. The equation of motion of a particle along a straight line varies in accordance with $s = t^2 + 3t$ ft. What is the velocity of the particle when (a) $t = 0.1$, and (b) $t = 0.5$ sec?
16. The charge transferred in a circuit varies with time as $q = 3/t$ coul. What current flows in the circuit when (a) $t = 0.5$ sec and (b) $t = 1.5$ sec?

3
Differentiation Formulas and Further Applications

3.1 INTRODUCTION

In the previous chapter we defined the derivative of a function after observing that the same limit appeared in several problems that were quite different in application. In this chapter we shall show some additional applications of the derivative. Before we do that, however, we shall develop formulas for the derivatives of certain types of functions. Our interest at this time will be restricted to power functions and polynomials, although some of the formulas that we shall develop apply to a much larger class of functions.

It is convenient at this time to introduce another widely used notation for the derivative of $f(x)$. The symbol d/dx is called the derivative operator; it indicates that the function written after it is to be differentiated with respect to x. Thus

$$\frac{d}{dx}[f(x)] = f'(x) \tag{1}$$

Notice also that the expression on the left is sometimes written df/dx with the understanding that f is a function of x.

3.2 LINEARITY PROPERTY

The first of our results is a very general property called the linearity property.

THEOREM 3.1

Let du/dx and dv/dx exist over an interval $[a, b]$. Then,

(a) $\dfrac{d}{dx}[cu(x)] = c\dfrac{d}{dx}[u(x)] = c\dfrac{du}{dx}$ \hfill (2)

(b) $\dfrac{d}{dx}[u(x) + v(x)] = \dfrac{d}{dx}[u(x)] + \dfrac{d}{dx}[v(x)] = \dfrac{du}{dx} + \dfrac{dv}{dx}$ \hfill (3)

56

We shall prove this theorem, not because we wish you to memorize proofs of theorems, but because we wish you to see the generality of the statement of the theorem. The letters u and v represent differentiable functions, although we may know nothing about them.

Proof
(a) Let $f(x) = cu(x)$. Then

$$f(x+\Delta x) - f(x) = cu(x+\Delta x) - cu(x) = c[u(x+\Delta x) - u(x)]$$

Since

$$f'(x) = \lim_{\Delta x \to 0} \frac{f(x+\Delta x) - f(x)}{\Delta x}$$

we have

$$= \lim_{\Delta x \to 0} \frac{c[u(x+\Delta x) - u(x)]}{\Delta x}$$

$$= c\frac{du}{dx}$$

(b) Let $f(x) = u(x) + v(x)$. Then

$$f(x+\Delta x) - f(x) = u(x+\Delta x) + v(x+\Delta x) - u(x) - v(x)$$

and, therefore

$$f'(x) = \lim_{\Delta x \to 0} \frac{u(x+\Delta x) - u(x)}{\Delta x} + \lim_{\Delta x \to 0} \frac{v(x+\Delta x) - v(x)}{\Delta x}$$

$$= \frac{du}{dx} + \frac{dv}{dx}$$

Perhaps you think the result of Theorem 3.1 is a natural consequence of something that we would have expected before actually proving it. You may even fail to see its importance until you think of some instances in other parts of mathematics when things have not been quite so fortunate. For example, does the function $\sin x$ or $\log x$ have the linearity property?

The linearity property allows us to narrow a given differentiation problem down to manageable size.

Example 1.   Find

$$\frac{d}{dx}[3x^5 + 7x^2 + 2]$$

Solution.   In this case

$$f(x) = 3x^5 + 7x^2 + 2$$

and

$$f'(x) = 3\frac{d}{dx}[x^5] + 7\frac{d}{dx}[x^2] + \frac{d}{dx}[2]$$

We could complete this problem if we knew the derivatives involved; that is, we need to know $d/dx\ [x^n]$ and $d/dx\ [c]$.

Example 2.   Find $f'(x)$ if $f(x) = 2\sin x + \log x$.

Solution.   Applying the linearity principle, we obtain

$$f'(x) = 2\frac{d}{dx}[\sin x] + \frac{d}{dx}[\log x]$$

and here again we need only know the *specific* formulas for

$$\frac{d}{dx}[\sin x] \text{ and } \frac{d}{dx}[\log x]$$

Both of these examples demonstrate the need for specific differentiation formulas to complement the general statement of linearity. We shall derive differentiation formulas for polynomials in this chapter, but we shall delay until a later chapter the derivation of differentiation formulas like those in Example 2.

3.3   DIFFERENTIATION OF POLYNOMIALS

In view of the linearity property, the following theorems will enable us to differentiate any polynomial.

THEOREM 3.2

$$\frac{d}{dx}[c] = 0 \tag{4}$$

Proof
This statement is immediately obvious from our geometric interpretation of the derivative as the slope of a curve at a point. The graph of $y = c$ is a straight line parallel to the x-axis (see Figure 3.1). Its slope

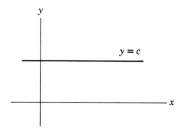

Figure 3.1

is therefore zero. While this is not a rigorous proof, it is an important type of intuitive thinking to which you should become accustomed. See if you can do the proof by direct appeal to the definition of the derivative.

THEOREM 3.3

$$\frac{d}{dx}[x^n] = nx^{n-1}$$

where n is a positive integer.

Proof
To show

$$\frac{d}{dx}[x^n] = nx^{n-1}$$

we proceed "inductively," that is, we prove it first for $n = 1$, which means that we wish to prove that $(d/dx)[x] = 1$. If we examine the sketch of $y = x$ (see Figure 3.2), we see that the slope is *always* 1,

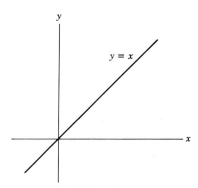

Figure 3.2

which is a geometric confirmation of the fact that $(d/dx)[x] = 1$. To prove the theorem for $n = 2$, we must show that

$$\frac{d}{dx}[x^2] = 2x$$

Let $f(x) = x^2$. Then

$$f(x + \Delta x) = (x + \Delta x)^2$$

and

$$\frac{d}{dx}[x^2] = \lim_{\Delta x \to 0} \frac{x^2 + 2x(\Delta x) + (\Delta x)^2 - x^2}{\Delta x}$$

$$= \lim_{\Delta x \to 0} (2x + \Delta x)$$

$$= 2x$$

Similarly, we can prove the theorem for $n = 3$ by showing that

$$\frac{d}{dx}[x^3] = 3x^2$$

Thus, if $f(x) = x^3$; $f(x + \Delta x) = (x + \Delta x)^3$ and

$$\frac{d}{dx}[x^3] = \lim_{\Delta x \to 0} \frac{x^3 + 3x^2(\Delta x) + 3x(\Delta x)^2 + (\Delta x)^3 - x^3}{\Delta x}$$

$$= \lim_{\Delta x \to 0} [3x^2 + 3x(\Delta x) + (\Delta x)^2]$$

$$= 3x^2$$

We could continue like this for $n = 4, 5, 6$, etc., but eventually we would have to prove the theorem for the general case. The details of such a proof involve the use of the binomial theorem – a detail which we choose to omit. For the present, we shall accept the theorem on the basis of the three specific proofs given above.

One further remark is in order before we turn to a few examples. The formula $(d/dx)[x^n] = nx^{n-1}$ was established for positive integral powers only. *The formula holds, however, for all real exponents.* For example, $(d/dx)[x^\pi] = \pi x^{\pi-1}$. The formula is *not* true if the exponent is variable as in the case of the function $f(x) = x^x$. (Indeed, you may ask: What *is* the derivative of x^x, if it is *not* $x \cdot x^{x-1}$? The answer to this question is given in Chapter 9.)

Example 3.   Differentiate $y = x^4$.

Solution.   Our basic formula with $n = 4$ is

$$\frac{d}{dx}[y] = \frac{d}{dx}[x^4]$$

Therefore

$$\frac{dy}{dx} = 4x^3$$

Example 4.   Complete Example 1.

Solution.   Applying the basic formula to the three terms, we obtain

$$\frac{dy}{dx} = 15\,x^4 + 14x$$

Example 5.   Find $\frac{dy}{dx}$ if $y = \sqrt{x}$

Solution.   If we write $y = x^{1/2}$, we can apply our formula with $n = \frac{1}{2}$.

$$\frac{dy}{dx} = \tfrac{1}{2}x^{-1/2} = \frac{1}{2\sqrt{x}}$$

Example 6.   Differentiate

$$y = \frac{1}{x^5}$$

Solution.   We write $y = x^{-5}$ and apply our basic formula with $n = -5$.

Then

$$\frac{dy}{dx} = -5x^{-6} = \frac{-5}{x^6}$$

Example 7.   Differentiate $s = \sqrt{3t^3}$ with respect to t.

Solution.   We write $s = \sqrt{3}t^{3/2}$ and apply the linearity principle along with our basic formula with $n = \frac{3}{2}$. Then

$$\frac{ds}{dt} = \sqrt{3}\,[\tfrac{3}{2}t^{1/2}]$$

$$= \frac{3^{3/2}}{2}\sqrt{t}$$

Example 8.   Find $f'(x)$ if $f(x) = ax^2 + bx^{1/3}$. Note, a and b are constants.

Solution.

$$f'(x) = \frac{d}{dx}[ax^2 + bx^{1/3}]$$

$$= a\frac{d}{dx}[x^2] + b\frac{d}{dx}[x^{1/3}] \text{ (use of linearity principle)}$$

$$= a[2x] + b[\tfrac{1}{3}x^{-2/3}]$$

$$= 2ax + \frac{b}{3x^{2/3}}$$

Example 9.   Find the equation of the straight line drawn tangent to the graph of $y = x^3/3$ at $(2,\tfrac{8}{3})$.

Solution.   Recall that to write the equation of a line, we need to know the slope of the line (which is constant) and any point through which the line passes. In this case we know the line passes through the point with coordinates $(2,\tfrac{8}{3})$. To obtain the slope, we differentiate the function and evaluate f' at $x = 2$. Since $y = x^3/3$ and $dy/dx = x^2$. Therefore, $f'(2) = 4$. Using the point slope form of the line (Figure 3.3), we obtain

$$y - y_1 = m(x - x_1)$$
$$y - \tfrac{8}{3} = 4(x - 2)$$

or

$$3y - 12x + 16 = 0$$

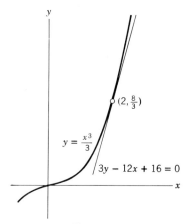

Figure 3.3

EXERCISES

Differentiate the following functions.

1. $y = 3x + 5$

2. $y = ax^2 + bx^3$

3. $\theta = 3\phi + 4\phi^2$

4. $g(s) = \sqrt{s} - \sqrt[3]{s}$

5. $y = 1/x^3$

6. $y = (2/x)^4 + (2/x)^2$

7. $z = \sqrt{t^3} - 5$

8. $f(x) = 3/4x$

9. $p = \sqrt{3t}$

10. $h = \sqrt[3]{4k^2}$

11. $v = at^{5/2} - bt^{3/5}$

12. $y = 1/\sqrt{x^7}$

13. $r = 4/\sqrt[3]{s} + 5/\sqrt{s}$

14. $s = 10/t$

15. $y(x) = x^4 - 2x^3 + 5x^2 - 7x - 1$

16. $y = \sqrt[5]{w} - 6\sqrt[3]{w^2}$

17. $p = 1/\sqrt{2t^3}$

18. $m = \sqrt{4k^6}$

19. $y = \sqrt{x^2 + 2x + 1}$

20. $v = \sqrt[3]{\sqrt{x}}$

21. $y = (x - 2)^2$

22. $r = (s^{1/3} + 2)^2$

23. $w = 3t^{1/2} - 4t^{1/3} + 2$

24. $f(x) = (2x^3 - 3x^2 + 5x)/x^3$

25. $i = (3 + t^2)/\sqrt{t}$

Find the equation of the line drawn tangent to the function at the indicated point.

26. $y = x^3 - 3x^2 - 2x$, at $x = 3$

27. $y = x^2 + 1/x^2$, at $x = \frac{1}{2}$

28. $y = x^{3/2} - x^{4/3}$, at $x = 1$

29. $y = x + 3x^{-1}$, at $x = -1$

In the following problems, find the values of x for which the graph has a slope of zero.

30. $y = x^2 - 6x + 13$

31. $y = x^3 + 9x^2 + 7$

32. $y = x^3 + 4x^2 + 4x$

33. $y = x - 8/x$

3.4 HIGHER ORDER DERIVATIVES

We observed in Chapter 2 that if we compute the limit of the difference quotient at each x, we arrive at a correspondence between x and the value of this limit at each point. This correspondence gives us another function denoted by $f'(x)$.

This process may be repeated to give us the "derivative of the derivative of f," or as it is usually called, the *second derivative of f*, that is,

$$\frac{d}{dx}[f'(x)] = f''(x) \tag{6}$$

Another very widespread notation of $f''(x)$ is

$$\frac{d^2}{dx^2}[f(x)]$$

Notice that neither use of the 2 in this notation has anything to do with "squaring."

We have seen several important applications of the first derivative in the previous chapter, in particular, its geometric interpretation as the slope of the curve. The second derivative, too, has some important applications that will come to light as we proceed. Here we shall learn the mechanics of finding $f''(x)$. Later we shall discuss the actual use of $f''(x)$ to describe geometric or physical phenomena.

Higher derivatives than the second are defined in analogous fashion, for example, the *third derivative of f* is

$$\frac{d}{dx}[f''(x)] = \frac{d^3}{dx^3}[f(x)] = f'''(x) \tag{7}$$

and the *fourth derivative of f* is

$$\frac{d}{dx}[f'''(x)] = \frac{d^4}{dx^4}[f(x)] = f^{iv}(x) \tag{8}$$

By far the most important derivatives for our use will be the first and second. Exercises are also provided for other higher orders.

Example 10.   Find the first, second, and third derivatives of

$$y = x^2 + \sqrt{x}$$

Solution

$$y = x^2 + x^{1/2}$$

$$y' = 2x + \frac{1}{2\sqrt{x}}$$

$$y''(x) = 2 - \frac{1}{4\sqrt{x^3}}$$

$$y'''(x) = \frac{3}{8\sqrt{x^5}}$$

Example 11.   Find the value of $f''(2)$ if $f(x) = x^3 - 3x^2 - 9x + 2$. Check the result by drawing the graph of $f'(x)$ and then measuring the slope at $x = 2$.

Solution

$$f(x) = x^3 - 3x^2 - 9x + 2$$
$$f'(x) = 3x^2 - 6x - 9$$
$$f''(x) = 6x - 6$$
$$f''(2) = 6$$

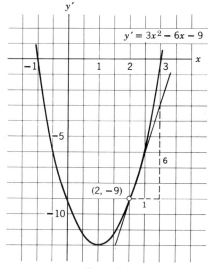

Figure 3.4

The graph of $f'(x)$ is shown in Figure 3.4. You can now measure the slope of the tangent line at $x = 2$ and show it is 6. (In Section 3.6 we shall see an easy way to sketch a figure such as the one above.)

EXERCISES

In the following problems, find the first three derivatives.

1. $y = x^3 - 3x^2 + 5$

2. $f(x) = x^{1/3} - x^{1/4}$

3. $z = 1/t^2 + 3t^2$

4. $g(x) = 1/\sqrt{x}$

5. $S = -k^{-3/5}$

6. $V = 3w^4 - w^3 - 10w^2$

7. $p = t + 1/t$

8. $h(n) = (n + 2)(n^2 - 3)$

9. $A(x) = (x^3 - 3)/\sqrt{x}$

10. $c = 3d^5(d^2 - 4)$

Find the value of the second derivative of the given function at the indicated point. Check your answer graphically, as shown in Example 11.

11. $y(x) = x^2 - 3x + 12$; find $y''(3)$.

12. $g(x) = x^3 - \frac{3}{2}x^2 - 6x + 7$; find $g''(0)$.

13. $h(z) = 6z^2 - 2z^3$; find $h''(1)$.

3.5   APPLICATION OF THE DERIVATIVE TO MECHANICS PROBLEMS

The concepts of velocity, acceleration, and power are examples of a large set of rate problems that we study in mechanics. For our purposes, when we say "rate of change," we shall mean instantaneous rate of change—a quantity easily obtained by differentiation.

Velocity.   We introduced the velocity problem in Chapter 2, and we found that it could be solved by evaluating $\lim_{\Delta t \to 0} \Delta s/\Delta t$. We can just as well, in fact more easily, obtain the answer by recognizing that the velocity is equal to derivative of the distance-time equation. Then we can use the formulas of Section 3.3.

Example 12.   Find the velocity of a solenoid actuated plunger when $t = 0.01$ sec, if the displacement of the plunger is given by $s = 150\sqrt{t}$ cm.

Solution.   The equation for the velocity of the plunger at any time t is

$$v = \frac{ds}{dt} = \frac{d}{dt}[150t^{1/2}] = 150\frac{d}{dt}[t^{1/2}] = \frac{75}{\sqrt{t}}$$

To find the velocity at $t = 0.01$ sec we write

$$v = \frac{75}{\sqrt{0.01}} = \frac{75}{0.1} = 750 \text{ cm/sec}$$

Acceleration.   The *acceleration* of an object is defined in mechanics as the time rate of change of velocity. This means that acceleration is the derivative of the velocity-time equation, which is the same thing as the second derivative of the distance-time equation. The acceleration is then represented symbolically by

$$a = \frac{dv}{dt} = \frac{d^2s}{dt^2} \tag{9}$$

Example 13.   Find the equations of the velocity and acceleration of a rocket having $s = 3t^3 + 5t$ as its equation of motion. Also, find the velocity and acceleration of the projectile after 1 sec. Assume s is measured in feet.

Solution.   The equation for the velocity is found by differentiating the given equation with respect to time. Thus

$$v = \frac{ds}{dt} = 9t^2 + 5$$

The equation for the acceleration is now found by differentiating the velocity equation with respect to time.

$$a = \frac{dv}{dt} = 18t$$

It is then a simple matter to find the values of the velocity and acceleration at $t = 1$ sec.

$$v = 9(1)^2 + 5 = 14 \text{ ft/sec}$$
$$a = 18(1) \quad\;\; = 18 \text{ ft/sec}^2$$

In some practical situations it is easier to obtain a graphical representation of the distance-time relationship than it is an equation. In these cases, it is important to recall that the first derivative is just the slope of the graph at every point and, hence, the velocity of an object at time t_1 is numerically equal to the slope of the distance-time graph at time t_1. Similarly, the acceleration of an object at time t_1 is numerically equal to the slope of the velocity-time graph at time t_1.

Example 14.   Find the velocity of an object when $t = 1.0$ sec, if the equation of motion has the indicated graph (Figure 3.5).

t	s
0	0
0.5	2.0
1.0	3.3
1.5	3.9
2.0	3.8
2.5	2.3
3.0	1.5
3.5	1.9
4.0	1.3
4.5	1.0

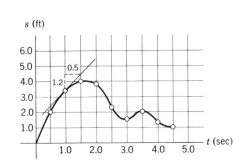

Figure 3.5

Solution.   First we make a sketch of the distance-time relationship as shown in Figure 3.5. Then drawing the tangent line as carefully as we can at (1.0,3.3), we can estimate the velocity of the object at $t = 1.0$ sec. Our estimate of the slope of the tangent line is $(1.2/0.5) = 2.4$. Hence, the velocity at $t = 1.0$ sec is $v = 2.4$ ft/sec.

The problem being considered here is that of *linear* motion, that is, the object is assumed to be traveling in a straight line. The distance-time graph does not represent the path traveled by the object, but rather $s(t)$ represents the distance of the particle away from some fixed reference point.

Power.   *Power* is defined as the time rate of change of doing work. Power, therefore, measures a *rate* and, as such, can be calculated by using the first derivative. Under the assumption that the work expended is given as a function of time, the power is the derivative of the work-time equation. This is written symbolically as

$$p = \frac{dw}{dt} \tag{10}$$

The units of power are, therefore, foot-pounds/second.

Example 15. Find the expression for the power of a mechanical system if the system has the capacity to work according to $w = t - 3t^4$. Work is in foot-pounds, and time is in seconds.

Solution

$$p = \frac{dw}{dt}$$

$$= \frac{d}{dt}[t - 3t^4] = 1 - 12t^3 \frac{\text{ft-lb}}{\text{sec}}$$

EXERCISES

1. The displacement of a projectile varies with time according to $s = 16t^2 - 4t + 5$ ft. Find the velocity-time equation for the projectile.
2. What is the projectile's velocity after: (a) 1 sec, (b) 2 sec, (c) 3 sec?
3. Show that the acceleration of the projectile in Exercise 1 is constant.
4. Find the velocity of a missile 3 sec after launch if its displacement, in feet, is given by $s = 3t^3 - 4t$.
5. The plunger of a solenoid has a velocity $v = 3t^{1/2}$ centimeters per second. Find the acceleration of the plunger when $t = 0.01$ sec.
6. If the distance-time equation of a dragster is $s = 2t^3 + t^2$ ft, what is the velocity of the dragster after 6 sec? Give your answer in miles per hour. (*Note.* 88 ft/sec $= 60$ mi/hr.)
7. An electric motor does work according to $w = 3t^{-2}$ ft-lb. What is the expression for the power developed by the motor?
8. What power is being developed by the motor after 1 sec?
9. A cannon is fired vertically upward. If the altitude of the projectile varies with time from launch according to $h = 100 + 640t - 16t^2$ ft, what is the maximum altitude obtained? (*Hint.* The velocity of the projectile is zero when it reaches its maximum altitude.)
10. In the metric system, the common unit of work is the joule. The unit of power is then the joule/sec. If a machine works according to $w = 3t^4 - 4t$ joules, what power is being developed by the machine when $t = 1.1$ sec? (*Note.* 1 joule/sec $= 1$ watt.)
11. A particle is observed in a laboratory experiment and its displacement-time coordinates are recorded as follows.

t(sec)	0	1	2	3	4	5
s(cm)	0	0.1	0.3	0.6	1.0	1.5

Estimate the velocity of the particle at $t = 3$ sec from the graph obtained by connecting the points with a smooth curve.

12. The following data

t(sec)	0	1	2	3	4
R(km)	5.75	5.79	5.85	5.93	6.05

were obtained from the radar tracking of a weather balloon, where R is the slant range. At what rate is the balloon moving away from the radar when $t = 2$ sec?

3.6 APPLICATION OF THE DERIVATIVE TO ELECTRICITY PROBLEMS

Just as it does in mechanics, the derivative has many important applications in the field of electricity. We defined *electric current* in Chapter 2 as the $\lim_{\Delta t \to 0} \Delta q / \Delta t$, where q is the quantity of electrical charge. Therefore, the current, i, is the derivative of the charge with respect to time. Symbolically, this is written

$$i = \frac{dq}{dt}$$

(11)

If the charge is measured in coulombs and the time in seconds, the unit of electric current is the ampere.

Example 16. Find the current when $t = 0.1$ sec if the charge varies according to $q = 2t - 3t^2$.

Solution. The equation for the current is $i = (dq/dt) = 2 - 6t$. Therefore, the current when $t = 0.1$ sec is

$$i = 2 - 6(0.1) = 2 - 0.6 = 1.4 \text{ amp}$$

Voltage Induced in a Coil. There are other quantities in electricity besides current that involve the concept of rate of change. One of these is the voltage induced in a coil of wire by a change in current in the coil. If the inductance of the coil in henrys is denoted by L, it is one of the laws of electricity that

$$v = -L \frac{di}{dt}$$

(12)

where the minus sign indicates that the induced voltage will oppose the current that is inducing it.

Example 17. Find the equation for the voltage induced in a coil if $L = 10$ henrys, and the current varies according to $i = 3t^{3/2}$.

Solution

$$v = -L\frac{di}{dt}$$

$$= -10\frac{d}{dt}[3t^{3/2}] = -10\left[\frac{9}{2}t^{1/2}\right] = -45t^{1/2} \text{ volts}$$

Current in a Capacitor.   Another relation that involves rates in electricity is the current in a capacitor. Again, it is a part of empirical knowledge that if the capacitance is given by C, then the current is given by

$$i = C\frac{dv}{dt} \tag{13}$$

In this equation, the current is in amperes, the capacitance is in farads, and the voltage is in volts.

Example 18.   Find the equation for the current in a capacitor if $C = 10^{-6}$ farads, and the voltage varies as $v = 2 \times 10^6(t^3 + 4t)$.

Solution.   Since

$$i = C\frac{dv}{dt}$$

We have

$$i = 10^{-6}\frac{d}{dt}[2 \times 10^6(t^3 + 4t)]$$

$$= 10^{-6} \times 2 \times 10^6\frac{d}{dt}[t^3 + 4t]$$

$$= 2(3t^2 + 4) \text{ amp}$$

In each of the preceding articles we have been concerned with finding rates of change. The physical interpretation of the rate of change is dependent upon what quantities we are considering, but the process of finding the rate of change is not. We should always remember: to find the time rate of change of a given physical quantity, we simply take the derivative of that quantity with respect to time.

EXERCISES

1. Find the current in a resistor when $t = 0.2$ sec if the charge varies according to $q = 4t^{4/3}$ coul.
2. If the charge transferred is $q = 3t - 2t^2$ coul, what charge has been transferred by the time the current becomes equal to zero?

3. A coil of copper wire has an inductance $L = 15$ henrys. What is the expression for the voltage induced in the coil if the current varies as $i = 4t^3 + 5$ amp?

4. Find the equation for the voltage induced in a coil if $L = 50$ henrys, and the current varies according to $i = 0.01t^2$ amp.

5. What is the induced voltage in the coil in Exercise 4 when $t = 2$ sec?

6. The electrical power loss in a resistor of R ohms when a current of i amperes flows in the resistor is given by $p = i^2R$. Find the power loss after 4 sec in a 1000-ohm resistor if the charge varies according to $q = t^{3/2}$ coul. The unit of power is the watt when R is in ohms and i is in amperes.

7. Find the electrical power loss in a 200-ohm resistor when $t = 8$ sec if the charge transferred is $q = 3t^{4/3}$ coul.

8. A capacitor of capacitance $C = 2$ farads has a voltage of $v = 25 + t^2$ volts applied to its terminals. What is the expression for the current in the capacitor?

9. In Exercise 8, what is the magnitude of the current when $t = 0.1$ sec?

10. A coil of wire has a current of $i = 4t^2 - 6t$ passing through it. If the induced voltage is 0.4 volts when $t = 2$ sec, what is the inductance of the coil?

11. The charge transferred in an inductance coil is $q = 0.5t^3$ coul. If the inductance of the coil is 7 henrys, what is the expression for the induced voltage?

3.7 CURVE SKETCHING

In Chapter 1 we discussed the graphing of functions without the use of calculus. In many cases, the tools introduced there, such as the finding of intercepts, the determination of symmetry, and the location of asymptotes, will be sufficient for your purposes. However, as we shall see, the knowledge of the sign of the first two derivatives can be of significant assistance.

Sign of $f(x)$.   For the sake of completeness, we will give a short review as to how the sign of the function itself affects the graph of $f(x)$. If $f(x) > 0$, the graph is above the x-axis; if $f(x) = 0$, the graph is on the x-axis (these points are also called intercepts); and if $f(x) < 0$, the graph is below the x-axis. It is often true that the information about the sign of $f(x)$ is sufficient to give us a graph which will satisfy our needs.

Example 19.   Sketch $f(x) = x^2 - 6x + 8 = (x-4)(x-2)$.

Solution.   To find the values of x for which $f(x) > 0$, we set $(x-4)(x-2) > 0$ and, solving this inequality, we obtain $x > 4$ or $x < 2$. It is then obvious that $f(x) < 0$ for $2 < x < 4$. Further, the intercepts are easily seen to be $x = 2$ and $x = 4$. With this much information we can make a sketch of the graph. Since the information is limited, our sketch might look like any of the three graphs shown in Figure 3.6. In each sketch $f(x) > 0$ for $x > 4$ and $x < 2$; $f(x) < 0$ for $2 < x < 4$; $f(x) = 0$ for $x = 2$ and $x = 4$. Knowledge of the geometric interpretation of the first and second derivatives of the function will enable us to improve our sketch.

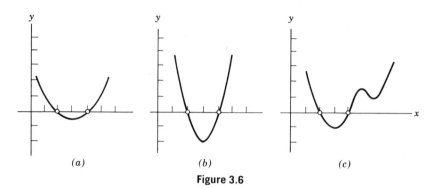

(a)   *(b)*   *(c)*

Figure 3.6

Sign of $f'(x)$. Figure 3.7 shows the graph of a rather typical continuous function of x, say $y = f(x)$, on which we have indicated certain points. We make the following observations concerning the graph of $f(x)$ and the numerical sign of its first derivative.

(a) At point A, the curve is rising (when viewed from left to right). The slope of a line drawn tangent to the curve at this point is positive; therefore $f'(x_A) > 0$. In fact, $f'(x) > 0$ for all values of x between A and B.

(b) At the point B, the slope of the curve is zero so that $f'(x_B) = 0$.

(c) At point C, in fact over the whole segment of the curve between B and D, the curve is falling as we move from left to right. The slope of the curve is, therefore, negative and $f'(x_C) < 0$.

(d) At the point D, the slope of the curve is again zero, and $f'(x_D) = 0$.

(e) To the right of D, the curve is rising and $f'(x_E) > 0$.

Thus we see how knowledge of the graph of a function can tell us where its derivative is positive, negative, and zero. Conversely, we can use this knowledge of the sign of $f'(x)$ to assist us in sketching the graph of a function.

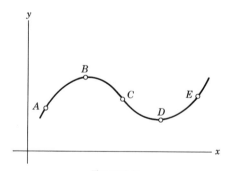

Figure 3.7

Example 20.   Sketch the graph of $f(x) = x^2 - 6x + 8$.

Solution.   This, you will recall, is the same function we discussed in Example 19 with reference to the sign of $f(x)$. To obtain the desired graph, we analyze $f'(x)$ with respect to its numerical sign. The derivative of $f(x)$ is

$$f'(x) = 2x - 6 = 2(x - 3)$$

The regions where $f'(x)$ is positive, negative, and zero are

$$f'(x) > 0 \text{ for } x > 3$$
$$f'(x) < 0 \text{ for } x < 3$$
$$f'(x) = 0 \text{ for } x = 3$$

Thus, the curve is falling for $x < 3$ and rising for $x > 3$. The slope is zero when $x = 3$. Computing the value of the function for $x = 3$, we find $f(3) = -1$. Using information about the signs of $f(x)$ (See Example 19) and $f'(x)$, we sketch the graph in Figure 3.8.

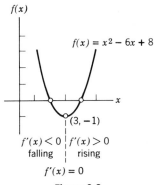

Figure 3.8

Example 21.   Using information about the sign of $f'(x)$, sketch the graph of $f(x) = \dfrac{x^3}{3} - 2x^2 + 3x + 1$.

Solution.   In this problem it is not convenient to find the regions for which $f(x)$ is positive and negative, so we base our sketch largely on the sign of $f'(x)$. Here

$$f'(x) = x^2 - 4x + 3 = (x - 1)(x - 3)$$

The regions where the graph is rising are $x < 1$ and $x > 3$, since $f'(x) > 0$ in these regions. Similarly, the graph is falling in the region $1 < x < 3$,

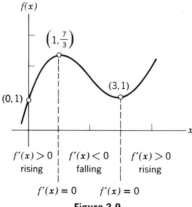

Figure 3.9

since $f'(x) < 0$. The slope is zero when $x = 1$ and $x = 3$. Computing the values of the function for $x = 0$. 1. and 3, we have $f(0) = 1$, $f(1) = \frac{7}{3}$, and $f(3) = 1$. Figure 3.9 is the graph of this function.

Example 22.   Sketch the graph of $y = x^{2/3}$.

Solution.   Notice that y is never negative, and $y = 0$ when $x = 0$. Calculating y', we have

$$y' = \frac{2}{3\sqrt[3]{x}}$$

We see that $y' > 0$ when $x > 0$, $y' < 0$ when $x < 0$, and y' does not exist for $x = 0$. Further, y' becomes arbitrarily large in the neighborhood of the origin. This means that the tangent approaches the vertical near the origin. We describe this phenomena by saying that the graph has a vertical tangent at the origin. Using this information and plotting a few additional points, we have the graph shown in Figure 3.10.

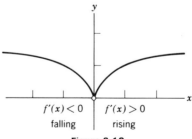

Figure 3.10

EXERCISES

In Problems 1–8, indicate the values of x for which y is positive.

1. $y = 6x - 12$
2. $y = x^2 + 1$
3. $y = 4 - x^2$
4. $y = x^3$
5. $y = x^2 - 3x$
6. $y = 2x - 2x^2$
7. $y = x^2 + x - 6$
8. $y = 3x^2 - 6x + 3$

Use the information about the sign of $f(x)$ and $f'(x)$ to sketch the graph of the indicated functions.

9. $f(x) = 8 - x^2$
10. $f(x) = 3x^2 - 27$
11. $f(x) = x^2 - 4x$
12. $f(x) = x^2 + 2x$
13. $f(x) = x^2 + x - 2$
14. $f(x) = -x^2 - 8x - 7$
15. $f(x) = x^3 + 3x^2$
16. $f(x) = x^3 - 4x^2$
17. $f(x) = x^3/3 - 3x^2/2 + 2x + 1$
18. $f(x) = x^3 - 6x^2 + 7$
19. $f(x) = x^{1/3}$
20. $f(x) = x^{2/3} + 1$
21. $f(x) = x^{4/3}$
22. $f(x) = -x^{2/3}$

Sign of $f''(x)$. In this section we examine the significance of the sign of the second derivative in graphing a function.

Let us first consider the case where $f''(x)$ is positive over some interval. This means we are assuming that the function $f'(x)$ has a positive derivative over that interval; that is, the derivative is an increasing function. Therefore the slope of the graph of $f(x)$ increases throughout the interval. The situation will look something like the graph in Figure 3.11.

On I_1 the derivative is negative, but increasing as x increases; on I_2 the derivative is positive, and increasing as x increases. Thus, on the complete interval, the derivative is increasing from negative values to positive values as x increases.

The curve is shaped like the inside of a cup and is described by a "sense of concavity." We say the curve is *concave-up* over the interval. The

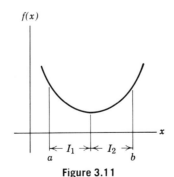

Figure 3.11

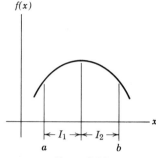

Figure 3.12

foregoing discussion is summarized in the form of a theorem (that can be proved by analytic methods).

THEOREM 3.4

If $f''(x) > 0$ on $[a,b]$, then the graph of $f(x)$ is concave-up over $[a,b]$.

A curve that is shaped like a cup with its open-end down is called *concave-down*. Referring to Figure 3.12, notice that the first derivative is positive but decreasing on I_1 as x increases, and negative and decreasing on I_2 as x increases. Thus, on the complete interval, the derivative is decreasing from positive to negative as x increases. The second derivative on an interval that is concave-down is, therefore, negative; we obtain the following theorem.

THEOREM 3.5

If $f''(x) < 0$ on $[a,b]$, then the graph of $f(x)$ is concave-down on $[a,b]$.

Many curves are composed of concave-up portions and concave-down portions. The sign of $f''(x)$ will tell us the sense of concavity of the graph of $f(x)$, as the next example illustrates.

Example 23.   Discuss the concavity of $y = 1 - x^3$.

Solution.   Here
$$y' = -3x^2$$
$$y'' = -6x$$

We can now make the following observations about the sign of the second derivative.
$$\text{If } x < 0, y'' > 0$$
$$\text{If } x > 0, y'' < 0$$

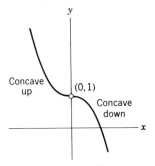

Figure 3.13

Therefore, at $x = 0$, the sense of concavity changes from concave-up, to the left of $(0,1)$, to concave-down to the right of this point. We can make a rough sketch of the graph of this function using only the above information (See Figure 3.13). At the point $(0,1)$ the sense of concavity changes as x passes through this point. Such points are called *points of inflection* of the curve. It is useful to note that $f''(x) = 0$ at the point of inflection. This fact is frequently used to locate points of inflection. However, as the next example demonstrates, there are points of inflection for which $f''(x) \neq 0$.

Example 24. Sketch $y = \sqrt[3]{x}$.

Solution. Here

$$y = x^{1/3}$$

$$y' = \frac{1}{3}x^{-2/3}$$

$$y'' = -\frac{2}{9}x^{-5/3} = -\frac{2}{9x^{5/3}}$$

We observe that

$$\text{If } x < 0, y'' > 0$$

$$\text{If } x > 0, y'' < 0$$

$$\text{If } x = 0, y'' \text{ is not defined}$$

The point $(0,0)$ is an inflection point, even though $f''(x)$ is not defined at this point, because the graph changes from concave-up to concave-down at this point (Figure 3.14).

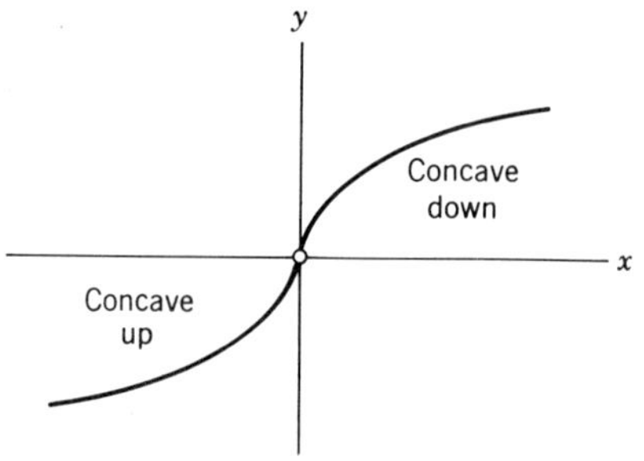

Figure 3.14

We now summarize the last few articles by means of the following table.

	Positive	Zero	Negative
$f(x)$	Curve is above x-axis	Intercept	Curve is below x-axis
$f'(x)$	Curve is rising	Horizontal tangent	Curve is falling
$f''(x)$	Concave-up	Possible inflection point	Concave-down

You should attempt to use the methods at your disposal when sketching the graph of a function instead of resorting to the time-consuming method of plotting points. You are still called upon to plot certain points (intercepts, critical points, points of inflection), but a much deeper meaning can now be attached to the graph of a function.

Example 25. Sketch $y = x^3 - 6x^2 = x^2(x-6)$.

Solution

$$y'(x) = 3x^2 - 12x = 3x(x-4)$$
$$y''(x) = 6x - 12 = 6(x-2)$$

We observe that:

(a) $y = 0$ at $x = 0$ and $x = 6$; $y > 0$ for $x > 6$, and $y < 0$ for $x < 6$.
(b) $y' > 0$ for $x < 0$ or $x > 4$; $y' < 0$ for $0 < x < 4$; $y' = 0$ at $x = 0$ and $x = 4$. Therefore, the tangent to the curve is horizontal at $(0,0)$ and $(4,-32)$; the curve is rising for $x < 0$ or $x > 4$ and falling for $0 < x < 4$.

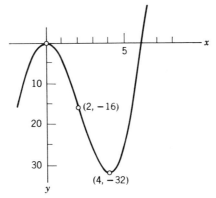

Figure 3.15

(c) $y'' < 0$ for $x < 2$, and $y'' > 0$ for $x > 2$. Therefore, the curve is concave-down for $x < 2$ and concave-up for $x > 2$, and the point $(2,-16)$
is a point of inflection.
(d) The y-intercept is the origin.
The graph of this function is shown in Figure 3.15.

The following summary will be useful as a guide to the student in sketching the graph of a function.

SUMMARY OF STEPS USED IN SKETCHING CURVES

1. Calculate the first and second derivatives.
2. Determine where $f(x)$, $f'(x)$, and $f''(x)$ are positive, negative, and
    zero. All three of these may not be possible or convenient, but you
    should do as many as possible.
3. Determine points of inflection.
4. Locate points where $f(x)$, $f'(x)$, and $f''(x)$ are not defined. Observe
    the trend in the value of the function near these points.
5. Locate vertical and horizontal asymptotes.
6. Observe the trend in the functional value as x becomes large.

Example 26.   Sketch the graph of $y = x + 1/x$

Solution
(a) $y' = 1 - 1/x^2$; $y'' = 2/x^3$
(b) $y = (x^2 + 1)/x$ and, since the numerator is always positive, the sign of
    y is the same as the denominator, that is, $y < 0$ for $x < 0$, and $y > 0$
    for $x > 0$. $y' = (x^2 - 1)/(x^2)$ and the sign of y' is determined by the
    numerator. $y'(x) = 0$ for $x = \pm 1$, $y'(x) > 0$ for $x < -1$ or $x > 1$,
    and is negative for x between -1 and 1. $y''(x) = 2/x^3$ and is thus nega-
    tive for $x < 0$ and positive for $x > 0$.

(c) Since $y(0)$ is undefined and $y''(x)$ is never zero, there are no points of inflection, even though the sense of concavity is different for $x < 0$ and $x > 0$.

(d) $y(x)$ is undefined at $x = 0$ and, indeed, is very large positively for x a small positive number and very large negatively when x is a small negative number.

(e) $x = 0$ is a vertical asymptote.

(f) As x becomes large, the term $1/x$ becomes small. Therefore, y behaves approximately as x for large x.

Figure 3.16 is a sketch of this curve.

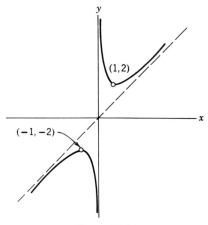

Figure 3.16

EXERCISES

Using information about the sign of $f''(x)$, determine where the graph of $f(x)$ is concave-up and concave-down?

1. $y = x^3$
2. $f(x) = 4 - 2x^3$
3. $f(x) = x^2 + 3x + 2$
4. $y = x^3 + 9x^2$
5. $y = x^{1/3}$
6. $f(x) = 6x - 3x^2$
7. $y = x^3 - 9x^2 + 24x - 10$
8. $y = x^4 - 4x^3$

9–16. Using information about the signs of $f(x)$, $f'(x)$, and $f''(x)$, sketch the graphs of the functions in Exercises 1–8.

In the remaining problems, use all of the methods at your disposal to sketch the graphs of the indicated functions.

17. $y = 4 + 5/x$
18. $y = x - 1/x$
19. $f(x) = x + 1/x^2$
20. $f(x) = 2x^5 - 5x^2$

3.8 MAXIMA AND MINIMA

Of considerable importance in some applications is the determination of the largest and/or smallest value of a function. For example, one may wish to maximize power, minimize heat loss, maximize profit, etc. We begin by defining the general concepts of extremums.

Definition
$f(x)$ is said to have a relative maximum at x_1 if $f(x) < f(x_1)$ for all x in the immediate neighborhood of x_1.

The concept of relative minimum is defined analogously.

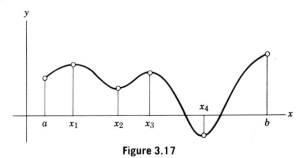

Figure 3.17

Figure 3.17 shows the graph of a function with a relative maximum at x_1, x_3 and b. Relative minimums are seen to be at a, x_2, and x_4.

Definition
The largest value of $f(x)$ on the interval $[a,b]$ is called the absolute maximum of $f(x)$ on $[a,b]$. The absolute maximum is obviously found by determining the largest relative maximum.

The concept of absolute minimum is defined analogously.

In Figure 3.17, $f(b)$ is the absolute maximum, and $f(x_4)$ is the absolute minimum for the function on the interval $[a,b]$. We shall be interested primarily in finding the absolute maximum or minimum of a function.

We may now ask: How do we proceed analytically to determine the extremes of a function? The answer to this question is based on the following theorem.

THEOREM 3.6

Let f be a differentiable function on the interval $a \leqslant x \leqslant b$. If a relative maximum (or minimum) value of the function occurs at a point X_m which is not an end point of the interval, then $f'(X_m)$ must equal 0.

This theorem is the basis of the procedure for finding the extremes of a differentiable function, because it tells us that potential extreme points can be found by setting $f'(x)$ equal to zero and solving for x. The points where $f'(x) = 0$, and the end points of the domain, are called *critical points*.

It is important to realize that the converse of Theorem 3.6 is not true. As an illustration, the function $f(x)$ considered in Example 23 has a zero derivative at $x = 0$, since $f'(x) = -3x^2$. Referring to Figure 3.13, notice that the curve $y = 1 - x^3$ has a point of inflection with a horizontal tangent at $(0, 1)$. We conclude from this that we need a procedure for distinguishing between maximum and minimum points. Recalling our discussion of the use of the first derivative in curve sketching, we state the following more-or-less obvious principles that together are called the *first derivative test*.

The First Derivative Test.   Assume that $f'(X_m) = 0$. Then

(a) $f(X_m)$ is a relative maximum if $f'(x)$ changes from $+$ to $-$ as x increases through X_m.

(b) $f(X_m)$ is a relative minimum if $f'(x)$ changes from $-$ to $+$ as x increases through X_m.

(c) If $f'(x)$ does not change sign as x increases through X_m, then $f(X_m)$ is neither a maximum nor a minimum.

Often the determination of the change in sign can be made just by checking the value of the derivative slightly to the left and then slightly to the right of the point; sophisticated methods are not necessary.

Example 27.   Show that $x^3 - 3x^2 - 45x$ has a relative maximum at $x = -3$ and a relative minimum at $x = 5$.

Solution.   The derivative of $y = x^3 - 3x^2 - 45x$ is

$$y' = 3x^2 - 6x - 45$$
$$= 3(x-5)(x+3)$$

Since the function is everywhere continuous, and there are no end point considerations, the only critical points are obtained by setting $y'(x) = 0$.

$$3(x-5)(x+3) = 0$$
$$x = -3, +5$$

Therefore, the critical values are -3 and $+5$. To test $x = -3$, substitute values into $y'(x)$ to the left of -3.

$$y'(x) = 3(x-5)(x+3) = 3(-)(-)$$

To the right of -3, the factor $(x+3)$ becomes positive and everything else remains unchanged; therefore, we have a sign change from $+$ to $-$ and $y(-3)$ is a relative maximum. You may easily check that $y(-3) = 81$. To test $x = 5$, we note that to the left of 5 the derivative is negative, and to the right, positive. Hence, $y(5)$ is a relative minimum, and $y(5) = -175$.

Whereas the procedure described and used in Example 27 for examining critical points for maxima and minima is usually adequate, there is another test frequently used, especially when the task of finding the second derivative is easy. This test follows immediately from our previous discussion of concavity. The bottom of a concave-up portion of a curve is obviously a relative minimum, and the top of a concave-down portion is obviously a relative maximum. In terms of values of $f''(x)$, we have

The Second Derivative Test

If $f'(X_m) = 0$ and $f''(X_m) > 0$, $f(X_m)$ is a relative minimum.
If $f'(X_m) = 0$ and $f''(X_m) < 0$, $f(X_m)$ is a relative maximum.

The so-called "second derivative test" fails if $f''(X_m) = 0$, or if the second derivative does not exist at X_m.

Example 28. Test $y = x^2 + 128/x$ for maxima and minima.

Solution. The given function may be rewritten as $y = x^2 + 128x^{-1}$, from which we obtain $dy/dx = 2x - 128x^{-2} = 2x - 128/x^2$. Setting $dy/dx = 0$, we obtain

$$2x - \frac{128}{x^2} = 0$$

or

$$2x^3 - 128 = 0$$
$$x^3 = 64$$
$$x = 4$$

To test this value by the second derivative test, we calculate the second derivative

$$\frac{d^2y}{dx^2} = 2 + 256x^{-3} = 2 + \frac{256}{x^3}$$

Evaluating the second derivative at $x = 4$, we obtain

$$\frac{d^2y}{dx^2} = 2 + \frac{256}{4^3} = 2 + 4 = 6$$

which is positive. Therefore, $f(4)$ is a relative minimum. It is easy to show $f(4) = 48$.

When finding absolute extrema, it is necessary to compute all relative extrema and determine the absolute extrema by comparison.

Example 29.   Find the absolute minima of

$$f(x) = \frac{x^4}{4} - \frac{5x^3}{3} + 3x^2$$

Solution

$$f'(x) = x^3 - 5x^2 + 6x = x(x-2)(x-3)$$
$$f''(x) = 3x^2 - 10x + 6$$

$f'(x)$ is equal to zero for $x = 0, 2$, and 3. Furthermore, $f''(0) = 6; f''(2) = -2; f''(3) = 3$; and thus, by the second derivative test, there are relative minimums at $x = 0$ and $x = 3$. Since $f(0) = 0$ and $f(3) = \frac{9}{4}$, the absolute minima of this function occurs at $x = 0$, and is $f(x) = 0$ (Figure 3.18).

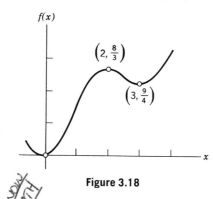

Figure 3.18

EXERCISES

Use the first derivative test to locate any relative maxima and minima of the following functions. Find the numerical value of each maxima and minima.

1. $y = x^2 - 6x + 5$
2. $s = 3t^2 + 12t$
3. $i = t^3 - 3t^2 + 2$
4. $f(x) = x^4 - 8x^2$

5. $y = x^3 + 4$
6. $v = 6t^2 - t^3$
7. $p = x^3 - 3x$
8. $g(s) = s^4 + 10s^2$

Use the second derivative test to locate any relative maxima and minima of the following functions. Find the numerical value of each maxima and minima.

9. $y = 3x^2 + 24x - 30$

10. $y = 4x - x^2$

11. $m = 3r^4 - 4r^3 + 1$

12. $s = 2t^3 - 18t^2 + 48t - 20$

13. $w = t^3 + 3t^2 + 3t + 5$

14. $f(x) = 2 + 5x - x^5$

15. $y = x + 10/x$

16. $v = x^2 + 16/x$

3.9 APPLIED MAXIMA AND MINIMA PROBLEMS

In applied problems we are interested primarily in finding the absolute maximum or minimum of a function. The technique is illustrated by the following examples.

Example 30. The work done by the plunger of a solenoid varies as $w = 6t^2 - t^4$ ft-lb. What is the greatest power produced by the solenoid?

Solution. Power is the rate of change of work with respect to time; therefore

$$p = \frac{dw}{dt} = 12t - 4t^3$$

We now wish to find the maximum power. The critical values are found by letting $dp/dt = 0$.

$$\frac{dp}{dt} = 12 - 12t^2$$

$$12 - 12t^2 = 0$$
$$t = \pm 1$$

We can eliminate the critical value $t = -1$ since negative times are not feasible. Taking d^2p/dt^2, we have $d^2p/dt^2 = -24t$. When $t = 1$, $d^2p/dx^2 = -24$ and, therefore, the power is a maximum. The power at $t = 1$ sec is

$$p = 12(1) - 4(1)^3 = 8 \, \frac{\text{ft-lb}}{\text{sec}}$$

Example 31. The ABC Company is going to construct a cylindrical oil can that will have a volume of 1000 in.^{3}. In order to minimize cost of producing the can, it is desirable to use as little material as possible. What are the dimensions of the can having minimum surface area for the given volume?

Solution. To determine the minimum surface area, we must know the relationship between the volume V, the surface area A, and the radius r. The volume V of a cylinder of radius r, and height h, is

$$V = \pi r^2 h \qquad (1)$$

The surface area of a cylinder, including the top, is

$$A = 2\pi r^2 + 2\pi rh \tag{2}$$

Equation 2 contains both r and h as does Equation 1. Since the volume is given, Equation 1 can be written as

$$1000 = \pi r^2 h$$

or,

$$h = \frac{1000}{\pi r^2} \tag{3}$$

This relation may now be substituted into Equation 2 to yield A as a function only of the radius r.

$$A = 2\pi r^2 + 2\pi r \left(\frac{1000}{\pi r^2}\right)$$

$$= 2\pi r^2 + \frac{2000}{r}$$

$$= 2\pi r^2 + 2000r^{-1} \tag{4}$$

The derivative of this function with respect to the radius r is

$$\frac{dA}{dr} = 4\pi r - 2000r^{-2}$$

The critical values are found by letting $dA/dr = 0$

$$4\pi r - 2000r^{-2} = 0$$

This can be solved for r by multiplying the equation by r^2. Thus

$$4\pi r^3 - 2000 = 0$$

$$r^3 = \frac{2000}{4\pi}$$

$$r = 5.4 \text{ in.}$$

This value may be checked using the second derivative test with

$$\frac{d^2A}{dr^2} = 4\pi + 4000r^{-3}$$

The second derivative is positive for all positive values of r and, therefore, we conclude the area will be a minimum for $r = 5.4$ in. The height can be found by substituting this value of r into Equation 3.

$$h = \frac{1000}{\pi (5.4)^2} = 10.9 \text{ in.}$$

The minimum area is 554 in.², which is found by substituting $r = 5.4$ in. and $h = 10.9$ in. into Equation 2.

It is important that you note the steps involved in finding maxima and minima in problems like Example 31.

1. Study the problem and determine what relationships are known. In most cases you will be able to determine equations involving the variables in the problem.

2. Determine which variable is to be either maximized or minimized. Using the equations found in step 1, write the variable of interest as a function of only one of the other variables. In Example 31, the surface area of the cylinder was expressed as a function of only the radius by using the given information and Equations 1 and 2.

3. Find the critical values by setting the derivative of the function to be maximized or minimized equal to zero.

4. Test the critical values to establish maxima and minima.

EXERCISES

1. The velocity of a rocket varies according to $v = 4t - t^2$ miles per hour. At what time does the rocket reach its maximum velocity? What is the magnitude of the velocity at this time?

2. The displacement of a cam-operated push-rod is given by $s = 1 + 3t^2 - t^3$, where t is the elapsed time and s is in inches. What is the maximum displacement of the push-rod?

3. The displacement of an electron in a magnetic field varies with time according to $s = t + 3t^2 - 4t^3$ inches. What is the maximum velocity that the electron achieves?

4. The energy in an inductance coil varies with time as $w = 3 + 4t - 3t^2$ joules. What is the maximum energy of the coil?

5. Find two integers whose sum is 30, and whose product is a maximum.

6. Find two integers whose sum = 12, such that the product of one number by the square of the other, is a maximum.

7. A box is to be constructed with a volume of 32 cu ft. The box is to have a square base. If the top is not included, what are the dimensions of the box that will have a minimum surface area?

8. A storage bin is to be constructed with a square base and no top. What is the maximum volume of the storage bin if only 432 sq ft of material can be used?

9. A box without a top is to be constructed from a 10 in. × 16 in. metal sheet. The box is to be formed by cutting equal squares from each corner of the sheet and then bending up the sides. What size square should be cut to give the maximum volume?

10. Find the dimensions of an open cylindrical oil drum having a minimum surface area that will hold 64 cu ft of oil.

11. A rectangular field is to be enclosed by 1000 ft of fence. What is the maximum area that can be enclosed?

12. A battery has an electromotive force E and an internal resistance r. When a current I is delivered by the battery, its power output is given by $P = EI - rI^2$. For what current is the power output a maximum?

13. The strength of a beam having a rectangular cross section varies directly as the breadth and the square of the depth. What is the strongest beam that can be cut from a log whose diameter is 10 in.

14. The Ace Safe Company sells n safes per month at S dollars each. The cost of manufacturing n safes varies in accordance with $C = 600 + 10n + n^2$ dollars/month. The selling price of each safe is $S = 70$ dollars. How many safes should be made each month to maximize the profit? (*Hint*. Profit equals total income minus cost.)

15. A capacitor is to be sealed in a cylindrical container having a volume of 2000 cm³. Find the dimensions of the container requiring the least amount of material.

4

Differentials

4.1 INTRODUCTION

When we introduced the symbol dy/dx to designate the derivative, we indicated that it was to be thought of as $\lim_{\Delta x \to 0} (\Delta y/\Delta x)$ and not as a fraction formed by dividing dy by dx. However the fractional interpretation is sometimes convenient in calculus, especially in making approximations.

In this chapter we are going to show that the derivative dy/dx may be interpreted as a fraction if the proper meaning is given to dy and dx.

4.2 DIFFERENTIALS

The symbol dx when considered by itself is called the *differential of x*. Similarly, dy is called the *differential of y*. Before we investigate the geometrical interpretation of differentials, let us note the relationship that exists between dx and dy.

We define the *differential dx* to be equal to any arbitrarily chosen increment of x. This means that we may choose dx as large or small as we please. In this sense, dx is an independent variable. Now we shall define the differential of the dependent variable in terms of the differential dx.

The *differential dy is the product of the derivative of the function times the differential dx*. This is written as

$$dy = f'(x) \, dx \qquad (1)$$

where $f'(x)$ is the derivative of $f(x)$.

Obviously, the definition in (1) can be divided by dx to yield the familiar statement

$$\frac{dy}{dx} = f'(x) \qquad (2)$$

The importance of this result is that the left side of Equation 2 was obtained by dividing the differential dy by the differential dx. Consequently, when dx and dy are defined as above, we may interpret dy/dx as a fraction.

The differential of a function can be found by multiplying the derivative of the function by the differential of the independent variable.

89

Example 1.   Find the differential of $y = 4x^3 + 5x$.

Solution

$$\frac{dy}{dx} = 12x^2 + 5$$

$$dy = (12x^2 + 5)\, dx$$

We use the letter d as a symbol for the *differential operator*. When written before a function, it indicates that we are to find the differential of the given function. Specifically, the differential of a function $f(x)$ is denoted

$$d[f(x)] = f'(x)\, dx$$

where $f'(x)$ is the derivative of $f(x)$.

Example 2.   Evaluate $d[\sqrt{z}]$.

Solution

$$d[\sqrt{z}] = d[z^{1/2}] = \frac{1}{2} z^{-1/2}\, dz$$

EXERCISES

Find the differential of each of the following expressions.

1. $y = 4x^2$
2. $p = t^3 + 4t$
3. $z = 1/\sqrt{x}$
4. $y = (x^3 - x)/\sqrt{x}$
5. $s = \sqrt{3t}$

6. $\theta = (\phi + 4)^2$
7. $a = 4\sqrt[3]{b^2}$
8. $y = t^4 - 1/t^2$
9. $m = p^{-3} - p^{-2} - 4p$
10. $i = 4\pi t^2$

Evaluate the indicated differentials.

11. $d[3x - x^2]$
12. $d[\sqrt{q} - \sqrt[3]{q}]$
13. $d[s^{5/3} + 3s^{2/3}]$
14. $d[(\sqrt{x} - x)^2]$
15. $d\left[\frac{1}{3} t^{-5}\right]$

4.3   GEOMETRIC INTERPRETATION OF DIFFERENTIALS

The physical meaning of the differentials dx and dy can best be explained by comparing them graphically with the increments Δx and Δy, respectively. To this end we have drawn two identical curves in Figure 4.1. Both curves are drawn through the points P and P_1.

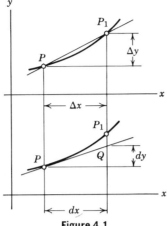

Figure 4.1

It is customary to let dx equal Δx, as we have done in Figure 4.1. This helps to emphasize the difference between dy and Δy.

For purposes of comparing Δy and dy, we have drawn a line through PP_1 in the upper curve and a line tangent to the lower curve at P. The increment Δy is represented in the upper curve, and the differential dy in the lower. We see from this that Δy and dy represent very different quantities. The increment Δy represents the change in the functional value as a point moves along the curve. In contrast, the differential dy represents the change in y that would occur if the point moved along the tangent line. Algebraically, Δy and dy are related by the equation

$$\Delta y = dy + P_1 Q$$

where $P_1 Q$ is the vertical distance from the indicated tangent line to the curve at the point P_1. We observe that dy will always differ from Δy by an amount $P_1 Q$. However, it should be noted that dy is not necessarily less than Δy. Had we drawn the curve between P and P_1 to be concave-down, dy would have been greater than Δy.

4.4   APPROXIMATION

There are times when we wish to compute the change in the value of a function caused by a small change in the independent variable, but the complexity of the manipulation makes it very difficult to do so. In problems of this type, an estimate of the change in the value of the function can be obtained by using the concept of a differential. As we pointed out in the preceding article, dy and Δy are very different quantities. We note,

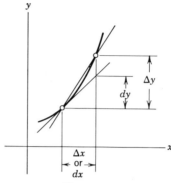

Figure 4.2

however, that when Δx is very small, dy and Δy are nearly equal. This fact is illustrated in Figure 4.2.

We denote the relationship between Δy and dy by

$$\Delta y \approx dy \qquad \text{(if } \Delta x \text{ is very small)} \tag{3}$$

or replacing dy with $f'(x)\, dx$

$$\Delta y \approx f'(x)\, dx \tag{4}$$

Equations 3 and 4 together show how the differential dy is used to approximate the increment Δy. To illustrate the use of this concept, consider the following examples.

Example 3.   Given the function $y = x^{3/2}$, what is the approximate change in y when x changes from 9 to 9.01?

Solution.   The differential dy is found as indicated in Example 1.

$$\frac{dy}{dx} = \frac{3}{2} x^{1/2}$$

$$dy = \frac{3}{2} x^{1/2}\, dx$$

The numerical value of dy may be found by letting $x = 9.00$ and $dx = 0.01$ in this equation. Thus

$$dy = \frac{3}{2}(9.00)^{1/2}(0.01) = 0.045$$

You can find the exact change in y by subtracting $(9.00)^{3/2}$ from $(9.01)^{3/2}$. As you can see, this is a more laborious task than finding dy.

Example 4.   A square sheet of steel has an initial side length of 10 inches. After the sheet is heated, the side length is 10.02 inches: a. Find the approximate change in area; and b. Find the exact change in area.

Solution.   The functional relationship between the area A, and the side length s, can be written $A = s^2$.
a.   The derivative of area A with respect to the side length s yields

$$\frac{dA}{ds} = 2s$$

and so

$$dA = 2s\,ds$$

Letting $s = 10.00$, and $ds = 0.02$

$$dA = 2(10.00)(0.02) = 0.4 \text{ in.}^2$$

b.   The exact change, ΔA, is found by

$$\Delta A = (10.02)^2 - 10^2$$
$$= 100.4004 - 100.0000 = 0.4004 \text{ in.}^2$$

In this case, our approximation is within 0.0004 in.^{2} of the actual value.

Example 5.   The work done by a particle in an electric field is given by $W = s^3$, where s is the displacement. What is the work done in moving the object from 1.99 to 2.01 meters. Work is measured in joules if displacement is given in meters.

Solution.   The approximate work done is given by

$$dW = 3s^2\,ds$$

Here we choose $s = 2.00$, since it is easier to manipulate than either 1.99 or 2.01. Remember that dW is only an approximation to ΔW and, therefore, we assign to s the most convenient value. Letting $ds = 0.02$, we have

$$dW = 3(2.00)^2(0.02) = 0.24 \text{ joules}$$

EXERCISES

Find the differential of the given function. Using the differential, find the approximate change in the function for the indicated change in the independent variable.

1. $y = 4x^3$ for $x = 5.0$ to $x = 5.1$
2. $z = 3t^{4/3}$ for $t = 8.00$ to $t = 8.04$
3. $s = \sqrt{t}$ for $t = 8.95$ to $t = 9.00$
4. $p = r^4 - 3r^2$ for $r = 3.98$ to $r = 4.01$
5. $y = (x+2)^2$ for $x = 15.0$ to $x = 14.8$

6. The horsepower of an internal combustion engine is given by $P = nd^2$, where n is the number of cylinders and d is the bore of each cylinder. What is the approximate increase in horsepower for an eight cylinder engine if the cylinders are rebored from 3.500 in. to 3.525 in.?

7. The current in a circuit changes according to $i = 0.1t^4 + 5$. What is the approximate change in current from $t = 4.00$ sec to $t = 4.02$ sec?

8. The acceleration of an object is $a = (t+2)/\sqrt{t}$ ft/sec². What is the approximate change in velocity from $t = 4.00$ to 4.08 sec?

9. The current in a circuit varies according to $i = (t+1)^{1/3}$ amp. What is the approximate charge transferred between $t = 0.500$ and $t = 0.501$ sec?

10. The power in a resistor is given by $p = ri^2$, where r is the resistance and i is the current. If the current in a 750 ohm resistor changes from 3.000 to 3.005 amp, what is the approximate change in power?

11. The heat radiation R from an incandescent lamp is $R = \sigma T^4$, where σ is a constant, and T is the operating temperature. What is the approximate change in radiation if the temperature changes from $T = 2500$ to $T = 2510$ degrees? Assume that $\sigma = 5 \times 10^{-10}$.

12. A circular disk is supposed to have a radius of 2.00 in. During the machining operation, the radius is cut to 1.98 in. What is the approximate loss in area?

13. What is the approximate error in the volume of a cubical box if the sides are made 25.10 cm, instead of 25.00 cm?

14. A 15 in. square plate is to be constructed so that its area is not in error by more than 0.03 in.². What is the approximate tolerance that must be placed on the side length?

15. A spherical oil tank is to be constructed with a radius of 25 ft. Determine the approximate error in the volume of the tank if it is constructed with a radius of 24 ft.

16. The force of attraction between two unlike magnetic poles is given by $F = 10/s^2$, where F is the force, in dynes, and s is the distance, in cm, separating the two poles. If the poles are initially 20 cm apart, what is the approximate change in the force of attraction when $ds = 0.5$ cm?

17. A circular ring has an inside radius of 5 in. and an outside radius of 10 in. What change in the inside radius will decrease the area of the ring by 1 in.²?

18. In the preceding problem, what change in outside radius will decrease the area by 1 in.²?

19. Referring again to Exercise 17, assume that the inside and outside radius change simultaneously by equal amounts. What equal changes in the radii will decrease the area by 1 in.²?

5

The Antiderivative and its Applications

5.1 INTRODUCTION

In differential calculus we study methods of finding the derivative of a function. Perhaps you have wondered if it is of practical value to reverse the differentiation process and reconstruct a function if we know its derivative. The answer to this is "yes." The process by which we do this is called *antidifferentiation*, or sometimes (erroneously) it is called integration. Differentiation and antidifferentiation are inverse operations in the same sense that exponentiation and extraction of roots are inverse operations.

5.2 THE ANTIDERIVATIVE

Antidifferentiation is quite simply the reverse of differentiation. Thus, antidifferentiation problems are essentially equivalent to the following: given $f(x)$, find a function F such that $F'(x) = f(x)$. There are two commonly used notations for the antiderivative of $f(x)$, namely

$$D_x^{-1} f(x) \tag{1}$$

and

$$\int f(x) \, dx \tag{2}$$

Of the two notations, the latter is by far the most popular although probably the less desirable. We shall use $\int f(x) \, dx$ as the symbol for the antiderivative of $f(x)$ because of its wide acceptance in the field of engineering. We point out in passing that (2) is frequently called the "integral of $f(x)$." The danger in using the word "integral" to denote the antiderivative of $f(x)$ will become clear in the next chapter.

From Chapter 3 and the definition of the antiderivative, it follows that

Since $d/dx \, [x^3 + 1] = 3x^2$,   we have $\int 3x^2 \, dx = x^3 + 1.$

Since $d/dx \, [x^3 - 2] = 3x^2$,   we have $\int 3x^2 \, dx = x^3 - 2.$

Since $d/dx \, [x^3 + 1599] = 3x^2$,   we have $\int 3x^2 \, dx = x^3 + 1599.$

We note here that $\int 3x^2 \, dx$ has three correct answers depending upon what constant term appeared in the original function. This fact could have been anticipated since the derivative of a constant is zero. In cases where we have no previous knowledge of the constant, we must still allow for its existence. Therefore, we shall write the antiderivative of $3x^2$ as

$$\int 3x^2 \, dx = x^3 + C$$

where C may be any constant. A constant arising in this way is called an *arbitrary constant* or a *constant of antidifferentiation*.

Just as we have formulas for differentiation, we would hope to have formulas for antidifferentiation. At present our best source of formulas to tell us how to antidifferentiate is a differentiation formula. Since

$$\frac{d}{dx}\left[\frac{x^{n+1}}{n+1}\right] = \frac{(n+1)x^{(n+1)-1}}{n+1} = x^n$$

for all n except $n = -1$, we have the antidifferentiation formula

$$\int x^n \, dx = \frac{x^{n+1}}{n+1} + C, \qquad n \neq -1 \tag{3}$$

The exceptional case, $n = -1$, is quite important; we shall study it at a later time.

Example 1.   Find $\int \sqrt{x} \, dx$.

Solution

$$\int \sqrt{x} \, dx = \int x^{1/2} \, dx = \frac{x^{3/2}}{3/2} + C = \frac{2}{3}x^{3/2} + C$$

The following properties of the antiderivative can be proved by appealing directly to its definition as the inverse of the derivative.

Property 1.   If c is a constant, then $\int cf(x) \, dx = c \int f(x) \, dx$.

Property 2. $\int [f(x) + g(x)] \, dx = \int f(x) \, dx + \int g(x) \, dx$

Properties 1 and 2, taken together, are often called the linearity property of antiderivatives. Although both properties are extremely useful, the first one causes beginning students the most difficulty because they sometimes forget that c must be a constant.

Example 2.   Find $\int 6ds/s^3$.

Solution

$$\int \frac{6ds}{s^3} = 6 \int s^{-3}ds = 6\frac{s^{-2}}{-2}+C = -3s^{-2}+C$$

Example 3.   Find $\int (3x^2+5x+6) \, dx$.

Solution

$$\int (3x^2+5x+6) \, dx = 3 \int x^2 \, dx + 5 \int x \, dx + 6 \int dx$$

$$= x^3+\frac{5x^2}{2}+6x+C$$

Only one constant is written in this solution, because the sum of three constants is, in turn, a constant.

Many antidifferentiation problems arise which are not immediately in the form $\int x^n \, dx$. In these instances it may be possible to rearrange the function into the desired form by algebraic manipulation.

Example 4.   Evaluate $\int (x-1)^2 \, dx$.

Solution.   The antidifferentiation of $(x-1)^2$ can be effected by first expanding the binomial and then using the linearity property of anti-derivatives. Thus

$$\int (x-1)^2 \, dx = \int (x^2-2x+1) \, dx$$

$$= \int x^2 dx - 2 \int x \, dx + \int dx$$

$$= \frac{x^3}{3} - x^2 + x + C$$

Example 5.   Evaluate $\int x^{1/3}(x^2-x^{-1}) \, dx$.

Solution.   This antiderivative may be evaluated if we first multiply each term in the parenthesis by $x^{1/3}$. Then

$$\int x^{1/3}(x^2-x^{-1}) \, dx = \int (x^{7/3}-x^{-2/3}) \, dx$$

$$= \int x^{7/3} \, dx - \int x^{-2/3} \, dx$$

$$= \frac{x^{10/3}}{10/3} - \frac{x^{1/3}}{1/3} + C$$

$$= \frac{3}{10}x^{10/3} - 3x^{1/3} + C$$

EXERCISES

Evaluate the following antiderivatives.

1. $\int 3\,dx$

2. $\int y\,dy$

3. $\int 3z\,dz$

4. $\int (z^2 - 3)\,dz$

5. $\int (x^3 + x^2)\,dx$

6. $\int 5x^4\,dx$

7. $\int (3x^7 + 6x^9)\,dx$

8. $\int \dfrac{x^3}{2}\,dx$

9. $\int p^{1/3}\,dp$

10. $\int 4r^{4/5}\,dr$ $20 r^{\frac{9}{5}} + C \over 6$

11. $\int \sqrt{3\phi}\,d\phi$

12. $\int q^{-3}\,dq$ $\dfrac{q^{-2}}{-2} + C$

13. $\int \dfrac{dx}{x^5}$

   $\int y^{\frac{1}{2}}\,dy = \dfrac{y^{\frac{3}{2}}}{\frac{3}{2}} + C = 2y^{\frac{3}{2}} + C$

14. $\int \dfrac{dy}{\sqrt{y}}$ $2y^{\frac{1}{2}} + C$

15. $\int x(x - x^3)\,dx$

16. $\int x\sqrt{x}\,dx$ $\dfrac{x \cdot x^{\frac{5}{2}}}{\frac{5}{2}}$

   $= \int x \cdot x^{1/2}\,dx$ $= x^{\frac{3}{2}}\,dx = x^{\frac{5}{2}}$
   $= \dfrac{x^2 \cdot x^{\frac{3}{2}}}{\frac{3}{2}} \div \dfrac{x^3}{3}$ $\dfrac{5}{\frac{5}{2}}$ $\dfrac{2x^{\frac{5}{2}}}{5} + C$

17. $\int \dfrac{\theta^2 - 2\theta^3}{\theta^5}\,d\theta$

18. $\int \dfrac{dt}{2\sqrt{t^3}}$

19. $\int (x - 3)^2\,dx$

20. $\int x(3x + 2)^2\,dx$

21. $\int \dfrac{(x^2 + 1)^2}{x^2}\,dx$

22. $\int (\sqrt{x} - 3)^2\,dx$

23. $\int \dfrac{2x^3 - 3x^2 + x}{\sqrt[3]{x}}\,dx$

24. $\int \dfrac{3\,dx}{\sqrt[3]{x^2}}$

25. $\int \left(x^3 + \dfrac{1}{x^3}\right)\,dx$

26. $\int (\sqrt[3]{x} - x)\sqrt{x}\,dx$

27. $\int [t^3]^4\,dt$ $= t^{12}\,dt = \dfrac{t^{13}}{13} + C$

28. $\int [\sqrt{t^5}]^{1/3}\,dt$

29. $\int [\sqrt[3]{\phi} + \sqrt[4]{\phi}]^2\,d\phi$

30. $\int (x^3 - \sqrt{x})(x - 4)\,dx$

31. $\int p^{1/8}(1 - p^{-1/3})\,dp$

32. $\int (x - 2)(x + 1)^2\,dx$

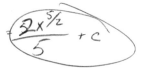

5.3   CONSTANT OF ANTIDIFFERENTIATION

Geometrically, curves of the form

$$y = x^2 + C$$

all have the same shape. From our study of curve plotting we realize that the constant term gives the y-intercept; it does not alter the shape of the graph. Therefore, $y = x^2 + C$ actually represents many curves, one curve for each value of C that we choose. Each curve has an identical shape but a different y-intercept. When we have a collection of curves differing only in their location in the plane, we call this collection a *family of curves*. Figure 5.1 shows several members of the family of curves of the equation $y = x^2 + C$. Our reason for introducing the idea of a family of curves is to show that the antiderivative of a function represents such a family of curves all of which have the same derivative.

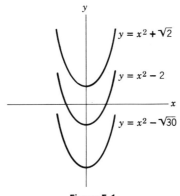

Figure 5.1

Example 6.   Find the equation of the family of curves whose slope is given by x^2.

Solution.   The slope of a curve at any point is given by its derivative, which means that

$$\frac{dy}{dx} = x^2$$

The equation of the family of curves having this slope (Figure 5.2) is then

$$y = \int x^2 \, dx = \frac{x^3}{3} + C$$

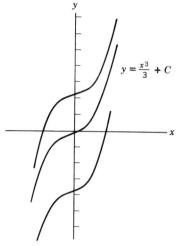

Figure 5.2

In some problems a knowledge of the family of functions having a given derivative is sufficient, but in most cases it is necessary to find a particular member of the family. In order to determine such a member of the family, information must be given to determine a particular value of C. Conditions of this type are referred to as *boundary conditions*. A problem in which the derivative and a boundary condition are given is called a *boundary value problem*. The remainder of this chapter is concerned with this kind of problem.

Example 7.   Find the equation of the graph passing through the point $(2,5)$ if the slope of the curve is given by $3x$.

Solution.   The slope of a curve is given by its derivative, so

$$f'(x) = 3x$$

The family of functions having this derivative is

$$f(x) = \int 3x\,dx = \frac{3}{2}x^2 + C \tag{4}$$

Since the curve passes through $(2,5)$, the coordinates of this point must satisfy Equation 4. Thus,

$$5 = \frac{3}{2}(2)^2 + C$$

or

$$C = -1$$

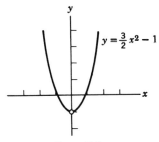

$$y = \frac{3}{2}x^2 - 1$$

Figure 5.3

The desired equation is then $y = \frac{3}{2}x^2 - 1$; its graph is shown in Figure 5.3.

In some cases it may be necessary to antidifferentiate more than once to obtain the relationship between x and y. As a case in point, suppose we wish to find the function whose second derivative is given. To do this requires that we find two separate antiderivatives. The antiderivative of the second derivative is the first derivative of the function, and the antiderivative of this is, in turn, the desired function. Two arbitrary constants are required since we must find two antiderivatives. In order to determine the values of both constants, we must be given two boundary conditions.

Example 8.   Given that $(d^2y)/(dx^2) = x$, find the equation relating x and y if $dy/dx = 2$, when $x = 1$ and $y = 2$, when $x = -3$.

Solution.   The first derivative is given by the antiderivative of the second derivative and, therefore,

$$\frac{dy}{dx} = \int x\,dx = \frac{x^2}{2} + B$$

We may now evaluate the constant B by using the boundary condition: $dy/dx = 2$, when $x = 1$. Thus

$$2 = \frac{1}{2}(1)^2 + B$$

or

$$B = \frac{3}{2}$$

The first derivative of the desired function is then

$$\frac{dy}{dx} = \frac{1}{2}x^2 + \frac{3}{2}$$

The antiderivative of this is the family

$$y = \int \left(\frac{1}{2}x^2 + \frac{3}{2} \right) dx = \frac{1}{6}x^3 + \frac{3}{2}x + C$$

Substituting the boundary condition $y = 2$ when $x = -3$ into this equation, we have

$$2 = \frac{1}{6}(-3)^3 + \frac{3}{2}(-3) + C$$

or

$$C = 11$$

Thus, the desired equation is

$$y = \frac{1}{6}x^3 + \frac{3}{2}x + 11$$

EXERCISES

In Exercises 1–10, determine the function having the given derivative or differential and sketch two or three members of each family.

1. $\dfrac{dy}{dx} = 5$ $y = 5x + C$

6. $F'(s) = \sqrt{s}$

2. $\dfrac{dp}{dt} = -4$

7. $f'(x) = x^3$ $y = \dfrac{x^4}{4} + C$

3. $\dfrac{ds}{dt} = 3t$ $s = \dfrac{3t^2}{2} + C$

8. $dy = 3dx$

4. $\dfrac{dy}{dx} = -4x$

9. $dy = (1 - 3x)\,dx$ $y = x - \dfrac{3x^2}{2} + C$

5. $\dfrac{di}{dt} = t^2$ $i = \dfrac{t^3}{3} + C$

10. $dy = x\,dx$

Determine the equation of the graph having the given slope and passing through the indicated point.

Slope	Boundary Condition

11. $\dfrac{dy}{dx} = 3x$ (1,2) $c = \frac{1}{2}$ $y = \dfrac{3x^2}{2} + \dfrac{1}{2}$

$y = \dfrac{3x^2}{2} + C$

$2 = \dfrac{3}{2} + C$

12. $\dfrac{dq}{dt} = 4t^2 + 5$ (0,3)

(handwritten) $q = \dfrac{4t^3}{3} + 5t + C$ $\left(q = \dfrac{4t^3}{3} + 5t + 3\right)$

$3 = 0 + 0 + C$ $\boxed{C = 3}$ or $\dfrac{102}{34}$

13. $\dfrac{ds}{dr} = \sqrt[3]{r} \;\Rightarrow\; r^{\frac{2}{3}}$ (8,−2)

(handwritten) $s = \dfrac{3r^{\frac{5}{3}}}{5} - 9.905$ $s = 3r^{\frac{5}{3}}$ ... $\dfrac{5}{5}$ ... $8 = \dfrac{3(-2)^{5/3}}{5} + C$ $C = -9.905$

14. $\dfrac{dy}{dx} = x^3 - 4x$ (2,0)

(handwritten) $\dfrac{3(-3.175)}{5} + C$

15. $\dfrac{dp}{dt} = \sqrt{t} + \dfrac{1}{\sqrt{t}} = t^{\frac{1}{2}} + t^{-\frac{1}{2}}$ (4,1)

(handwritten) $= \dfrac{2t^{3/2}}{3} + 2t^{1/2}$ $1 = \dfrac{16}{3} + 4 + C$ $8 = -1.905 + C$

$1 = 6.667 + C$ $p = \dfrac{2t^{3/2} + 2t^{1/2}}{3} - 8.33t$

In each of the following problems the second derivative is given with the indicated boundary conditions. Find the function satisfying these conditions.

16. $\dfrac{d^2s}{dt^2} = t$; $s = 2$ and $\dfrac{ds}{dt} = -1$ when $t = 0$

17. $\dfrac{d^2y}{dx^2} = x$; $y = 0$ when $x = 4$, and $\dfrac{dy}{dx} = 2$ when $x = 0$

18. $\dfrac{d^2m}{dp^2} = 1 - p$; $m = 1$, and $\dfrac{dm}{dp} = 6$ when $p = 2$

(handwritten) do

19. $f''(z) = 3z^2$; $f(0) = 4$ and $f'(-1) = -3$

20. $g''(t) = \dfrac{1}{t^3}$; $g(3) = 1$ and $g'\left(\dfrac{1}{2}\right) = 4$

5.4 KINEMATICS

Kinematics is the study of the motion of objects without regard to the forces causing the motion. The motion of an object is described by its displacement, velocity, and acceleration. We have already seen how we can find the equations for velocity and acceleration from the distance-time equation. We now reverse this process and find the velocity equation if we know the acceleration, and the displacement equation if we know the velocity. Stating this symbolically, it follows that since $v = ds/dt$,

$$s = \int v\,dt \qquad (5)$$

Equation 5 says that the displacement-time equation is the antiderivative of the velocity-time equation. Similarly, it follows that the velocity-time equation is the antiderivative of the acceleration-time equation; that is, since $a = dv/dt$,

$$v = \int a\, dt \tag{6}$$

The following examples will help to explain the use of Equations 5 and 6 in solving kinematic problems.

Example 9.   A rocket is fired vertically upward from its launching pad at a time designated $t = 0$. If the velocity, in ft/sec, of the rocket is given by the equation $v = 2t^3$, how high above the ground is the rocket after 10 sec?

Solution.   The distance-time equation is the antiderivative of the velocity-time equation, so that

$$s = \int v\, dt = \int 2t^3 dt = \frac{1}{2}t^4 + C$$

In order to find the value of C, we must use the implied condition that $s = 0$ when $t = 0$. Hence

$$0 = \frac{1}{2}(0)^4 + C$$

or

$$C = 0$$

The distance-time equation for this rocket is then

$$s = \frac{1}{2}t^4$$

The altitude of the rocket after 10 sec can now be found by substituting $t = 10$ into this equation. Thus

$$s = \frac{10^4}{2} = 5000 \text{ ft}$$

Example 10.   The acceleration of a particle is given by $a = 6t + 10$ cm/sec². If the initial velocity of the particle is 3 cm/sec and the initial displacement is zero, how far has the particle moved after 2 sec, and what is its velocity at this time?

Solution.   The equation for the velocity is

$$v = \int a\, dt = \int (6t + 10)\, dt = 3t^2 + 10t + K$$

The constant K is evaluated by using the boundary conditions $v = 3$ when $t = 0$ in this equation. Thus

$$3 = 3(0)^2 + 10(0) + K$$

or

$$K = 3$$

and

$$v = 3t^2 + 10t + 3$$

We now use the velocity-time equation to find the distance-time equation; that is,

$$s = \int v\,dt = \int (3t^2 + 10t + 3)\,dt = t^3 + 5t^2 + 3t + K_1$$

Substituting the boundary conditions $s = 0$ when $t = 0$, we have

$$K_1 = 0$$

and, therefore

$$s = t^3 + 5t^2 + 3t$$

The distance traveled by the particle in 2 sec is then

$$s = (2)^3 + 5(2)^2 + 3(2) = 34 \text{ cm}$$

and its velocity at that time is

$$v = 3(2)^2 + 10(2) + 3 = 35 \text{ cm/sec}$$

Before giving the next example, let us consider some facts about objects moving through the atmosphere without some propelling force, such as, a thrown ball or a bullet after it leaves the gun barrel. Objects of this type are called *projectiles*. First, note that all projectiles have a constant acceleration toward the earth of approximately 32 ft/sec². If the projectile is falling, the acceleration is considered positive, and if it is rising the acceleration is considered negative. Secondly, if a projectile is thrown upward it will be decelerated 32 ft/sec² until its velocity is zero; at that instant it starts to fall. It is then accelerated 32 ft/sec² until it strikes the ground. When it strikes the ground its speed is the same as it was when it was launched upward.

Example 11.   A man fires a bullet vertically upward from the roof of a building which is 75 ft high. The bullet has an initial velocity of 500 ft/sec.

(a) What is the velocity of the bullet after 10 seconds?
(b) How long does it take the bullet to reach its maximum altitude?
(c) How high is the bullet above the ground when it reaches maximum altitude?

Solution

(a) The acceleration-time equation is

$$a = -32$$

The negative sign indicates that the bullet will initially be decelerated. The equation for the velocity is then given by

$$v = \int a\,dt = \int (-32)\,dt = -32t + C_1$$

The constant of antidifferentiation can be evaluated using $t = 0$ and $v = 500$. Thus

$$500 = -32(0) + C_1$$
$$C_1 = 500$$

and

$$v = -32t + 500 \tag{7}$$

The velocity when $t = 10$ sec is then

$$v = -32(10) + 500 = 180 \text{ ft/sec}$$

(b) The time required for the bullet to reach maximum altitude is found by letting $v = 0$ in Equation 7 and solving for t. We let $v = 0$, because the velocity at the instant that the bullet quits rising is zero. Thus

$$0 = -32t + 500$$

and

$$t = \frac{-500}{-32} = 15.6 \text{ sec}$$

(c) In order to find its maximum altitude above the ground, we need to know the equation that relates displacement to time. This can be found by taking the antiderivative of the velocity-time equation; that is,

$$s = \int v \, dt = \int (-32t + 500) \, dt = -16t^2 \mid 500t + C_2$$

To evaluate C_2 we note that we want the altitude above the ground and, therefore, we use $s = 75$ when $t = 0$. (The bullet is already 75 ft above the ground when the gun is fired.)

$$75 = -16(0)^2 + 500(0) + C_2$$

$$C_2 = 75$$

Hence

$$s = -16t^2 + 500t + 75$$

Now substituting $t = 15.6$ sec into this equation, we get

$$s = -16(15.6)^2 + 500(15.6) + 75 = 3980 \text{ ft}$$

EXERCISES

Solve each of the following problems by the use of antidifferentiation.

1. The acceleration of a rocket sled starting from rest is $a = \sqrt{t}$, where t is the time. What is the equation for the velocity of the sled?
2. Find the equation for the displacement of the sled in Exercise 1.
3. A car starts from rest and accelerates 8 ft/sec². What is the elapsed time to 88 ft/sec? (88 ft/sec = 60 mi/hr)
4. In Exercise 3, how far has the car traveled from the starting line by the time it reaches a velocity of 100 ft/sec?
5. An airplane lands with a velocity of 200 ft/sec. At the instant of touchdown the airplane decelerates -10 ft/sec² until it comes to rest. How far does the plane taxi before coming to rest?
6. A new braking system has been designed to decelerate an airplane -4 ft/sec². What is the minimum runway necessary for planes landing with a velocity of 200 ft/sec?
7. The velocity of a particle is equal to the reciprocal of the displacement. Find the equation for the displacement of the particle, if $t = 2$, $s = 3$ is a boundary condition.
8. A stone is dropped from the top of a building. If the building is 50 ft high, how long will it take the stone to reach the ground?
9. How long will it take the stone to reach the ground if it is thrown downward with a velocity of 5 ft/sec?
10. The velocity of a particle is $v = t^{4/3}$ ft/sec. Find the acceleration of the particle when $t = 27$ sec.

11. A baseball is thrown vertically upward with a velocity of 96 ft/sec. Find the velocity of the ball after 2 sec.

12. In Exercise 11, determine the maximum altitude reached by the ball. How long does it take to reach this height?

13. The kinetic energy of a moving object is the capacity of the object to do work by virtue of its motion. The formula for kinetic energy is $K.E. = Wv^2/2g$, where W is the weight, v is the velocity, and g is the acceleration of gravity. If a 2000 lb car starts from rest and accelerates according to $a = \frac{1}{4}t^{-1/2}$ ft/sec², what is its $K.E.$ when $t = 2$ sec?

14. What is the $K.E.$ of a 2 lb rocket 3 sec after it is launched, if its acceleration is given by $a = 3t^3$ ft/sec²?

15. The Mach number N of an object moving through the atmosphere is defined by $N = v/v_s$, where v is the velocity of the object and v_s is the speed of sound. Under standard conditions $v_s = 1110$ ft/sec. A bomb is dropped from an altitude of 10,000 ft. Find the Mach number of the bomb 20 sec after it is released if its acceleration is $a = (32 - 0.1\, t)$ ft/sec².

16. A SAM missile accelerates from rest according to $a = 20t + 100$ ft/sec². How long does it take the missile to reach Mach 1?

5.5   SERIES AND PARALLEL CIRCUITS

The many electronic devices on the market today are made up of various series and parallel combinations of resistors, inductors, and capacitors. The relationship that exists between current and voltage in these RLC combinations is of primary importance in the design of electronic equipment. (The letters RLC are used to denote the combination of a resistor, an inductor, and a capacitor.) To determine the current in either a series or a parallel RLC circuit when a voltage is applied to the combination, it is necessary to use Kirchhoff's two laws. These laws are basic to the study of series and parallel circuits. In this article we are going to describe the use of Kirchhoff's laws, but first let us extend the electrical concepts that we introduced in Section 3.6.

The electric charge transferred past a point can be found from the definition of current. It will be recalled that current is defined by $i = dq/dt$. The equation for the charge is then found by taking the antiderivative of the current-time equation. Symbolically, this is written

$$q = \int i\,dt \tag{8}$$

Example 12.   The current in a resistor is given by $i = t^{1/2} + 2$ amp. What is the equation for the charge if $q = 0$ when $t = 0$?

Solution.   Using $i = t^{1/2} + 2$ in Equation 8, we have

$$q = \int i\,dt = \int (t^{1/2} + 2)\,dt = \frac{2t^{3/2}}{3} + 2t + K$$

We use K for the constant of antidifferentiation, because C is the symbol for capacitance. Letting q and t equal zero, we have

$$0 = \frac{2(0)^{3/2}}{3} + 2(0) + K$$

so

$$K = 0$$

and the equation for the charge is

$$q = \frac{2t^{3/2}}{3} + 2t$$

Another relationship that we recall from Chapter 3 is

$$v = -L\frac{di}{dt}$$

which gives the voltage induced in coil of inductance L by a given current. The voltage drop v_L across the inductor is the negative of the induced voltage. Thus,

$$\frac{di}{dt} = \frac{1}{L}v_L$$

it follows that the current required to cause a voltage drop v_L is

$$i = \int \frac{1}{L}v_L dt = \frac{1}{L}\int v_L dt \qquad (9)$$

Also, we have the current in a capacitor given by $i = C(dv/dt)$ and, therefore, the equation for the voltage across the plates of the capacitor can be written

$$v = \frac{1}{C}\int i dt \qquad (10)$$

Now we are ready to discuss Kirchhoff's two laws. These laws are basic to the analysis of series and parallel electrical circuits.

Kirchhoff's first law states that, at any instant, the algebraic sum of the currents flowing toward any point in a circuit is equal to zero. This law can be written as the equation

$$\Sigma \text{ currents} = 0 \qquad (11)$$

In mathematics and science, the Greek capital letter sigma (Σ) is used to represent the algebraic summation of quantities. In the next chapter we

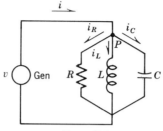

Figure 5.4

shall use this notation frequently. At that time we shall extend our understanding of the symbol.

To illustrate this law, consider the circuit in Figure 5.4. This figure shows a generator connected to a parallel *RLC* circuit. If the generator supplies a voltage v to the parallel circuit, a current will flow in the exterior circuit and in each branch of the parallel arrangement. Kirchhoff's first law indicates that the algebraic sum of currents at point P will equal zero. For convenience, we let currents moving toward the point be positive, and currents moving away from the point be negative. Using this convention in Figure 5.4, we obtain

$$\sum \text{currents} = i - i_R - i_L - i_C = 0 \qquad (12)$$

In addition, the voltage drop across each branch of the parallel circuit will be equal to the voltage drop across the terminals of the generator. This means that the current in each branch can be expressed in terms of the applied voltage. Therefore, the current in the resistor is given by Ohm's law to be $i_R = v/R$; the current in the inductance coil by $i_L = 1/L \int v dt$; and the current in the capacitor by $i_C = C(dv/dt)$. Making these substitutions in Equation 12, we have

$$i - \frac{v}{R} - \frac{1}{L} \int v dt - C \frac{dv}{dt} = 0 \qquad (13)$$

The total current can then be found if we know the equation for the voltage being applied and the numerical values of R, L, and C.

Example 13. The voltage being applied across a parallel *RLC* circuit is given by $v = t^4 + 10$. Find the equation for the total current if $R = 2$ ohms, $L = 5$ henrys, and $C = 10$ farads*. The current is initially zero.

*In practice, a capacitance of 10 farads is extremely large. Capacitors are generally of the order of 10^{-6} farads. We have ignored this in this exercise so that the numerical calculations will be more simple.

Solution.   This solution is obtained by a direct substitution of the given values into Equation 13.

$$i = \frac{v}{R} + C\frac{dv}{dt} + \frac{1}{L}\int v\,dt$$

$$= \frac{(t^4+10)}{2} + 10\frac{d}{dt}(t^4+10) + \frac{1}{5}\int (t^4+10)\,dt$$

$$= \frac{t^4}{2} + 5 + 40t^3 + \frac{t^5}{25} + 2t + K$$

$$i = 0.04t^5 + 0.5t^4 + 40t^3 + 2t + 5 + K$$

The value of K can be found by letting $i = 0$ and $t = 0$ in this equation. Thus

$$K = -5$$

Consequently, the equation for the total current is

$$i = 0.04t^5 + 0.5t^4 + 40t^3 + 2t$$

Kirchhoff's second law is concerned with voltage drops around the closed loop. It states that, at any instant, the algebraic sum of the voltage drops around the circuit is zero. This is denoted by

$$\Sigma \text{ voltage drops} = 0 \tag{14}$$

Kirchhoff's second law can be used to find the voltage being supplied to a series RLC circuit like the one shown in Figure 5.5 if we know the equation for the current.

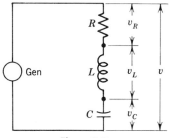

Figure 5.5

Here we consider the applied voltage, v, to be positive and the voltage drops across the individual components to be negative. Using this convention and applying Kirchhoff's second law, we have

$$\Sigma \text{ voltage drops} = v - v_R - v_L - v_C = 0 \tag{15}$$

Since the components are connected in series, the current in each component is the same as the total current. This fact allows us to write Equation 15 as

$$v - iR - L\frac{di}{dt} - \frac{1}{C}\int idt = 0 \tag{16}$$

Example 14.   A series RLC circuit has $R = 200$ ohm, $L = 80$ henrys, and $C = 0.5$ farads. If the current in the circuit is given by $i = t^{-2}$, find the voltage being applied when $t = 2$ sec. Assume $v = 38$ volts when $t = 1$ sec.

Solution.   The equation for the voltage is found by substituting into Equation 16 and solving it for v.

$$v = 200\,(t^{-2}) + 80\frac{d}{dt}\,(t^{-2}) + \frac{1}{0.5}\int t^{-2}\,dt$$

$$= 200\,t^{-2} - 160t^{-3} - 2t^{-1} + K$$

Using the boundary condition $v = 38$ when $t = 1$, the value of K is

$$K = 0$$

The equation for the applied voltage is then

$$v = 200t^{-2} - 160t^{-3} - 2t^{-1}$$

Substituting $t = 2$, we have

$$v = 200\left(\frac{1}{4}\right) - 160\left(\frac{1}{8}\right) - 2\left(\frac{1}{2}\right) = 29 \text{ volts}$$

.EXERCISES

1. The current in a resistor varies according to $i = 0.1t$ amp. Find the charge transferred after 1 sec, if the initial charge is 0.25 coul.
2. A capacitor contains an initial charge of 0.5 coul. The terminals of the capacitor are then connected to a circuit which delivers a current of $i = 0.5t^4 - 0.4t$ amp. What is the charge on the plates of the capacitor after 2 sec?

3. The voltage induced in a coil of wire is $v = 3t^{1/2}$ volts. If the current in the coil is initially $\frac{1}{2}$ amp and the inductance is 4 henrys, what is the equation for the current in the coil?

4. What is the current in the coil of Exercise 3 after 0.1 sec?

5. The voltage across the plates of a capacitor is initially 6 volts. What is the equation for the voltage across the plates of the capacitor, if the current in the capacitor is $i = (t+3)^2$ amp? Assume that the capacitance of the capacitor is 0.5 farads.

6. Find the voltage across the plates of the capacitor in Exercise 5 when $t = 0.2$ sec.

7. The power in a resistor is given by $p = ri^2$. Find the energy dissipated by 100 ohm resistor after 8 sec, if the current being supplied to the resistor is $i = \sqrt[3]{t}$. Assume that no current flows in the resistor initially.

8. Another expression for the power in an electrical circuit is $p = vi$, where v is the voltage and i is the current. If the voltage is $v = 3t + 2$ volts, and current is $i = \sqrt{t}$ amp, what is the energy delivered to the circuit in 10 sec? (Let $v = 2$ and $i = 0$ when $t = 0$.)

9. A 25 ohm resistor and a 5 henry inductance coil are connected in parallel and a voltage $v = 7t^3 + 5$ volts is applied to the combination. What is the equation of the resulting current, if $i = 2$ when $t = 0$?

10. The current in a series RL circuit is $i = 15\sqrt{t}$ amp. Find the voltage across the combination when $t = 9$ sec, if $R = 12$ ohms and $L = 2$ henrys.

11. Find the formula for the voltage across a series RC circuit required to deliver a current of $i = 5t^3 - 4t + 3$ amp, if $R = 50$ and $C = 10$. Assume that the initial voltage across the circuit is 1 volt.

12. Find the voltage in the circuit in Exercise 11 when $t = 1$ sec.

13. If the voltage across a 200 ohm resistor and a 0.1 farad capacitor connected in parallel is $v = 1 - t$ (volts), what is the current in the combination after 2 sec?

14. Consider a parallel RLC circuit with $R = 20$, $L = 0.5$, and $C = 5$. Let the applied voltage be $v = t^2 + 4$. Find the formula for the current if $i = 1.5$ when $t = 0$.

15. The current in a series RLC circuit is to vary according to $i = t^2$ amp. In this circuit, $R = 20$, $L = 500$, and $C = 10^{-2}$. Assuming that $v = 0$ when $t = 0$, find the formula for the applied voltage.

16. Use the expression for the voltage obtained in the Exercise 15 to calculate the voltage drop across the circuit when $t = 0.1$ sec.

17. If $R = 10$, $L = 100$, and $C = 10^{-2}$, what voltage must be applied to a series RLC circuit to yield a current of $i = 3t - 2$ amp? Assume $v = 5$ volts, when $t = 1$. What is the voltage when $t = 0.1$ sec?

18. In a parallel RLC circuit, $R = 10$, $L = 0.25$, and $C = 2$. Find the formula for the total current if the applied voltage is $v = t^{5/3}$ volts. Assume $i = 0$ when $t = 0$.

19. Consider a parallel RLC circuit with $R = 5$, $L = 0.1$ and $C = 3$. If $v = t + 4t^2$ volts is applied to the circuit, what is the rate of change of current with respect to time?

6

The Definite Integral

6.1 INTRODUCTION

In the preceding chapters we pursued one of the two main branches of the calculus, namely, differential calculus. We now consider the other main branch which is called integral calculus. For the sake of simplicity, we shall utilize geometric and physical illustrations to introduce the concept of the *definite integral*, which is the basic notion in the integral calculus.

6.2 THE SUMMATION NOTATION

In the integral calculus we are frequently concerned with writing sums of quantities such as the sum of the first n integers or the sum of the first n integers squared. To facilitate the writing of such sums, we introduce a compact notation (called summation notation) for which we give the following definition.

Definition

Let $u_1, u_2, \ldots, u_n$ be a set of n numbers. Then

$$\sum_{k=1}^{n} u_k = u_1 + u_2 + \cdots + u_n$$

The Greek letter sigma (Σ) is used to denote summation. The subscript k on the sigma is called the *index* of summation. The choice of the letter k for the index was arbitrary — any other symbol will serve as well. The 1 and the n indicate the range of the index; that is, k begins at 1 and increases in steps of 1 until it becomes equal to n. The following examples help to clarify the use of summation notation.

Example 1

$$\sum_{k=1}^{n} k = 1 + 2 + 3 + \cdots + n$$

Example 2

$$\sum_{i=1}^{k} i^2 = 1 + 2^2 + 3^2 + \cdots + k^2$$

114

Example 3

$$\sum_{r=1}^{4} x^r = x^1 + x^2 + x^3 + x^4$$

It is easy to show that the summation symbol has the two basic linearity properties.

Property 1

$$\sum_{k=1}^{n} ca_k = c \sum_{k=1}^{n} a_k$$

Property 2

$$\sum_{k=1}^{n} (a_k + b_k) = \sum_{k=1}^{n} a_k + \sum_{k=1}^{n} b_k$$

You should also be aware of the fact that a sum may be represented by more than one summation expression. One such instance is given in the following example.

Example 4

$$\sum_{k=1}^{n} \frac{1}{k} = \sum_{k=0}^{n-1} \frac{1}{k+1}$$

The fact that these two sums are equal is merely a matter of writing them out. The index on the left summation ranges from 1 to n, and the other ranges from 0 to $n-1$. This is commonly called a "shift" of index.

EXERCISES

Write the terms of the indicated summations.

1. $\sum_{i=1}^{5} x_i$ **5.** $\sum_{m=-1}^{4} \frac{1}{2+m}$

2. $\sum_{P=0}^{4} \frac{1-P}{1+P}$ **6.** $\sum_{n=1}^{5} (-1)^n (3n)$

3. $\sum_{n=1}^{3} n^2$ **7.** $\sum_{q=0}^{3} \left(-\frac{1}{2}\right)^q$

4. $\sum_{j=1}^{5} 2j$

Represent each of the following summations by the sigma notation.

8. $1+2+3+4$

9. $x_0 + x_1 + x_2 + x_3 + x_4$

10. $1+4+9+16$

11. $\frac{1}{2} + \frac{1}{4} + \frac{1}{8}$

12. $-3+9-27+81-243$

6.3   THE AREA PROBLEM

We introduce the concept of the *definite integral* in much the same way that we introduced the concept of the derivative in the differential calculus; that is, by considering the solutions to three seemingly unrelated but mathematically similar problems. The first problem is that of finding the area bounded by a curve $y = g(x)$, and the x-axis over an interval $a \leqslant x \leqslant b$. For simplicity we assume the graph of $g(x)$ is above the x-axis over the given interval. Figure 6.1 depicts such an area. The area so formed is frequently called the *area under the curve*.

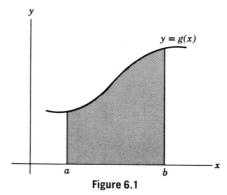

Figure 6.1

We begin our discussion of the area problem by describing a means of estimating the area bounded by the curve $y = x^2$, and the x-axis over the interval $0 \leqslant x \leqslant 2$. We can obtain an estimate of this area by constructing a series of rectangles as shown in Figure 6.2. The sum of these rectangular areas is then an estimate of the precise area under the curve. (We are, of course, assuming that we know how to find the area of a rectangle.)

The rectangles in Figure 6.2 are constructed by first dividing the interval $0 \leqslant x \leqslant 2$ into four equal subintervals of $\frac{1}{2}$ unit each. A rectangle is then drawn above each of the subintervals such that its height is that of the curve $y = x^2$ above the righthand end point of each subinterval. Referring to Figure 6.2, we see that our approximating sum is equal to

rt. hand
endpts

$$A_{\text{est}} = \left(\frac{1}{2}\right)^2\left(\frac{1}{2}\right) + \left(1\right)^2\left(\frac{1}{2}\right) + \left(\frac{3}{2}\right)^2\left(\frac{1}{2}\right) + \left(2\right)^2\left(\frac{1}{2}\right)$$

$$= \frac{1}{2}\left[\frac{1}{4} + 1 + \frac{9}{4} + 4\right]$$

$$= \frac{1}{2}\left[\frac{15}{2}\right] = \frac{15}{4} = 3.75$$

$$A = h\left(y_1 + y_2 + y_3 + \cdots y_N\right)$$

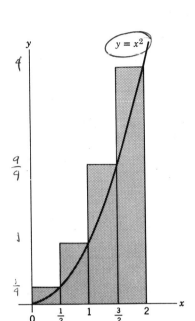

Figure 6.2

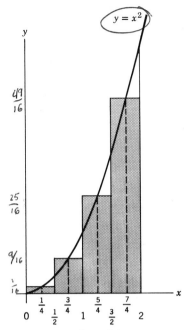

Figure 6.3

A little investigation reveals that a better estimate of this area can be obtained by constructing a series of rectangles as shown in Figure 6.3.

The rectangles in Figure 6.3 are constructed by first dividing the interval $0 \leq x \leq 2$ into four equal subintervals of $\frac{1}{2}$ units each. A rectangle is then drawn above each of these subintervals such that its height is that of the curve $y = x^2$ above the *midpoint* of each subinterval. Referring to Figure 6.3, we see that our approximating sum in this case is equal to

$$A_{est} = \left(\frac{1}{4}\right)^2\left(\frac{1}{2}\right) + \left(\frac{3}{4}\right)^2\left(\frac{1}{2}\right) + \left(\frac{5}{4}\right)^2\left(\frac{1}{2}\right) + \left(\frac{7}{4}\right)^2\left(\frac{1}{2}\right)$$

$$= \frac{1}{2}\left[\frac{1}{16} + \frac{9}{16} + \frac{25}{16} + \frac{49}{16}\right]$$

$$= \frac{1}{2}\left[\frac{84}{16}\right] = \frac{21}{8} = 2.625$$

We have an intuitive feeling that this estimate is closer to the precise area than that obtained in Figure 6.2; however, at this point we do not know how good either of these estimates are. The important thing is that you understand the means by which we formed our estimates. This process is of immense practical value as well as a prelude to finding the precise answer to the problem.

We might ask if this approach can be used to estimate the area under any continuous curve, such as the one shown in Figure 6.4. The answer to this question is *yes*. To obtain an estimate of the area bounded by the curve $y = g(x)$ and the x-axis over the interval $a \leqslant x \leqslant b$, we proceed as follows.

1. We subdivide the interval $a \leqslant x \leqslant b$ into n sections of width $\Delta x = (b - a)/n$.

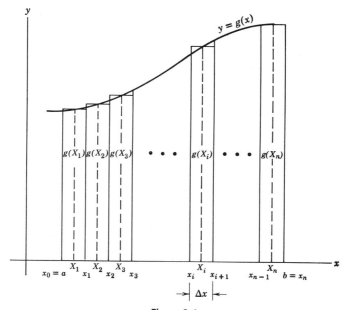

Figure 6.4

2. We pick a point in each of the n sections at which to evaluate the function. (The point we pick may be the midpoint, or it may be some other point.) We designate the point chosen in the first section by X_1, in the second by X_2, and so on until we reach the nth section. We designate the point chosen in a typical section by X_i where i is an integer from 1 to n.

3. We draw a rectangle above each section such that its height is that of the curve $y = g(x)$ above the point chosen in each section. (See Figure 6.4.) The area of a typical rectangle is then $g(X_i)\Delta x$.

4. We form the approximating sum

$$A_n = g(X_1)\Delta x + g(X_2)\Delta x + \cdots + g(X_n)\Delta x$$

$$= \sum_{i=1}^{n} g(X_i)\Delta x$$

While $A_n = \sum_{i=1}^{n} g(X_i)\Delta x$ can be used to estimate the area under any continuous curve $y = g(x)$, we choose to restrict its use in this section to the area under $y = x^2$ from 0 to 2.

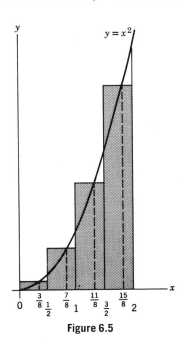

Figure 6.5

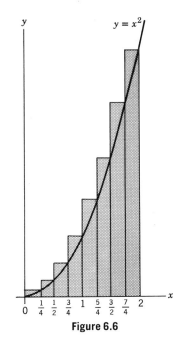

Figure 6.6

Example 5. Estimate the area bounded by $y = x^2$ and the x-axis over the interval $0 \le x \le 2$ by using four subintervals and selecting X_i to be one-quarter of the distance from the righthand end point of each subinterval (Figure 6.5).

Solution.   The end points of the subintervals are $x_0 = 0$, $x_1 = \frac{1}{2}$, $x_2 = 1$, $x_3 = 3/2$, $x_4 = 2$. One quarter of a subinterval is $(1/4)(1/2) = 1/8$ so that $X_1 = 1/2 - 1/8 = 3/8$. Likewise, X_2 is $1/8$ to the left of x_2 or $X_2 = 7/8$; X_3 is $1/8$ to the left of x_3 or $X_3 = 11/8$; X_4 is $1/8$ to the left of x_4

or $X_4 = 15/8$. The approximating sum is then

$$A = \sum_{i=1}^{4} g(X_i)\Delta x = \sum_{i=1}^{4} X_i^2 \Delta x$$

$$= \left(\frac{3}{8}\right)^2\left(\frac{1}{2}\right) + \left(\frac{7}{8}\right)^2\left(\frac{1}{2}\right) + \left(\frac{11}{8}\right)^2\left(\frac{1}{2}\right) + \left(\frac{15}{8}\right)^2\left(\frac{1}{2}\right)$$

$$= \frac{1}{2}\left[\frac{9}{64} + \frac{49}{64} + \frac{121}{64} + \frac{225}{64}\right]$$

$$= \frac{1}{2}\left[\frac{404}{64}\right] = \frac{101}{32} = 3.16$$

Example 6.   Estimate the area in the preceding example by using eight subintervals and selecting X_i to be the righthand end point of each subinterval (Figure 6.6).

Solution.   The width of each subinterval is

$$\Delta x = \frac{2-0}{8} = \frac{1}{4}$$

The end points of each subinterval are then $x_0 = 0$, $x_1 = 1/4$, $x_2 = 1/2$, $x_3 = 3/4$, $x_4 = 1$, $x_5 = 5/4$, $x_6 = 3/2$, $x_7 = 7/4$, $x_8 = 2$. Using the righthand end points of each subinterval to construct our rectangles, the approximating sum is

$$A_{est} = \sum_{i=1}^{8} g(X_i)\Delta x$$

$$= g\left(\frac{1}{4}\right)\cdot\frac{1}{4} + g\left(\frac{1}{2}\right)\cdot\frac{1}{4} + g\left(\frac{3}{4}\right)\cdot\frac{1}{4} + g(1)\cdot\frac{1}{4} + g\left(\frac{5}{4}\right)\cdot\frac{1}{4} + g\left(\frac{3}{2}\right)\cdot\frac{1}{4}$$

$$+ g\left(\frac{7}{4}\right)\cdot\frac{1}{4} + g(2)\cdot\frac{1}{4}$$

$$= \frac{1}{4}\left[\left(\frac{1}{4}\right)^2 + \left(\frac{1}{2}\right)^2 + \left(\frac{3}{4}\right)^2 + (1)^2 + \left(\frac{5}{4}\right)^2 + \left(\frac{3}{2}\right)^2 + \left(\frac{7}{4}\right)^2 + (2)^2\right]$$

$$= \frac{1}{4}\left[\frac{1}{16} + \frac{1}{4} + \frac{9}{16} + 1 + \frac{25}{16} + \frac{9}{4} + \frac{49}{16} + 4\right]$$

$$= \frac{1}{4}\left[\frac{204}{16}\right] = \frac{51}{16} = 3.185$$

EXERCISES

1. Estimate the area under the curve $y = x^2$ from 0 to 2 by using 4 partitions and selecting X_i to be the lefthand end point of each partition. Sketch the graph and show the indicated rectangles.

2. Estimate the area in Exercise 1 by selecting X_i to be one-quarter of the distance from the lefthand end point of each partition. Sketch the graph and show the indicated rectangles.

3. Estimate the area in Exercise 1 using 8 partitions. (a) Select X_i to be the lefthand end point of each partition. (b) Select X_i to be the righthand end point. (c) Select X_i to be the midpoint of each partition. Sketch each case.

4. From Exercise 3 we know that the area under the curve $y = x^2$ from 0 to 2 is greater than _____ and less than _____.

5. Estimate the area under the curve $y = \frac{1}{4}x^2$ from 0 to 2 using 4 partitions. Select X_i to be: (a) the lefthand end point, (b) the righthand end point, and (c) the midpoint. Sketch each case.

6. From Exercise 5 we know that the area under the curve $y = \frac{1}{4}x^2$ from 0 to 2 is greater than _____ and less than _____.

★7. Estimate the area under the curve $y = x^2$ from 1 to 3 using 4 partitions. Select X_i to be: (a) the lefthand end point, (b) the righthand end point, and (c) the midpoint of each partition. Sketch each area and show the indicated rectangles.

8. From Exercise 7 we know that the area under the curve $y = x^2$ from 1 to 3 is greater than _____ and less than _____.

9. Estimate the area under the curve $y = 4 - x^2$ from $x = 0$ to $x = 2$ using four partitions. (a) Select X_i to be the lefthand endpoint of each partition. (b) Select X_i to be the righthand endpoint. Sketch each case showing the approximating rectangles.

6.4   THE DISPLACEMENT PROBLEM

The second problem we shall consider is that of finding the displacement of a moving object when its velocity is known. We know that the distance traveled by an object which is moving with a constant velocity is given by

$$s = v \cdot t \qquad (1)$$

where v is the velocity and t is the elapsed time. This formula cannot be used if the velocity changes during the time interval; however, it can be used to estimate the displacement of an object whose velocity varies with time. To illustrate this, let us consider a rocket sled whose velocity during the first two seconds of propulsion varies with time according to $v = 25t^2$ ft/sec (Figure 6.7). We see that the velocity of the sled varies from $25(0)^2 = 0$ to $25(2)^2 = 100$ ft/sec during this time period. Clearly the displacement obtained by using either one of these two extreme values in Equation 1 would be incorrect. You may notice that an estimate of the

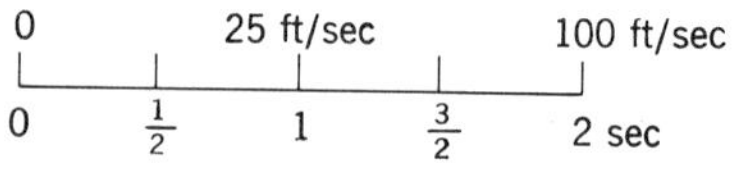

```
0              25 ft/sec        100 ft/sec
└──────┴──────┴──────┴──────┘
0      1/2      1      3/2      2 sec
```

Figure 6.7

distance traveled by the sled during this time interval can be obtained by using an arithmetic average of the two extreme velocities in Equation 1. However, this would be a very rough estimate. We can obtain a better estimate of the sled's displacement by using velocity averages over shorter time intervals in Equation 1, and then summing these values. Let us subdivide the 2 sec duration into four equal intervals, each $\frac{1}{2}$-sec long, and use the velocity of the sled at the midpoint of each interval to estimate the distance covered during each of these shorter time periods.

To estimate the distance the sled travels during the first time interval, we multiply its velocity at $\frac{1}{4}$ sec by the duration, $\frac{1}{2}$ sec. Thus, during the first time period, the sled's displacement is about $25(\frac{1}{4})^2(\frac{1}{2})$ ft. By making similar estimates for the remaining time intervals, the estimated distance traveled by the sled is

$$S_{est} = 25\left(\frac{1}{4}\right)^2\left(\frac{1}{2}\right) + 25\left(\frac{3}{4}\right)^2\left(\frac{1}{2}\right) + 25\left(\frac{5}{4}\right)^2\left(\frac{1}{2}\right) + 25\left(\frac{7}{4}\right)^2\left(\frac{1}{2}\right)$$

$$= \frac{25}{2}\left[\frac{1}{16} + \frac{9}{16} + \frac{25}{16} + \frac{49}{16}\right]$$

$$= \frac{25}{2}\left[\frac{21}{4}\right] = \frac{525}{8} = 65.6 \text{ ft}$$

Here we have an estimate of the precise distance traveled. As in the area problem, this approach can be extended to include motion problems in which the velocity of the object varies with time as, say $v = \mathcal{V}(t)$. Let us write the approximating sum for the displacement of an object over the interval $a \leqslant t \leqslant b$ when the velocity varies according to $v = \mathcal{V}(t)$ (Figure 6.8). First we subdivide the interval $a \leqslant t \leqslant b$ into n equal subintervals of duration $\Delta t = (b-a)/n$. Next we pick a point within each subinterval, either the midpoint or some other point, and denote the

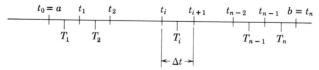

Figure 6.8

point chosen in a typical subinterval by T_i, where i is an integer from 1 to n. Each term of the approximating sum is determined by evaluating the velocity of the object at the point T_i of each subinterval and multiplying this value by Δt. The approximating sum is then formed by adding these products. Thus

$$S_n = \mathscr{V}(T_1)\Delta t + \mathscr{V}(T_2)\Delta t + \cdots + \mathscr{V}(T_n)\Delta t$$

$$= \sum_{i=1}^{n} \mathscr{V}(T_i)\Delta t$$

Example 7. Estimate the distance traveled by the rocket sled using four subintervals and selecting T_i to be the righthand end point of each subinterval.

Solution.   The duration of each subinterval is

$$\Delta t = \frac{2-0}{4} = \frac{1}{2}$$

Consequently, the end points of the subintervals are $t_0 = 0$, $t_1 = \frac{1}{2}$, $t_2 = 1$, $t_3 = \frac{3}{2}$, $t_4 = 2$. Using the righthand end point of each subinterval for T_i, we have the following estimate of the displacement of the sled during the first 2 sec.

$$S_{\text{est}} = \sum_{i=1}^{n} \mathscr{V}(T_i)\Delta t$$

$$= \mathscr{V}\left(\frac{1}{2}\right) \cdot \frac{1}{2} + \mathscr{V}(1) \cdot \frac{1}{2} + \mathscr{V}\left(\frac{3}{2}\right) \cdot \frac{1}{2} + \mathscr{V}(2) \cdot \frac{1}{2}$$

$$= 25\left(\frac{1}{2}\right)^2\left(\frac{1}{2}\right) + 25(1)^2\left(\frac{1}{2}\right) + 25\left(\frac{3}{2}\right)^2\left(\frac{1}{2}\right) + 25(2)^2\left(\frac{1}{2}\right)$$

$$= \frac{25}{2}\left[\frac{1}{4} + 1 + \frac{9}{4} + 4\right]$$

$$= \frac{25}{2}\left[\frac{30}{4}\right] = \frac{375}{4} = 93.75 \text{ ft}$$

6.5   THE WORK PROBLEM

The work done by a constant force F in moving an object through a distance S is the product FS. We wish to estimate the work done in moving a charged particle a distance of 2 cm in an electrical field if the force exerted on the particle by the electric field varies with the displacement of the particle according to $F = 0.01s^2$ dynes.

Let us divide the distance traveled by the particle into four equal lengths. The force varies less in each interval than it does for the whole distance. To obtain an estimate of the work done in each interval, we multiply the force being exerted on the particle at the midpoint of each interval by the length of each interval, $\frac{1}{2}$ cm. The force on the particle at $\frac{1}{4}$ cm is $0.01(\frac{1}{4})^2$ dyn. The work done in the first interval is, therefore, approximated by $0.01(\frac{1}{4})^2(\frac{1}{2})$ dyn-cm. Similarly, the work done in the second interval is about $0.01(\frac{3}{4})^2(\frac{1}{2})$ dyn-cm. Extending this reasoning to the third and fourth interval, our approximating sum is

$$W_{est} = 0.01\left(\frac{1}{4}\right)^2\left(\frac{1}{2}\right) + 0.01\left(\frac{3}{4}\right)^2\left(\frac{1}{2}\right) + 0.01\left(\frac{5}{4}\right)^2\left(\frac{1}{2}\right) + 0.01\left(\frac{7}{4}\right)^2\left(\frac{1}{2}\right)$$

$$= \frac{0.01}{2}\left[\frac{1}{16} + \frac{9}{16} + \frac{25}{16} + \frac{49}{16}\right]$$

$$= 0.005\left[\frac{21}{4}\right] = 0.0262 \text{ dyn-cm}$$

As in the other two problems, the approximating sum for the work done on an object by a force whose magnitude varies with displacement according to $F = \mathscr{F}(s)$ can be written

$$W_n = \mathscr{F}(S_1)\Delta s + \mathscr{F}(S_2)\Delta s + \cdots + \mathscr{F}(S_n)\Delta s$$

$$= \sum_{i=1}^{n} \mathscr{F}(S_i)\Delta s$$

where S_i is a point chosen within a typical subinterval of displacement.

Example 8.   Estimate the work done on the electric charge by using four subintervals and selecting S_i to be a point one-quarter of the way from s_{i-1} to s_i.

Solution.   Since Δs is

$$\Delta s = \frac{2-0}{4} = \frac{1}{2}$$

the end points of the subintervals are $s_0 = 0$, $s_1 = \frac{1}{2}$, $s_2 = 1$, $s_3 = \frac{3}{2}$, $s_4 = 2$. S_1 is then one-quarter of the distance from s_0 to s_1; that is, $S_1 = \frac{1}{4}(\frac{1}{2} - 0) = \frac{1}{8}$. Likewise, S_2 is $\frac{1}{8}$ cm to the right of s_1 or $S_2 = \frac{5}{8}$; S_3 is $\frac{1}{8}$ cm to the right of

s_2 or $S_3 = \frac{9}{8}$; S_4 is $\frac{1}{8}$ cm to the right of s_3 or $S_4 = \frac{13}{8}$ (Figure 6.9). The approximating sum for this value is then

$$W_{est} = \sum_{i=1}^{4} \mathscr{F}(S_i)\Delta s$$

$$= 0.01\left(\frac{1}{8}\right)^2\left(\frac{1}{2}\right) + 0.01\left(\frac{5}{8}\right)^2\left(\frac{1}{2}\right) + 0.01\left(\frac{9}{8}\right)^2\left(\frac{1}{2}\right) + 0.01\left(\frac{13}{8}\right)^2\left(\frac{1}{2}\right)$$

$$= \frac{0.01}{2}\left[\frac{1}{64} + \frac{25}{64} + \frac{81}{64} + \frac{169}{64}\right]$$

$$= 0.005\left[\frac{276}{64}\right] = 0.0216 \text{ dyn-cm}$$

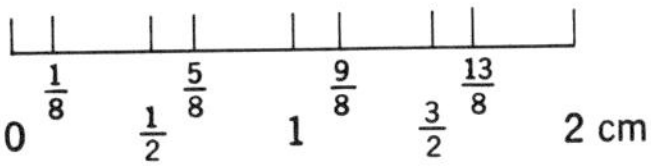

Figure 6.9

EXERCISES

1. Estimate the distance traveled by the rocket sled in Example 7 using four partitions and selecting T_i to be the lefthand end point of each partition.
2. From Example 7 and Exercise 1 we know that the distance traveled by the sled in 2 sec is greater than _____ and less than _____.
3. What is the velocity of the sled in Exercise 1 at T_1, T_2, T_3, and T_4?
4. Estimate the distance traveled by the rocket sled in 2 sec using eight subintervals, and selecting T_i to be: (a) the lefthand end point of each subinterval, and (b) the righthand end point of each subinterval.
5. From Exercise 4 we can refine our estimate of the distance traveled by the sled and say that it is greater than _____ and less than _____.
6. The force on a charged particle varies with the displacement of the particle according to $F = 0.01s^2$ dyn. Estimate the work done on the charged particle by the force F in moving the particle a distance of 2 cm. Choose four subintervals and select S_i to be: (a) the lefthand end point of each subinterval, (b) the righthand end point of each subinterval, and (c) a point one-quarter of the distance from the righthand end point of each subinterval.
7. From Exercise 6 we know that the work done in moving the charged particle 2 cm is greater than _____ and less than _____.
8. What force is being applied to the particle in Exercise 6c at S_1, S_2, S_3, and S_4?
9. Estimate the work done on the particle in Exercise 6 using eight subintervals and selecting S_i to be: (a) the lefthand end point of each subinterval, and (b) the righthand end point of each subinterval.
10. From Exercise 9 we know that the work done in moving the particle 2 cm is greater than _____ and less than _____. How does this compare with the result in Exercise 7?

$-M\alpha\!\!\curvearrowright$

6.6   PRECISE ANSWERS

We reiterate that we have not found the exact answer to any of the three problems; we have only estimated their values. In this Section we shall obtain the precise answers to all three problems. The scheme that we shall use is based on the one we used in the previous articles to obtain approximate solutions. We begin with the area problem.

The Area Problem.   To find the precise area bounded by the curve $y = x^2$ and the x-axis over the interval $0 \leqslant x \leqslant 2$, we note that we expect the approximating sum

$$A_n = \sum_{i=1}^{n} g(X_i)\Delta x \tag{2}$$

to approach the area under the curve as the number of subdivisions is increased. This remark is the key to the area problem and makes the following definition plausible. The area under a curve is the limit of the finite sum of rectangular areas as the number of partitions increases without bound; that is,

$$A = \lim_{n \to \infty} \sum_{i=1}^{n} g(X_i)\Delta x \tag{3}$$

The limiting value must be the same regardless of the choice of X_i in each subinterval if indeed we have a workable definition of area.

This definition is usable to give us the *exact* area under the curve only if the limit can be evaluated. The following example shows how this definition is used to solve our area problem.

Example 9.   Find the exact area under the curve $y = x^2$ from $x = 0$ to $x = 2$, selecting X_i to be the left-hand end point of each subinterval.

Solution.   To find the area under the curve by Equation 3 we first evaluate A_n. Dividing the interval $0 \leqslant x \leqslant 2$ into n equal parts, each of width Δx, the successive points of subdivision can be written

$$x_0 = 0$$
$$x_1 = \Delta x$$
$$x_2 = 2\Delta x$$
$$x_3 = 3\Delta x$$
$$\cdot$$
$$\cdot$$
$$\cdot$$
$$x_{n-1} = (n-1)\Delta x$$
$$x_n = n\Delta x$$

Selecting X_i to be the left-hand end point of each subinterval, the area of the successive rectangles is given by

$$f(X_1)\Delta x = (0)^2\Delta x = 0$$
$$f(X_2)\Delta x = (\Delta x)^2(\Delta x) = 1^2(\Delta x)^3$$
$$f(X_3)\Delta x = (2\Delta x)^2(\Delta x) = 2^2(\Delta x)^3$$
$$f(X_4)\Delta x = (3\Delta x)^2\Delta x = 3^2(\Delta x)^3$$
$$\vdots$$
$$f(X_n)\Delta x = [(n-1)\Delta x]^2\Delta x = (n-1)^2(\Delta x)^3$$

Summing these areas, we obtain

$$A_n = 0 + 1^2(\Delta x)^3 + 2^2(\Delta x)^3 + 3^2(\Delta x)^3 + \cdots + (n-1)^2(\Delta x)^3$$
$$= [1^2 + 2^2 + 3^2 + \cdots + (n-1)^2](\Delta x)^3$$
$$= \sum_{k=1}^{n-1} k^2(\Delta x)^3 = (\Delta x^3)\sum_{k=1}^{n-1} k^2$$

A not very well-known formula allows us to write

$$\sum_{k=1}^{n-1} k^2 = \frac{n(n-1)(2n-1)}{6}$$

In addition, we note that Δx can be written as

$$\Delta x = \frac{2-0}{n} = \frac{2}{n}$$

Therefore, A_n can be written

$$A_n = (\Delta x)^3 \sum_{k=1}^{n-1} k^2 = \left(\frac{2}{n}\right)^3 \left(\frac{n(n-1)(2n-1)}{6}\right)$$
$$= \frac{8(2n^3 - 3n^2 + n)}{6n^3} = \frac{8}{6}\left(2 - \frac{3}{n} + \frac{1}{n^2}\right)$$

The precise area under the curve is then

$$A = \lim_{n\to\infty} \sum_{k=1}^{n-1} k^2(\Delta x)^3 = \lim_{n\to\infty} \frac{8}{6}\left(2 - \frac{3}{n} + \frac{1}{n^2}\right) = \frac{8}{3}$$

A result that is used throughout the remainder of the chapters.

Remark 1.   You should be aware of the fact that the exact area under the curve can be found by selecting any X_i within each interval. In Exercise 1 (of the next Exercise Section) you will show that the same value is obtained for the area when the righthand end point of each sub-interval is chosen for X_i.

Remark 2.   You should not be alarmed at the complexity involved in finding exact areas. In the next chapter we will show you an easier method to get this number based upon the methods of calculus.

The Displacement Problem.   To find the exact displacement of the rocket sled, we observe that the approximating sum

$$S_n = \sum_{i=1}^{n} \mathscr{V}(T_i)\Delta t$$

will approach the length of the trip as the number of partitions increases without bound, a fact we can denote by

$$S = \lim_{n \to \infty} \sum_{i=1}^{n} \mathscr{V}(T_i)\Delta t$$

We could evaluate this limit as we did the limit in the area problem, but this is not necessary. By writing the expression for both the exact area and the exact distance traveled, we see that they are mathematically identical. Thus

$$A = \lim_{n \to \infty} \sum_{i=1}^{n} g(X_i)\Delta x = \lim_{n \to \infty} \sum_{i=1}^{n} X_i^2 \Delta x$$

$$S = \lim_{n \to \infty} \sum_{i=1}^{n} \mathscr{V}(T_i)\Delta t = \lim_{n \to \infty} \sum_{i=1}^{n} 25T_i^2 \Delta t$$

$$= 25 \left[\lim_{n \to \infty} \sum_{i=1}^{n} T_i^2 \Delta t \right]$$

The limit in the brackets is mathematically identical to the limit evaluated in the area problem. Therefore, the distance traveled by the sled is $S = 25[8/3] = 200/3$ ft.

The Work Problem.   We expect the approximating sum

$$W_n = \sum_{i=1}^{n} \mathscr{F}(S_i)\Delta s$$

to approach the work done in moving the charged particle a distance of

2 cm if we allow the number of partitions to increase without bound; that is,

$$W = \lim_{n \to \infty} \sum_{i=1}^{n} \mathscr{F}(S_i)\Delta s$$

The work done by a force $F = 0.01 \, s^2$ is then

$$W = \lim_{n \to \infty} \sum_{i=1}^{n} 0.01 \, S_i^2 \Delta s = 0.01 \left[\lim_{n \to \infty} \sum_{i=1}^{n} S_i^2 \Delta s \right]$$

Since the expression in the brackets is mathematically identical to the one evaluated in the area problem, we conclude that $W = 0.01[8/3] = 0.0267$ dyn-cm.

6.7 THE DEFINITE INTEGRAL

The fact that each of the preceding problems is mathematically identical suggests that it is to our advantage to generalize the discussion of the limit of a finite sum of products. Therefore, to avoid reference to specific applications, we denote the functional values $g(x)$, $\mathscr{V}(t)$, and $\mathscr{F}(s)$ by $f(x)$ and study the sum

$$\lim_{n \to \infty} \sum_{i=1}^{n} f(X_i)\Delta x$$

This sum, which is the focal point of the integral calculus, leads us to the following definition.

Definition
If f is a function defined on an interval $a \leqslant x \leqslant b$ and the

$$\lim_{n \to \infty} \sum_{i=1}^{n} f(X_i)\Delta x$$

approaches a number as n increases without bound regardless of the way in which X_i is chosen, we call that number the *definite integral* of the function f from a to b and denote it by $\int_a^b f(x) \, dx$.

We shall not prove the general conditions for which the definite integral exists; we only observe that if the function f is not too "wild," the definite integral exists. We shall say more about this in a later chapter.

The symbol $\int_a^b f(x) \, dx$ consists of four parts to which we have occasion to refer. The sign, $\int$ is called the *integral sign*. It comes from the S that originally was used to denote summation. The letters a and b stand for the extreme points of the interval $a \leqslant x \leqslant b$, and are called the *limits of integration*. There is no stipulation on the magnitudes of a and b, but we do insist that they be *real numbers*. The function f is known as the *integrand* and the dx suggests a small interval on the x-axis and indicates the variable of integration.

By virtue of its definition, the definite integral possesses the linearity properties; that is, if c is a constant then

$$\int_a^b cf(x)\, dx = c \int_a^b f(x)\, dx$$

and

$$\int_a^b [f(x) + g(x)]\, dx = \int_a^b f(x)\, dx + \int_a^b g(x)\, dx$$

As with antiderivatives, you must remember that the first property may be used only when c is a constant.

Example 10. Use the linearity property of definite integrals and previously established results to evaluate $\int_0^2 3x^2 dx$.

Solution.   The definite integral can be written

$$\int_0^2 3x^2 dx = 3 \int_0^2 x^2 dx = 3\left[\frac{8}{3}\right] = 8$$

Notice that the answer to the problem is known because we had previously calculated $\int_0^2 x^2 dx$.

It is important that you realize right from the beginning that $\int_a^b f(x)\, dx$ is a *number*. We get this number in a variety of ways, some of which you will learn in this book. It is also important for you to realize that area, displacement, and work are simply special applications of the definite integral. The following table emphasizes this point.

If $f(x)$ represents	$\sum_{i=1}^{n} f(X_i)\Delta x$ represents	$\int_a^b f(x)\, dx$ represents
Curve in the plane	Approximate area under curve	The exact area
Velocity	Approximate displacement	The exact displacement
Force	Approximate work done	The exact work done

The answers to the three special problems can now be written in terms of the definite integral. Thus, the area under the curve $y = x^2$ from 0 to 2 is given by

$$A = \int_0^2 x^2 dx = \frac{8}{3} = 2.667 \text{ sq units}$$

the displacement of the rocket sled by

$$S = \int_0^2 25t^2 dt = \frac{200}{3} = 66.67 \text{ ft}$$

and the work done on the charged particle by

$$W = \int_0^2 0.01s^2 ds = .02667 \text{ dyn-cm}$$

EXERCISES

1. Compute the area under the curve $y = x^2$ from 0 to 2 by selecting X_i to be the righthand end point of each partition. How does this answer compare with the answer obtained in Example 9 of Section 6.6?
2. Using only your present knowledge of definite integrals, evaluate

    (a) $\int_0^2 8x^2 dx$ (b) $\int_0^2 \pi x^2 dx$ (c) $\int_0^2 \frac{x^2}{2} dx$

3. Find the area under the curve $y = x$ from 0 to 2 by purely geometric considerations; that is, find the area by a known geometric formula.
4. Use the linearity property of the definite integral and the results from Exercise 3 to evaluate

    (a) $\int_0^2 3x dx$ (b) $\int_0^2 \frac{1}{2}t dt$ (c) $\int_0^2 100z dz$

5. Use the linearity property of the definite integral and the results from the foregoing exercises to evaluate

    (a) $\int_0^2 (2x^2 + 5x) dx$ (b) $\int_0^2 (s^2 - 7s) ds$ (c) $\int_0^2 (4x^2 + 2x + 3) dx$

    Hint for (c): What is the area under the curve $y = 3$ from $x = 0$ to $x = 2$?
6. The velocity of a solenoid actuated plunger is given by $v = 3t - t^2$ cm/sec. Using only the results from the foregoing exercises, find the distance traveled by the plunger in 2 sec.
7. The force on a proton varies with displacement as $F = 4s^2 + 100s$ dyn. How much work is done by the proton in moving 2 cm? Use only available information.

8. Find the area under the curve $y = 4 - x^2$ from 0 to 2 using the available information. Make a sketch of the area.
9. The force required to compress a certain spring is $F = 25s$ lb, where s is the displacement from the spring's free length. How much work is done in compressing the spring 2 in.?

6.8 PROPERTIES OF THE DEFINITE INTEGRAL

When a definite integral of a function f exists over an interval, we say that the function is *integrable*. In Chapter 7 we shall show some methods of calculating the definite integral of a function over some interval. In this chapter we wish to point out certain properties of the definite integral. These properties can be proven, but we will be content to explain them.

Property 1.

$$\int_a^b f(x)\, dx = -\int_b^a f(x)\, dx$$

Property 2.   If $a \leqslant X \leqslant b$, then

$$\int_a^b f(x)\, dx = \int_a^X f(x)\, dx + \int_X^b f(x)\, dx$$

The plausibility of Property 2 is established by using the area interpretation of the definite integral. Referring to Figure 6.10, we see that the area from a to b is equal to the area from a to X plus the area from X to b.

Property 3.   If f is continuous on $a \leqslant x \leqslant b$, then there is a number X on the interval such that

$$\int_a^b f(x)\, dx = f(X)(b - a)$$

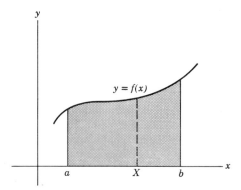

Figure 6.10

This property is referred to as the *mean value theorem for integrals*. It has an interesting geometric interpretation that is useful in many applications. Again, using the area interpretation of the definite integral, the mean value theorem asserts that there is a rectangle of width $b-a$ and height $f(X)$ that has the same area as the area under the curve $y=f(x)$. The area of the rectangle in Figure 6.11 is the same as the area under the curve.

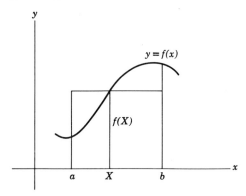

Figure 6.11

The height $f(X)$ is sometimes called the *average height of the curve* on the interval $[a,b]$. When Property 3 is used in this context it is commonly written

$$f(X) = \frac{1}{b-a} \int_a^b f(x)\, dx$$

Example 11.   Find the dimensions of the rectangle that has the same area as the region under the curve $y = x^2$ from 0 to 2.

Solution.   From our previous work we know that the area under the curve $y = x^2$ from 0 to 2 is

$$\int_0^2 x^2\, dx = \frac{8}{3}\text{ sq units}$$

By Property 3

$$\frac{8}{3} = f(X)(2-0)$$

Solving for $f(X)$, we have

$$f(X) = \frac{4}{3}$$

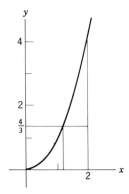

Figure 6.12

The dimensions of the desired rectangle are then $\frac{4}{3} \times 2$. This is shown geometrically in Figure 6.12.

Example 12. Find the average velocity of a particle whose velocity varies with time according to $v = 2t^2 + 5t$ mm/sec over the time interval $[0,2]$.

Solution.   The average velocity can be written

$$\bar{v} = \frac{1}{2-0} \int_0^2 (2t^2 + 5t) \, dt$$

From Problem 5a of the preceding assignment

$$\int_0^2 (2x^2 + 5x) \, dx = \frac{46}{3}$$

Consequently

$$\bar{v} = \frac{1}{2}\left(\frac{46}{3}\right) = \frac{23}{3} = 7.7 \text{ mm/sec}$$

EXERCISES

As in the preceding Exercise section use only the information that is presently available to solve the following exercises.
1. Find the average height of the curve $y = x + 2$ from $x = 0$ to $x = 2$. Draw the curve and indicate $f(X)$.
2. How does the answer obtained in Exercise 1 compare with your intuitive notion of the average height of the indicated straight line?

3. Find the average height of the curve $y = 3x^2$ over the interval $[0,2]$. At what point in $[0,2]$ does the height of the curve equal the average height of the curve?

4. Find the average height of the curve $y = 12x - 3x^2$ from 0 to 2.

5. During the first 2 sec of motion, the velocity of a rocket is given by $v = t^2 + 3t$ ft/sec. What is the rocket's (a) minimum velocity, (b) maximum velocity, and (c) average velocity during this time period?

6. Is the average velocity obtained in Exercise 6c equal to the arithmetic average of the two extreme velocities?

7. How far did the rocket in Exercise 5 travel during the first 2 sec?

8. What is the average force applied over a 2-foot interval if the force varies with displacement in accordance with $F = \frac{1}{2}s^2$ lb.

7
Integration

7.1 THE INDEFINITE INTEGRAL

From the discussion of the preceding chapter, it is clear that the letter denoting the variable of integration in a definite integral is immaterial to the value of the definite integral, just as the letter denoting the index of summation is immaterial to the value of the sum. In other words, $\int_a^b f(x)\, dx$ and $\int_a^b f(t)\, dt$ have the same value. You shall find this remark to be useful in the following discussion.

The value of the definite integral of a function ordinarily depends on the end points of the interval of integration. Thus, if the value of the upper limit of integration is changed, we would expect the value of the definite integral to be changed. This observation suggests that the definite integral can be used to define a function. In particular, by considering one of the end points of the interval of integration as a variable, we can define a function F on the interval $a \leqslant t \leqslant b$ such that

$$F(x) = \int_a^x f(t)\, dt$$

where x is an arbitrary point in the interval. We call $\int_a^x f(t)\, dt$ the *indefinite integral of f*, and we remark that this is the third use of the word "integral" that we have encountered: *the first was in connection with the antiderivative, the second in connection with the definite integral, and now the indefinite integral.* Unfortunately, in most engineering books, all three concepts are called "integrals" with only the context of the problem to tell you how the word is being used. It takes a thorough understanding of the definitions to keep the three ideas distinct.

Our interest in the indefinite integral is motivated by the fact that it leads to one of the most remarkable results in mathematics; a result which shows the close relationship between the derivative and the indefinite integral and is called the *fundamental theorem of calculus*. Instead of approaching the fundamental theorem of calculus in its most abstract form, we shall consider the case in which the indefinite integral is interpreted as the area under a curve; in this case it is called the "area" function. Denoting the area function by $A(x)$, we see that the area under the graph of $f(t)$ from $t = a$ to $t = x$ is given by

$$A(x) = \int_a^x f(t)\, dt$$

The fundamental theorem of calculus is now stated using this notation.

THEOREM 7.1

The Fundamental Theorem of Calculus (FTC) If f is a continuous function over the interval $a \leqslant x \leqslant b$ and $A(x)$ is defined by $A(x) = \int_a^x f(t)\, dt$, then

$$\frac{d}{dx}[A(x)] = \frac{d}{dx}\left[\int_a^x f(t)\, dt\right] = f(x)$$

In words, *the derivative of an indefinite integral is the integrand function evaluated at the upper limit.*

Proof

Figure 7.1 is a sketch of a continuous function f over the interval $a \leqslant x \leqslant b$. We want to investigate the manner in which the area A changes as x changes. Assume that the value of the independent variable changes from x to x_1, then the area under the curve will change by an amount ΔA. (This change in area is represented by the dark shading.) To find the relationship between Δx and ΔA, we construct two rectangles as shown in the figure. One rectangle is formed by drawing a line parallel to the x-axis from A to D. The other is formed by extending line AF and then drawing a line parallel to the x-axis from C to B.

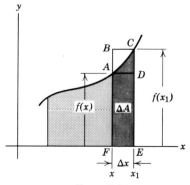

Figure 7.1

The smaller rectangle has an area of

$$\Delta A_s = f(x)\, \Delta x$$

and the area of the larger is

$$\Delta A_L = f(x_1)\, \Delta x$$

Obviously, ΔA is greater than ΔA_s and less than ΔA_L, a fact that may be written

$$f(x)\Delta x < \Delta A < f(x_1)\Delta x$$

Dividing the inequality by Δx, yields

$$f(x) < \frac{\Delta A}{\Delta x} < f(x_1)$$

What happens to this inequality if we let $\Delta x \to 0$? As $\Delta x \to 0$, we observe that the middle term becomes

$$\lim_{\Delta x \to 0} \frac{\Delta A}{\Delta x} = \frac{dA}{dx}$$

and, simultaneously, $f(x_1) \to f(x)$. But, if $f(x_1)$ approaches $f(x)$ as a limit and $\Delta A/\Delta x$ lies between $f(x_1)$ and $f(x)$, then $\Delta A/\Delta x$ must also approach $f(x)$. Hence,

$$\frac{dA}{dx} = f(x)$$

Recalling that $A = \int_a^x f(t)\, dt$, this completes the proof.

We have proven that the derivative of the indefinite integral is equal to the integrand evaluated at x; that is, if

$$F(x) = \int_a^x f(t)\, dt$$

then

$$F'(x) = f(x)$$

Example 1.   Let $F(x) = \int_1^x (1/t)\, dt$. Find $F'(2)$.

Solution. $dF/dx = 1/x$ and hence $F'(2) = \frac{1}{2}$.

Example 2.   Let $F(x) = \int_z^x t^3\, dt$. Find $F'(x)$.

Solution

$$dF/dx = x^3$$

Example 3.   Let $F(x) = \int_0^x \sin t\, dt$. Find $F'(\pi/2)$.

Solution

$$F'(x) = \sin x$$
$$F'(\pi/2) = 1$$

7.2   THE FUNDAMENTAL THEOREM AND THE DEFINITE INTEGRAL

The fundamental theorem of calculus demonstrates the essential unity of the subject of calculus. As a most profitable by-product, we are provided with a simple method for evaluating definite integrals as indicated in the next theorem.

THEOREM 7.2

Let G be any antiderivative of f, then

$$\int_a^b f(t)\, dt = G(b) - G(a)$$

In words, *the theorem states that the value of a definite integral is equal to the value of the antiderivative of the integrand function at the upper limit minus its value at the lower limit.*

Proof

Let

$$F(x) = \int_a^x f(t)\, dt \tag{1}$$

Then by the *FTC*, we have $F'(x) = f(x)$. Also by hypothesis $G'(x) = f(x)$. Since $G(x)$ and $F(x)$ have the same derivative, they differ at most by a constant; that is,

$$F(x) - G(x) + C \tag{2}$$

From (1) and (2) it follows that

$$F(b) = \int_a^b f(t)\, dt = G(b) + C$$

The value of C can be found by observing that

$$F(a) = \int_a^a f(t)\, dt = 0 = G(a) + C$$

or

$$C = -G(a)$$

Therefore

$$\int_a^b f(t)\, dt = G(b) - G(a)$$

which completes the proof.

When evaluating a definite integral it is customary to use the notation

$$\int_a^b f(x)\, dx = [G(x)]_a^b = G(b) - G(a) \tag{3}$$

where $G(x)$ represents *any* antiderivative of $f(x)$. The intermediate step shows an antiderivative of the integrand function, except for the arbitrary constant, with the limits of integration written to the right of the bracket.

Example 4. Evaluate

$$\int_1^8 x^{-1/3}\, dx$$

Solution

$$\int_1^8 x^{-1/3}\, dx = \left[\frac{3}{2} x^{2/3}\right]_1^8 = \left[\frac{3}{2} (8)^{2/3} - \frac{3}{2} (1)^{2/3}\right] = \frac{9}{2}$$

Example 5. Evaluate

$$\int_0^3 (x^2 + 4)\, dx$$

Solution

$$\int_0^3 (x^2 + 4)\, dx = \left[\frac{x^3}{3} + 4x\right]_0^3 = (9 + 12) - (0) = 21$$

Example 6. Show that $\int_1^2 3x^2\, dx = -\int_2^1 3x^2\, dx$.

Solution.   The integral on the left has the value

$$\int_1^2 3x^2\, dx = [x^3]_1^2 = 8 - 1 = 7$$

and the value of the integral on the right is

$$-\int_2^1 3x^2\, dx = -[x^3]_2^1 = -[1-8] = 7$$

Remark 1.   We now see the value of developing our skill in finding anti-derivatives. We see that if we can obtain *any* antiderivative for f, our problem of evaluating $\int_a^b f(x)\, dx$ is essentially finished. This is the reason some people (erroneously) equate the ideas of antidifferentiating and integrating. It is also the reason that the integral sign, without any indicated limits, is widely used to indicate an antiderivative.

Remark 2.   It is not at all uncommon to encounter a function that has no "elementary" function as its antiderivative. We should not say that such a function is "non-integrable" since schemes other than the previous theorem can be (and are) used to evaluate $\int_a^b f(x)\, dx$.

EXERCISES

In Exercises 1–4 find the indicated derivative.

1. $F(x) = \int_1^x 4t^2\, dt;$ find $F'(x)$

SKIP - these use these theorem. Just check in back

2. $G(x) = \int_2^x (t^2 - 1)\, dt;$ find $G'(x)$

3. $F(t) = \int_3^t \sqrt{4x}\, dx;$ find $F'(t)$

4. $H(z) = \int_4^z \cos x\, dx;$ find $H'(z)$

In Exercises 5–16 evaluate the given definite integrals.

5. $\int_0^1 5x^4\, dx$

10. $\int_{-1}^3 (v+2)^2\, dv$

6. $\int_0^3 \theta^3\, d\theta$

11. $\int_{-8}^0 (x^{1/3} + 2)x\, dx$

7. $\int_2^4 (3t^2 - 5)\, dt$

12. $\int_5^{10} \frac{dw}{w^2}$

8. $\int_1^4 z^2 \sqrt{z}\, dz$

13. $\int_{1/4}^{1/2} \frac{x^2\, dx}{2}$

9. $\int_{-2}^1 (q - q^3)\, dq$

14. $\int_{-3}^{-1} p^2\left(1 - \frac{1}{p^2}\right) dp$

15. $\int_2^4 \sqrt{8x}\, dx$

16. $\int_1^4 x^{1/2}(3x^2 - 5x)\, dx$

In each of the following problems show that the left side of the equation equals the right side.

17. $\int_1^3 x^3\, dx = -\int_3^1 x^3\, dx$

18. $\int_0^2 2x\, dx = -\int_2^0 2x\, dx$

19. $\int_{-1}^3 (x^3 - 5x)\, dx = \int_{-1}^2 (x^3 - 5x)\, dx + \int_2^3 (x^3 - 5x)\, dx$

20. $\int_0^4 s^{3/2}\, ds = \int_0^1 s^{3/2}\, ds + \int_1^4 s^{3/2}\, ds$

7.3   AREA PROBLEMS

The definite integral of a function over an interval is, as was pointed out, a process whereby one must eventually compute the limit of a finite sum. This process occurs in several widely different subject areas and, indeed, is the very reason the definite integral is considered to be so important. We now consider some additional facts about area problems.

In all of the area problems that we have considered, the curve has been *above* the x-axis. However, this is not a necessary restriction. The definite integral can also be used to compute areas of other type regions.

Example 7.   Find the area bounded by the curve $y = x^2 - 5x$ and the x-axis from $x = 1$ to $x = 4$ (Figure 7.2).

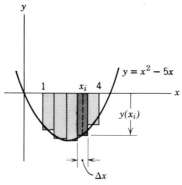

Figure 7.2

Solution. Dividing the interval $1 \leqslant x \leqslant 4$ into n equal parts, we conclude that the area of a typical element of area is given by $y(x_i)\Delta x = (x_i^2 - 5x_i)\Delta x$ and the desired area by

$$A = \lim_{n \to \infty} \sum_{i=1}^{n} y(x_i)\Delta x = \lim_{n \to \infty} \sum_{i=1}^{n} (x_i^2 - 5x_i)\,\Delta x$$

Then by definition of the definite integral and the FTC, we have

$$A = \int_1^4 (x^2 - 5x)\,dx = \left[\frac{x^3}{3} - \frac{5x^2}{2}\right]_1^4$$

$$= \left[\left(\frac{64}{3} - \frac{80}{2}\right) - \left(\frac{1}{3} - \frac{5}{2}\right)\right] = -\frac{33}{2}$$

Of course, we cannot have a negative area and so the negative sign must have some other interpretation. The negative sign simply indicates that the area under the curve lies below the x-axis.

We frequently encounter functions whose graphs lie alternately above and below the x-axis over the interval of integration as shown in Figure 7.3. To find the area under this curve from a to b it is necessary to realize that A_1 is positive and A_2 is negative. Since integration is a process of summation, the definite integral $\int_a^b f(x)\,dx$ will yield the algebraic sum of A_1 and A_2. In this sense, $\int_a^b f(x)\,dx$ is the net area above or below the x-axis. If we want the total area from a to b, regardless of numerical sign, we must find A_1 and A_2 separately and then add their absolute values; that is,

$$A_T = |A_1| + |A_2| = \left|\int_a^c f(x)\,dx\right| + \left|\int_c^b f(x)\,dx\right|$$

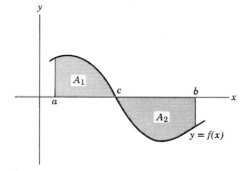

Figure 7.3

Example 8.   Consider the curve $y = x^2 - 4$ from $x = 0$ to $x = 5$. Find: (a) the net area under the curve, and (b) the total area under the curve (Figure 7.4).

Solution.   (a) The net area is found by simply evaluating the definite integral from $x = 0$ to $x = 5$.

$$A_N = \int_0^5 (x^2 - 4)\,dx = \left[\frac{x^3}{3} - 4x\right]_0^5$$

$$= \frac{125}{3} - 20 = \frac{65}{3}$$

(b) To find the total area, we first locate the points at which the curve crosses the x-axis. This is done by letting y equal zero and solving for x. Thus,

$$x^2 - 4 = 0$$
$$x = \pm 2$$

The curve has intercepts at both 2 and −2. We may ignore the −2 since it is not within the interval of integration. Hence,

$$A_1 = \int_0^2 (x^2 - 4)\,dx = \left[\frac{x^3}{3} - 4x\right]_0^2 = -\frac{16}{3}$$

$$A_2 = \int_2^5 (x^2 - 4)\,dx = \left[\frac{x^3}{3} - 4x\right]_2^5 = \left[\left(\frac{125}{3} - 20\right) - \left(\frac{8}{3} - 8\right)\right] = 27$$

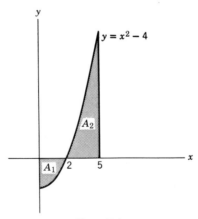

Figure 7.4

The total area is then

$$A_T = |A_1| + |A_2| = \left|-\frac{16}{3}\right| + |27| = \frac{16}{3} + 27 = \frac{97}{3}$$

To conclude this example, we show that the net area found in part (a) is truly the difference in A_1 and A_2. By adding A_1 and A_2 algebraically, we get

$$A_N = A_1 + A_2 = -\frac{16}{3} + 27 = \frac{65}{3}$$

Example 9.   Find the average height of the curve $y = \sqrt{x}$ on the interval [1,4]. (See Figure 7.5.)

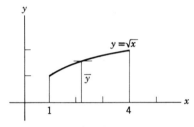

Figure 7.5

Solution.   The average height of a curve is

$$\bar{y} = \frac{1}{b-a} \int_a^b f(x)\, dx$$

Therefore, in the indicated problem

$$\bar{y} = \frac{1}{4-1} \int_1^4 x^{1/2} dx = \frac{1}{3}\left[\frac{2x^{3/2}}{3}\right]_1^4 = \frac{1}{3}\left[\frac{16}{3} - \frac{2}{3}\right] = \frac{14}{9}$$

The area bounded by a curve and the y-axis can also be stated as the limit of a finite sum. The next example shows how a problem of this type is solved.

Example 10.   Find the area bounded by the curve $y = x^{1/2}$, the y-axis and the lines $y - 1$ and $y - 3$. (See Figure 7.6.)

Solution.   To solve this problem, we divide the area into n horizontal strips each Δy units wide and construct an approximating rectangle in each case. Denoting the length of a typical rectangle by $x(y_i)$, we have

$$A_i = x(y_i)\Delta y$$

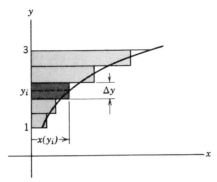

Figure 7.6

The desired area may then be written as

$$A = \lim_{n \to \infty} \sum_{i=1}^{n} x(y_i)\Delta y = \int_1^3 x(y)\,dy$$

Before we can evaluate this definite integral, we must replace $x(y)$ with its functional value in terms of y. Since the given equation can be expressed in the form $x = y^2$, we have

$$A = \int_1^3 x(y)\,dy = \int_1^3 y^2\,dy = \left[\frac{y^3}{3}\right]_1^3 = 9 - \frac{1}{3} = \frac{26}{3}$$

EXERCISES

In Exercises 1–8 find the *total area* bounded by the curve and the x-axis for the indicated interval on the x-axis. Sketch each figure.

1. $y = 2x - 6$; $x = 2$ to $x = 5$
2. $y = 1 - x^2$; $x = 0$ to $x = 3$
3. $y = x^3$; $[-1,2]$
4. $y = x^2 - 2x$; $[1,4]$
5. $y = x^3 + 1$; $-3 \leqslant x \leqslant 0$
6. $y = x^2 - 5x + 6$; $x = 0$ to $x = 4$
7. $y = x^2 + x - 2$; $x = -1$ to $x = 3$
8. $y = x^3 - 3x^2 + 2x$; $0 \leqslant x \leqslant 2$

In Exercises 9–12 find the average height of the curve above the x-axis over the indicated interval on the x-axis.

9. $y = \sqrt[3]{x}$; $[0,8]$
10. $y = x^2 - 4x$; $[0,2]$

11. $y = x^{-1/2}$; $4 \leqslant x \leqslant 9$

12. $y = \dfrac{1}{x^2}$; $\dfrac{1}{2} \leqslant x \leqslant 2$

In Exercises 13–18 find the area bounded by the curve, the y-axis, and the indicated interval on the y-axis. Sketch each figure.

13. $y = x$; $2 \leqslant y \leqslant 5$

14. $y^2 = x^3$; $1 \leqslant y \leqslant 8$

15. $y = 6x - 6$; $y = 3$ to $y = 6$

16. $y = x^{1/2} - 1$; $y = 0$ to $y = 3$

17. $y^3 = x^2$; $[1,4]$

18. $y = x^3$; $[0,8]$

7.4   THE AREA BETWEEN TWO CURVES

We continue our discussion of the area problem by considering areas bounded by two curves. We will illustrate the problem and the method of solution in the following examples.

Example 11.   Find the area bounded by the curves $y = 7x$ and $y = x^2$ and the lines $x = 2$ and $x = 5$. This area is depicted in Figure 7.7.

Solution.   We proceed as in the previous area problems by dividing the area into vertical rectangles of width Δx. Each rectangle is drawn so that it is bounded above by the curve $y_2 = 7x$ and below by the curve $y_1 = x^2$. The height of a typical element is then of the form

$$h_i = y_2(x_i) - y_1(x_i)$$

and the area of a typical rectangle is

$$A_i = [y_2(x_i) - y_1(x_i)]\Delta x$$

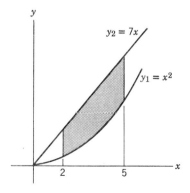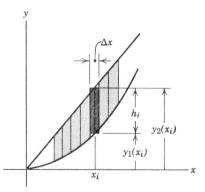

Figure 7.7

The area bounded by the two curves is then given by

$$A = \lim_{n \to \infty} \sum_{i=1}^{n} [y_2(x_i) - y_1(x_i)] \, \Delta x$$

or

$$A = \int_2^5 [y_2(x) - y_1(x)] \, dx$$

Since $y_2(x) = 7x$ and $y_1(x) = x^2$, we have

$$A = \int_2^5 (7x - x^2) \, dx = \left[\frac{7x^2}{2} - \frac{x^3}{3} \right]_2^5$$

$$= \left(\frac{175}{2} - \frac{125}{3} \right) - \left(\frac{28}{2} - \frac{8}{3} \right) = \frac{207}{6}$$

Example 12. Find the area bounded by the curves $y = x^3$ and $y^2 = x$. (See Figure 7.8.)

Solution. Dividing the area into vertical strips, the area can be written as

$$A = \lim_{n \to \infty} \sum_{i=1}^{n} [y_2(x_i) - y_1(x_i)] \, \Delta x$$

The next step is to express this sum as a definite integral; however, before we can do this, we have one more question to answer. What are the limits of integration? We see from the graph that we wish to sum the elements from the leftmost intersection of the two curves to the rightmost intersection. We find the points of intersection by solving the given equations simultaneously.

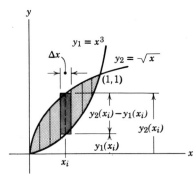

Figure 7.8

Thus
$$x^3 = x^{1/2}$$
$$x^6 = x$$
$$x^6 - x = 0$$
$$x(x^5 - 1) = 0$$
$$x = 0, 1$$

The limits of integration are, therefore, $x = 0$ to $x = 1$, and the definite integral is

$$A = \int_0^1 [y_2(x) - y_1(x)] \, dx$$

$$= \int_0^1 (x^{1/2} - x^3) \, dx$$

$$= \left[\frac{2x^{3/2}}{3} - \frac{x^4}{4} \right]_0^1 = \frac{5}{12}$$

In solving problems involving the application of the definite integral, you should give careful consideration to the choice of the elements to be summed. In most cases it is advisable to make a sketch of the problem before choosing the elements. As a case in point, consider the following problem.

Example 13.   Find the area formed by the intersection of the curves $y = x - 4$ and $y^2 = 2x$. (See Figure 7.9.)

Solution.   The solution to this problem is most readily obtained if we choose to take horizontal elements instead of vertical elements. The right

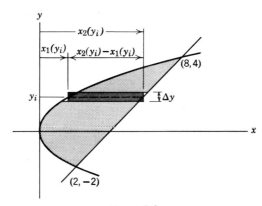

Figure 7.9

end of each element will then be on the curve $y = x - 4$ and the left end will be on $y^2 = 2x$. In this case, the area of each element given by

$$A_i = [x_2(y_i) - x_1(y_i)] \, \Delta y$$

and the area bounded by the two curves by

$$A = \lim_{n \to \infty} \sum_{i=1}^{n} [x_2(y_i) - x_1(y_i)] \, \Delta y$$

As in the previous example, we must find the points of intersection of the given curves before we can write the definite integral for this area. Solving the two given equations simultaneously, yields

$$(x - 4)^2 = 2x$$
$$x^2 - 8x + 16 = 2x$$
$$(x - 2)(x - 8) = 0$$
$$x = 2, y = -2$$
$$x = 8, y = 4$$

Since we are using horizontal elements, we want to sum these elements from $y = -2$ to $y = 4$. We then write the area as

$$A = \int_{-2}^{4} [x_2(y) - x_1(y)] \, dy$$

$$= \int_{-2}^{4} \left[(y + 4) - \frac{y^2}{2} \right] dy$$

$$= \left[\frac{y^2}{2} + 4y - \frac{y^3}{6} \right]_{-2}^{4}$$

$$= \left(8 + 16 - \frac{64}{6} \right) - \left(2 - 8 + \frac{8}{6} \right) = 18$$

Now let us consider the situation had we chosen vertical elements. Here we wish to sum the vertical elements from the left extremity of the area which is $x = 0$ to the right which is $x = 8$. The difficulty arises from the fact that the character of the elements change within the interval of integration. From Figure 7.10 we see that the elements between $x = 0$ and $x = 2$ have both ends on the curve $y_2^2 = 2x$. The height of a typical element in this region is then $2y_2(x_i)$, and the area from 0 to 2 is

$$A_{0 \to 2} = \lim_{n \to \infty} \sum_{i=1}^{n} 2y_2(x_i) \, \Delta x$$

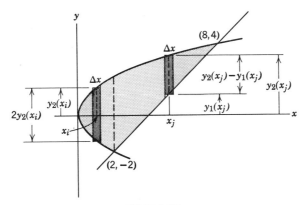

Figure 7.10

On the other hand, the elements between $x = 2$ and $x = 8$ have their upper end on the parabola and their lower end on the straight line so that the height of each element is given by $y_2(x_j) - y_1(x_j)$, and the area from 2 to 8 by

$$A_{2 \to 8} = \lim_{n \to \infty} \sum_{j=1}^{n} [y_2(x_j) - y_1(x_j)] \, \Delta x$$

Thus, the area between the curves is

$$A = \lim_{n \to \infty} \sum_{i=1}^{n} 2y_2(x_i) \, \Delta x + \lim_{n \to \infty} \sum_{j=1}^{n} [y_2(x_j) - y_1(x_j)] \, \Delta x$$

or

$$A = \int_0^2 2y_2(x) \, dx + \int_2^8 [y_2(x) - y_1(x)] \, dx$$

$$= \int_0^2 2\sqrt{2x} \, dx + \int_2^8 (\sqrt{2x} - x + 4) \, dx$$

$$= \left[\frac{4\sqrt{2}}{3} x^{3/2} \right]_0^2 + \left[\frac{2\sqrt{2}}{3} x^{3/2} - \frac{1}{2} x^2 + 4x \right]_2^8$$

$$= \left[\frac{16}{3} - 0 \right] + \left[\left(\frac{64}{3} - 32 + 32 \right) - \left(\frac{8}{3} - 2 + 8 \right) \right]$$

$$= 18$$

The solution can be obtained by either choice of elements as long as we take all of the information into consideration. It is obvious, however, that the choice of horizontal elements is the most expedient one in this case.

EXERCISES

1. Estimate the area bounded by $y = 2x$ and $y = x^2$ using four partitions and selecting X_i to be the midpoint of each partition. Draw the area and show each of the approximating rectangles. (Be sure to select a scale that is large enough for this purpose.)
2. Estimate the area bounded by $y = 4x - x^2$ and $y = x$ using three partitions and selecting X_i as the midpoint of each partition. Show the area and the approximating rectangles.

Find the area bounded by the given curves. Sketch the figure. Choose the most convenient element of area in solving each problem.

3. $y = x^4; y = x^2$
4. $y = 9x; y = x^3$; 1st quad. only
5. $y = x^2; y = 4$
6. $y^2 = x; x = 3$
7. $y = 2x - x^2; y = x^2$

8. $y^2 = 4x; y^2 = x^3$
9. $y = -x; y = x^2 - 5x$
10. $y = x^2 - 4; 3x + y = 0$
11. $3y + x - 4 = 0; y^2 = x$
12. $x = 4y - y^2; y^2 = 3x$

13. Find the area bounded by the curve $y = x^2 - 5$ and the straight line passing through the points $(-2,-1)$ and $(3,4)$.
14. Find the area bounded by $y^2 = x^3$ and the straight line passing through $(1,1)$ and $(4,8)$.
15. Using horizontal elements, find the area bounded by $y^2 = 4x$, and $y = 2x - 4$.
16. Solve Exercise 15 using vertical elements of area.
17. Find the area bounded by the straight lines $y = x$, $y = -3x + 4$, and $x = 0$. Use vertical elements of area.
18. Solve Exercise 17 using horizontal elements of area.
19. Find the area bounded by the curve $y = x^3 - 6x^2$ and the chord connecting the points $(1,-5)$ and $(6,0)$.

7.5   THE FIRST MOMENT OF A FORCE

Figure 7.11 shows a rod that is pivoted at point P with a force F applied to the rod. This applied force will cause the rod to rotate about P. By our knowledge of levers, we know that the tendency of a force to produce rotation is dependent upon the distance between the force and the axis of rotation. That is, the farther the line of action of F is from P, the greater will be the tendency of F to produce rotation. In mechanics, the tendency of a force to produce rotation is called torque, or moment. The quantitative measure of the moment of a force is defined to be the product

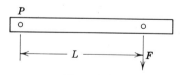

Figure 7.11

of the force times the perpendicular distance between the force and the axis of rotation. The perpendicular distance between the force and the axis of rotation is called the moment arm of the force. The definition of moment can be written symbolically as

$$M = F \cdot L \qquad (4)$$

The units of moment is the product of the units used for force and moment arm.

Example 14.   Compute the moment of a 10 lb force if the moment arm is (a) 5 ft,   (b) 12 ft,   (c) 3 in.

Solution.   (a) $M = 10 \cdot 5 = 50$ ft-lb
(b) $M = 10 \cdot 12 = 120$ ft-lb
(c) $M = 10 \cdot 3 = 30$ in.-lb

If we desire to express the answer to (c) in ft-lb, we write

$$M = 10\left(\frac{3}{12}\right) = 2.5 \text{ ft-lb}$$

7.6   CENTERS OF GRAVITY

A body is composed of a system of particles each of which is attracted by the gravitational force of the earth. The resultant of all of these forces acting on the particles making up the body is called the force of gravity, or weight of the body. The point in the body through which the weight may be considered acting is called the center of gravity.

In designing airplanes, the components that make up the airplane must be located so that their weights will balance about the center of gravity in order to maintain proper stability. The location of the center of gravity is also important in the design of automobiles. If the center of gravity is located too high above road level, the car may have a tendency to tip over when it turns a corner.

Let us consider centers of gravity of flat plates of uniform thickness. For certain shapes, the location of the center of gravity is obvious. The center of gravity of rectangles and circles coincides with their geometrical center. For more general shapes, we must employ calculus to determine the center of gravity.

7.7   CENTROIDS

When we talk about the center of gravity of an object, we imply that the object has weight. In the mechanics of materials we often consider cross-sectional areas of beams and columns. The point that corresponds to the center of gravity is called the *centroid* of the area. For example, if we consider a flat plate only as an area, the center of gravity becomes the centroid of the area. Conceptually, we regard the centroid as the point at which all of the area may be considered to be concentrated.

In Section 7.5 we defined the moment of a force. It is now convenient to define an analogous quantity for areas. If we assume that the area of a plane figure is concentrated at its centroid, the moment of area is defined as the product of the area times the perpendicular distance between the centroid and the axis of rotation. When an area A is located in the plane, as shown in Figure 7.12, the coordinates of the centroid are commonly designated by $(\bar{x}, \bar{y})$.

In all of the problems we shall deal with, the axis of rotation will be one of the coordinate axes. Referring to Figure 7.12, the moment of given area A about the y-axis is

$$M_y = A\bar{x} \tag{5}$$

and the moment of A about the x-axis is

$$M_x = A\bar{y} \tag{6}$$

where $\bar{x}$ and $\bar{y}$ are the coordinates of the centroid of A.

To determine the coordinates of the centroid of a given area, Equations 5 and 6 can be solved for $\bar{x}$ and $\bar{y}$, respectively. Thus,

$$\bar{x} = \frac{M_y}{A} \tag{7}$$

and

$$\bar{y} = \frac{M_x}{A} \tag{8}$$

We already know how to find the area of a given figure. We may use this same technique to evaluate the moments M_x and M_y. To find the moment

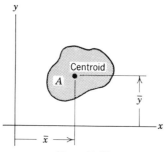

Figure 7.12

of a given area, divide the area into rectangular elements. The centroid of each of these rectangles is located at its geometrical center. The moment of each element may then be written as the product of the elemental area times the distance from the axis of rotation to the centroid of the element. Finally, we write the moment of the total area as the limit of the sum of the elemental moments. The technique that has just been outlined will be described in detail in the following examples.

Example 15. Find the centroid of the area bounded by the curve $y = x^2$ and the lines $x = 3$ and $y = 0$ (Figure 7.13).

Solution. Divide the area into rectangular elements as shown in Figure 7.13. The desired area is then given by

$$A = \lim_{n \to \infty} \sum_{i=1}^{n} y(x_i)\,\Delta x$$

or

$$A = \int_0^3 y(x)\,dx = \int_0^3 x^2 dx = \left[\frac{x^3}{3}\right]_0^3 = 9$$

Now let us calculate the moments M_x and M_y. We begin by considering the moment of an element of area about the y-axis. Referring to Figure 7.13, we see that x_i is the distance from the y-axis to the centroid of the element of area. Consequently, the moment of a typical element of area about the y-axis is given by

$$M_{y_i} = x_i A_i = x_i [y(x_i)\,\Delta x] = x_i y(x_i)\,\Delta x$$

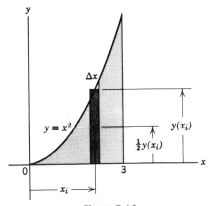

Figure 7.13

The moment of an area is the sum of the moments of the component areas. Therefore, the moment of the given area about the y-axis is

$$M_y = \lim_{n \to \infty} \sum_{i=1}^{n} x_i y(x_i) \, \Delta x$$

Writing this as a definite integral, we have

$$M_y = \int_0^3 x[y(x)] \, dx$$

Notice that the limits of integration are the same as those used to find the area in the first part of this problem. This points to the fact that the limits of integration are established by extremes of the elements being summed and not by the form of the elements. Here we sum the moments of vertical elements of area from $x = 0$ to $x = 3$, so the limits of integration are $x = 0$ and $x = 3$.

To evaluate the moment M_y we must replace $y(x)$ in the integrand with x^2, since $y = x^2$ is the given equation. Hence,

$$M_y = \int_0^3 x(x^2) \, dx = \int_0^3 x^3 dx = \left[\frac{x^4}{4} \right]_0^3 = \frac{81}{4}$$

The moment about the x-axis M_x is found by the same procedure used to find the M_y, the only difference being the expression for the moment arm of the centroid.

In Figure 7.13 we observe that the height of each elemental area is given by $y(x_i)$ and, therefore, the distance from the x-axis to the centroid is equal to $\frac{1}{2}y(x_i)$. (This follows from the fact that the centroid of a rectangle is located at its geometrical center.) The moment of each element about the x-axis is then

$$M_{x_i} = \frac{1}{2}y(x_i) \cdot A_i = \frac{1}{2}y(x_i) \cdot y(x_i) \, \Delta x = \frac{1}{2}[y(x_i)]^2 \Delta x$$

and the moment of the total area is

$$M_x = \lim_{n \to \infty} \sum_{i=1}^{n} \frac{1}{2}[y(x_i)]^2 \Delta x$$

As in the case of M_y, we are summing the vertical elements from $x = 0$ to $x = 3$ so that the limit may be evaluated by the integral

$$M_x = \frac{1}{2} \int_0^3 [y(x)]^2 \, dx$$

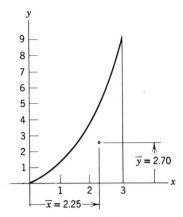

Figure 7.14

where $y(x)$ is replaced with the given function of x. Hence,

$$M_x = \frac{1}{2}\int_0^3 (x^2)^2\, dx = \frac{1}{2}\int_0^3 x^4\, dx = \frac{1}{2}\left[\frac{x^5}{5}\right]_0^3 = \frac{243}{10}$$

Finally, we substitute A, M_x, and M_y into Equations 7 and 8 (see Figure 7.14) to obtain

$$\bar{x} = \frac{M_y}{A} = \frac{81/4}{9} = \frac{9}{4} = 2.25$$

$$\bar{y} = \frac{M_x}{A} = \frac{243/10}{9} = \frac{27}{10} = 2.7$$

Example 16. Find the centroid of the area bounded by the curve $y = x^3$, the line $y = 10 - x$, and the y-axis (Figure 7.15).

Solution.   Choose a vertical element of area as shown above. The height of each of these elements is given by

$$h_i = y_2(x_i) - y_1(x_i)$$

since the upper end of the element lies on the line $y_2 = 10 - x$ and the lower end lies on the curve $y_1 = x^3$. Do not forget that y always represents the distance from the x-axis to the curve. The area of each element is then

$$A_i = [y_2(x_i) - y_1(x_i)]\,\Delta x$$

and the area is the sum

$$A = \lim_{n\to\infty} \sum_{i=1}^{n} [y_2(x_i) - y_1(x_i)]\,\Delta x$$

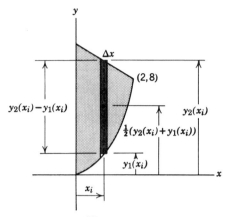

Figure 7.15

or

$$A = \int_0^2 [y_2(x) - y_1(x)]\, dx = \int_0^2 [(10 - x) - x^3]\, dx$$

$$= \left[10x - \frac{x^2}{2} - \frac{x^4}{4}\right]_0^2 = 14$$

Next we find the moments of this area about the respective axes. Referring to Figure 7.15, we see that the moment of each element about the y-axis is

$$M_{y_i} = x_i \cdot A_i = x_i [y_2(x_i) - y_1(x_i)]\, \Delta x$$

Thus,

$$M_y = \lim_{n \to \infty} \sum_{i=1}^{n} x_i [y_2(x_i) - y_1(x_i)]\, \Delta x$$

which can be written as

$$M_y = \int_0^2 x[y_2(x) - y_1(x)]\, dx$$

$$= \int_0^2 x[(10 - x) - x^3]\, dx$$

$$= \int_0^2 [10x - x^2 - x^4]\, dx = \left[5x^2 - \frac{x^3}{3} - \frac{x^5}{5}\right]_0^2 = \frac{164}{15}$$

The moment of area about the x-axis is found by a similar procedure. However, in this case, care must be used in evaluating the moment arm.

Analysis of Figure 7.15 reveals that the moment arm of each element about the x-axis is given by

$$L_i = \frac{1}{2}[y_2(x_i) + y_1(x_i)]$$

To explain this, note that the upper end of the ith element is $y_2(x_i)$ units above the x-axis and the lower end is $y_1(x_i)$ units above the x-axis. Since the centroid of each element is its midpoint, the centroid is located $\frac{1}{2}[y_2(x_i) + y_1(x_i)]$ units above the x-axis. This demonstrates an important fact: the centroid of each element must be located relative to the coordinate axes. The moment of a typical element about the x-axis is

$$M_{x_i} = L_i \cdot A_i = \frac{1}{2}[y_2(x_i) + y_1(x_i)][y_2(x_i) - y_1(x_i)] \Delta x$$

$$= \frac{1}{2}[(y_2(x_i))^2 - (y_1(x_i))^2] \Delta x$$

and the total moment about this axis is

$$M_x = \lim_{n \to \infty} \sum_{i=1}^{n} \frac{1}{2}[(y_2(x_i))^2 - (y_1(x_i))^2] \Delta x$$

Thus,

$$M_x = \frac{1}{2}\int_0^2 [(y_2(x))^2 - (y_1(x))^2] \, dx$$

$$= \frac{1}{2}\int_0^2 [(10-x)^2 - (x^3)^2] \, dx$$

$$= \frac{1}{2}\int_0^2 [x^2 - 20x + 100 - x^6] \, dx$$

$$= \frac{1}{2}\left[\frac{x^3}{3} - 10x^2 + 100x - \frac{x^7}{7}\right]_0^2$$

$$= \frac{1516}{21}$$

The coordinates of the centroid are then

$$\bar{x} = \frac{M_y}{A} = \frac{164/15}{14} = \frac{82}{105} = 0.78$$

$$\bar{y} = \frac{M_x}{A} = \frac{1516/21}{14} = \frac{758}{147} = 5.16$$

Example 17. Find the centroid of the area bounded by the curve $y^2 = -x + 4$ and the y-axis (Figure 7.16).

Solution. The most expedient approach to solving this problem is to choose a horizontal element of area. The area is then

$$A = \lim_{n \to \infty} \sum_{i=1}^{n} x(y_i) \, \Delta y$$

The horizontal elements are being summed from $y = -2$ to $y = 2$, so that

$$A = \int_{-2}^{2} x(y) \, dy$$

where $x(y) = 4 - y^2$. Since the area in question is symmetric with respect to the x-axis, this integral may be simplified by integrating from 0 to 2 and multiplying the integral by 2. Hence

$$A = 2 \int_{0}^{2} x(y) \, dy = 2 \int_{0}^{2} (4 - y^2) \, dy = 2 \left[4y - \frac{y^3}{3} \right]_{0}^{2} = \frac{32}{3}$$

The symmetry of the area also simplifies the computation of the centroid. We may immediately conclude that $\bar{y} = 0$ because the centroid must lie on the axis of symmetry.

The value of $\bar{x}$ is now computed. The moment arm of each element about the y-axis is $\frac{1}{2}x(y_i)$. Therefore, the moment of each element about the y-axis is of the form

$$M_{y_i} = \frac{1}{2} x(y_i) A_i = \frac{1}{2} x(y_i) [x(y_i) \, \Delta y] = \frac{1}{2} [x(y_i)]^2 \Delta y$$

and M_y is

$$M_y = \lim_{n \to \infty} \sum_{i=1}^{n} \frac{1}{2} [x(y_i)]^2 \Delta y$$

$$= \frac{1}{2} \int_{-2}^{2} [x(y)]^2 \, dy$$

We use symmetry of the integrand with respect to the x-axis to write

$$M_y = \int_{0}^{2} (4 - y^2)^2 \, dy = \int_{0}^{2} (16 - 8y^2 + y^4) \, dy$$

$$= \left[16y - \frac{8y^3}{3} + \frac{y^5}{5} \right]_{0}^{2} = \frac{256}{15}$$

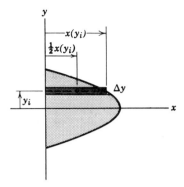

Figure 7.16

The value of $\bar{x}$ is then

$$\bar{x} = \frac{M_y}{A} = \frac{256/15}{32/3} = \frac{8}{5} = 1.6$$

4.8 ⃝

EXERCISES

Determine the location of the centroid of each of the given areas using the most convenient elemental moment. Sketch each figure.

1. $y = 2x$, $x = 3$, $y = 0$
2. $y = 5x$, $x = 2$, $y = 0$
3. $y - 3x = 0$, $x = 0$, $y = 4$
4. $2y + x = 6$, $x = 0$, $y = 0$
5. $y - x = 3$, $x = 0$, $x = 4$, $y = 0$
6. $2y + x - 4 = 0$, $x = 0$, $x = 3$, $y = 0$
7. $y = x^2$, $x = 4$, $y = 0$
8. $y = x^3$, $x = 1$, $y = 0$
9. $y = x^2$, $x = 0$, $y = 9$
10. $y = x^3$, $x = 0$, $y = 8$
11. $y = x^2 + 2x$, $y = -2x$
12. $y = x^2 - 8x$, $y = x$
13. $y = 4 - x^2$, $x = 2$, $y = 4$
14. $y^2 = 4x$, $y = x - 3$
15. $y - x = 0$, $3x + y - 4 = 0$, $x = 0$

7.8   VOLUMES OF REVOLUTION

Consider the segment of the curve $y = 2x$ from $x = 0$ to $x = 4$. Imagine this curve being revolved about the x-axis so that a cone is generated. The volume of this cone can be found by use of the definite integral if we proceed as follows. Divide the segment from $x = 0$ to $x = 4$ into n parts each Δx units wide and form rectangles of area as we have done previously. (See Figure 7.17b.) When each of these areas is revolved about the x-axis, they will generate cylinders similar to the one shown in Figure 7.17c. The volume of the cone can then be found by summing the volumes of these cylinders as $n \to \infty$.

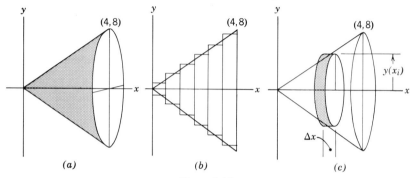

Figure 7.17

The volume of a cylinder is $\pi r^2 h$, where r is the radius and h is the width. Therefore, the volume of each of these cylinders is of the form

$$V_i = \pi [y(x_i)]^2 \Delta x$$

and the volume of the cone is

$$V = \lim_{n \to \infty} \pi \sum_{i=1}^{n} [y(x_i)]^2 \Delta x$$

or

$$V = \pi \int_0^4 [y(x)]^2 \, dx$$

Before integrating we replace $y(x)$ with its functional value $2x$, so

$$V = \pi \int_0^4 (2x)^2 \, dx$$

$$= 4\pi \int_0^4 x^2 \, dx$$

$$= 4\pi \left[\frac{x^3}{3} \right]_0^4 = \frac{256\pi}{3}$$

Example 18.   Find the volume generated by revolving the segment of the curve $y = x^{1/3}$ from $x = 0$ to $x = 8$ about the y-axis (Figure 7.18).

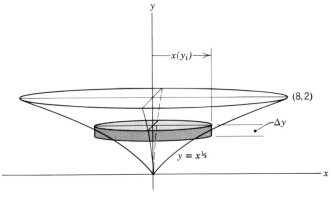

Figure 7.18

Solution.   Since the volume is generated about the y-axis, we divide the interval $y = 0$ to $y = 2$ into n horizontal elements of height Δy. The volume of this reservoir may then be expressed as

$$V = \lim_{n \to \infty} \pi \sum_{i=1}^{n} [x(y_i)]^2 \Delta y$$

and evaluated by

$$V = \pi \int_0^2 (y^3)^2 \, dy = \pi \int_0^2 y^6 dy$$

$$= \pi \left[\frac{y^7}{7} \right]_0^2 = \frac{128\pi}{7}$$

EXERCISES

Find the volume by revolving the given area about the x-axis.

1. $3y = x$, $x = 6$, $y = 0$ 5. $y = x^3$, $x = 2$, $y = 0$
2. $y^2 = x$, $x = 4$, $y = 0$ 6. $y = \frac{1}{2}x^2$, $x = 8$, $y = 0$
3. $y^2 = 9x$, $x = 4$, $y = 0$ 7. $y = 9 - x^2$, $y = 0$
4. $y^2 = x^3$, $x = 4$, $y = 0$ 8. $y = 8 - 2x^2$, $y = 0$

Find the volume generated by revolving the given area about the y-axis.

9. $y = \frac{1}{5}x$, $x = 0$, $y = 2$ 13. $y = x^3$, $x = 0$, $y = 1$
10. $y^2 = \frac{1}{2}x$, $x - 0$, $y - 2$ 14. $y = 3$, $y = 0$, $x = 0$, $x = 2$
11. $y^2 = x^3$, $x = 0$, $y = 8$ 15. $y^2 = x + 4$, $x = 0$
12. $y = x^2$, $x = 0$, $y = 9$ 16. $3y^2 = x + 27$, $x = 0$

Find the volume generated by revolving the given area about the indicated line. (The key is the expression that you write for the radius.)

17. $y^2 = x^3$, $x = 0$, $y = 8$; revolved about $y = 8$.

18. $y = x^2$, $x = 0$, $y = 9$; revolved about $y = 9$.
19. $y = x^2$, $x = 2$, $y = 0$; revolved about $x = 2$.
20. $y = x^3$, $x = 2$, $y = 0$; revolved about $x = 2$.

7.9 WORK DONE IN LIFTING LIQUIDS

The work done in lifting an object is

$$W = F \cdot S \tag{9}$$

where F is the weight of the object and S is the height through which it is lifted. This formula is true as long as F is constant. For example, the work done in lifting a 25 lb casting onto a 3 ft high bench is

$$W = 25 \cdot 3 = 75 \text{ ft-lb}$$

Equation 9 is not valid if either F or S is a variable quantity. In particular, the work done in pumping a liquid out of a container cannot be calculated by Equation 9 because as the liquid level is lowered, the height that the next layer must be raised is increased. Work problems that involve variable forces and distances can be solved by the use of the limit of a finite sum. To illustrate this, we consider the following example.

Example 19. Compute the work done in pumping the water out of a conical reservoir to a height of 3 ft above the reservoir. The vertical height of the reservoir is 2 ft and the diameter across its top is 8 ft. The density of water is 62.5 lb/ft³.

read through this example

Solution. Before we can solve the problem by the calculus, we must represent it by some model in the coordinate plane. We have chosen to locate the vertex of the cone at the origin, as shown in Figure 7.19.

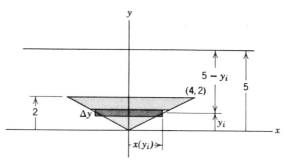

Figure 7.19

Consider the shape of the reservoir: the cone may be generated by revolving a straight line about the y-axis. Therefore, in order to represent the cone analytically, we must find the equation of the line which generates it.

The slope of this line is determined from the given dimensions as

$$m = \frac{\Delta y}{\Delta x} = \frac{2}{4} = \frac{1}{2}$$

The y-intercept is zero since the vertex of the cone is located at the origin. Using the slope-intercept form of a straight line, we have

$$y = \frac{1}{2}x$$

as the equation of the line of revolution.

We are now ready to compute the work required to lift the water to a height of 3 ft above the top of the reservoir. First divide the volume of the cone into cylindrical elements each of height Δy. The radius of a typical cylindrical element is $x(y_i)$ so that its volume is

$$V_i = \pi [x(y_i)]^2 \Delta y$$

The weight of each cylindrical element is the product of the density times the volume or

$$w_i = D\pi [x(y_i)]^2 \Delta y$$

Next we compute the work required to lift a typical cylindrical element to a height of 3 ft above the top of the reservoir. To do this we must determine how far each cylindrical element must be raised. Referring to Figure 7.19, we see that the ith element is y_i units above the x-axis. Therefore, each element must be raised $5 - y_i$ units. The work done in lifting the ith element will then be

$$\text{Work}_i = w_i(5 - y_i) = D\pi [x(y_i)]^2(5 - y_i)\,\Delta y$$

and the total work done in lifting the contents of the reservoir to a height of 5 ft is

$$\text{Work} = \lim_{n \to \infty} \sum_{i=1}^{n} D\pi [x(y_i)]^2(5 - y_i)\,\Delta y$$

In representing this summation by a definite integral, we use $y = 0$ and $y = 2$ as the limits of integration since the cylindrical elements that we are lifting lie between these two values. In general, the upper limit of the integral corresponds to the surface of the liquid in the reservoir and the lower limit corresponds to the bottom of the reservoir.

The work can then be evaluated by the integral

$$\text{Work} = D\pi \int_0^2 [x(y)]^2 (5-y) \, dy$$

where $x(y)$ must be expressed in terms of y from the equation of the line. Also, D must be replaced with the numerical value of the density of water. Solving the equation of the line for x, we have

$$x = 2y$$

and, therefore,

$$\text{Work} = 62.5\pi \int_0^2 (2y)^2 (5-y) \, dy$$

$$= 62.5\pi \int_0^2 (20y^2 - 4y^3) \, dy$$

$$= 62.5\pi \left[\frac{20y^3}{3} - y^4 \right]_0^2 = 7315 \text{ ft-lb}$$

The solution to the above problem was not dependent upon the location of the reservoir in the plane ... the problem can be solved no matter where we decide to locate the reservoir, as long as the proper coordinates are used. To illustrate this we now solve the above problem with the reservoir located differently.

Example 20.   Solve the problem described above, but locate the vertex of the cone at the point $(0, -2)$. (See Figure 7.20.)

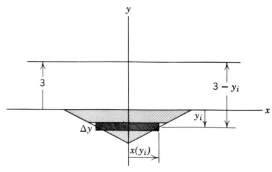

Figure 7.20

Solution. By locating the cone in this position, the line has a y-intercept of −2. Therefore, the equation is

$$y = \frac{1}{2}x - 2$$

Now we compute the work required to lift each of the elements to a height of 3 ft above the surface of the water. We see from Figure 7.20 that the cone has been located so that the surface of the water lies on the x-axis. Each of the elements then lie y_i units below the x-axis.

The distance that each element must be raised is equal to the distance between the line $y = 3$ and the line which generates the cone. Thus, the expression for h_i is

$$h_i = 3 - y_i$$

The work done on each element is the product

$$\text{Work}_i = (3 - y_i)(D\pi[x(y_i)]^2 \Delta y) = D\pi(3 - y_i)[x(y_i)]^2 \Delta y$$

The work done in emptying the conical reservoir is then the limit of the sum of these products, that is

$$\text{Work} = \lim_{n \to \infty} \sum_{i=1}^{n} D\pi(3 - y_i)[x(y_i)]^2 \Delta y$$

Using $y = -2$ and $y = 0$ as the limits of integration, we may write this sum as the definite integral

$$\text{Work} = D\pi \int_{-2}^{0} (3 - y)[x(y)]^2 \, dy$$

Before we can evaluate this integral, we must make the substitution $x(y) = 2y + 4$. Thus

$$\text{Work} = 62.5\pi \int_{-2}^{0} (3 - y)(2y + 4)^2 \, dy$$

$$= 62.5\pi \int_{-2}^{0} (-4y^3 - 4y^2 + 32y + 48) \, dy$$

$$= 62.5\pi \left[-y^4 - \frac{4y^3}{3} + 16y^2 + 48y \right]_{-2}^{0}$$

$$= 7315 \text{ ft-lb}$$

EXERCISES

1. A reservoir in the form of an inverted right circular cone is full of water. Find the work required to empty the reservoir if the altitude is 10 ft and the radius is 4 ft.
2. Solve Exercise 1 using an altitude of 4 ft and a diameter of 10 ft.
3. A right circular cylindrical tank has a diameter of 6 ft and an altitude of 10 ft. If the tank is half-full of gasoline, how much work is required to pump the gasoline to the top of the tank? (Use 60 lb/ft³ for the density of gasoline.)
4. A right circular cylindrical tank has a radius of 5 ft and an altitude of 2 ft. Find the work required to empty the tank if it is full of water.
5. A reservoir is formed by revolving the segment of the curve $y = \frac{1}{3}x^2$, from $x = 0$ to $x = 3$ ft, about the y-axis. If the reservoir is filled with oil (50 lb/ft³), find the work done in lifting the oil to a point 6 ft above the top.
6. The segment of the curve $y^2 = 4x$, from $x = 0$ to $x = 4$ ft, is revolved about the y-axis to form a reservoir. Find the work expended in emptying the reservoir, if it is full of water.
7. A tank is formed by revolving the straight line segment passing through $(2,0)$ and $(4,6)$ about the y-axis. Assume the coordinates are in feet. If the tank is full of water, how much work is done in emptying the tank to a depth of 3 ft?
8. Assume that the tank in Exercise 7 is initially empty and sitting on the ground. How much work is done in filling the tank with water, if the water is pumped over the top edge? Explain.
9. A right circular cylindrical tank 18 ft deep and 6 ft in diameter is filled with water. With the base of the tank on the x-axis, calculate the work done in pumping out the water.
10. Solve Exercise 9 when the top of the tank is located on the x-axis.
11. The inner surface of a tank has the form $y = x^3$; the diameter of the top of the tank is 4 ft. How much work is done in filling the tank with water if the water enters the tank at the bottom?
12. The curve of revolution of a tank is $y = x^2 - 4$. If the surface of the water in the tank is at the x-axis, find the work required to lift the water to a height 5 ft above the initial surface.
13. A tank is formed by revolving the curve $y^3 = x^2$ about the y-axis. If the tank is filled with water to a height of 4 ft, determine the work done in pumping all of the water to a point 3 ft above the initial surface.
14. A reservoir which is 8-ft deep is generated by the curve $y^2 = x$. Assume the reservoir is filled with water. Use four right circular cylinders to estimate the work done in emptying the reservoir. Take the radius of each cylinder to be the midpoint of each subinterval.
15. Find the work done in Exercise 14 by using the definite integral.

7.10   FURTHER APPLICATIONS OF THE DEFINITE INTEGRAL

As we have already observed, the definite integral is a useful tool in a wide range of subject areas. The following examples are intended to extend your comprehension of the definite integral and to show how physical problems other than those already presented can be solved using the definite integral.

Quantities that are expressible as the limit of a finite sum of product terms can be defined in terms of the definite integral.

Example 21. The rate at which heat is being absorbed by the thermostat in a cooling system is $dH/dt = (\frac{1}{5}t+1)$ cal/sec. What quantity of heat is absorbed by the thermostat from $t = 0$ to $t = 10$ sec?

Solution. Since the quantity of heat absorbed is equal to the rate times the elapsed time, the heat absorbed during a small time dt is given by

$$dH = \left(\frac{1}{5}t+1\right) dt$$

The total quantity of heat absorbed in 10 sec is then the limit of the sum of all such elements, or

$$H = \int_0^{10} \left(\frac{1}{5}t+1\right) dt = \left[\frac{1}{10}t^2 + t\right]_0^{10} = 20 \text{ cal}$$

Example 22. Over the time interval $t = 0$ to $t = 4$, the velocity in cm/sec of a cam-driven valve-lifter varies according to $v = t^3 - 6t^2 + 8t$. (a) How far from its rest position is the end of the lifter after 3 sec? (b) What was the total distance traveled by the lifter during this time? Note that the displacement of the lifter is equal to the area under the velocity curve. This is depicted in Figure 7.21.

Solution. (a) The cam initially pushes the lifter upward then it allows the lifter to return to its rest position. The distance that the end of the lifter will be from its rest position at any time t is the distance that it moved upward minus the distance that it returned. In other words, it is

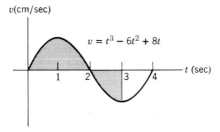

Figure 7.21

a *net* displacement and, therefore, may be found by evaluating

$$s = \int_0^3 (t^3 - 6t^2 + 8t)\, dt$$

$$= \left[\frac{t^4}{4} - 2t^3 + 4t^2\right]_0^3 = \frac{81}{4} - 54 + 36 = 2.25 \text{ cm}$$

(b) The *total* distance traveled is found by adding the absolute values of the distance moved upward and the distance returned. The point at which the lifter starts its return is the point at which its upward velocity becomes zero. We find the times at which the velocity is zero by setting $v = 0$. Thus

$$t^3 - 6t^2 + 8t = 0$$
$$t(t-2)(t-4) = 0$$
$$t = 0, 2, 4$$

From this we see that the lifter reaches its maximum travel when $t = 2$ sec. The total distance traveled in 3 sec will then be given by

$$s = \left|\int_0^2 (t^3 - 6t^2 + 8t)\, dt\right| + \left|\int_2^3 (t^3 - 6t^2 + 8t)\, dt\right|$$

$$= \left|\left[\frac{t^4}{4} - 2t^3 + 4t^2\right]_0^2\right| + \left|\left[\frac{t^4}{4} - 2t^3 + 4t^2\right]_2^3\right|$$

$$= |4| + \left|-\frac{7}{4}\right| = 5.75 \text{ cm}$$

Example 23.   The current in a resistor varies with time in accordance with $i = 3\sqrt{t}$ amp. Find the charge transferred from $t = 0$ to $t = 1$ sec.

Solution.   The charge Q transferred by a constant current is I in a time t is $Q = It$. Therefore, the approximate charge transferred by $i = 3\sqrt{t}$ in a time dt is

$$dq = i\,dt = 3t^{1/2}dt$$

The total charge transferred in 1 second is then

$$q = \int_0^1 3t^{1/2}dt = [2t^{3/2}]_0^1 = 2 \text{ coul}$$

Example 24.   An object is dropped from the top of a building. Find its average velocity during the first 5 sec of its fall.

Solution.   The acceleration of a free falling object is $a = 32$ ft/sec². To find the average velocity we need the equation for the velocity as a function of time. Since velocity is the antiderivative of the acceleration, we have

$$v = \int 32dt = 32t + C$$

The constant is zero, since $v = 0$ when $t = 0$ and, therefore

$$v = 32t$$

is the equation for the velocity. The average velocity is then found by

$$\bar{v} = \frac{1}{5-0} \int_0^5 32t\,dt$$

$$= \frac{1}{5}[16t^2]_0^5 = 80 \text{ ft/sec}$$

EXERCISES

1. A certain oil tank can be filled at a rate $R = 3t^2 + t$ gal/min. How many gallons are delivered in the first 10 min? How many are delivered in the next 10 min?

2. The rate at which heat is radiated by the plate of an electron tube is $dH/dt = 0.0006t^2$ cal/sec. Determine the quantity of heat radiated by the electron tube from $t = 0$ to $t = 10$ sec.

3. The rate of diffusion of helium gas through an orifice is $dG/dt = 1/t^4$ mole/sec. How many moles of helium pass through the orifice from $t = 2$ to $t = 4$ sec?

4. During the time interval $t = 0$ to $t = 0.1$ sec, the current in a circuit is $i = 0.3t^3$ amp. Find the charge transferred during this interval.

5. What is the average current in Exercise 4 during the indicated time interval?

6. During the first 10 sec after ignition, the thrust of a rocket is given by the formula $T = 4t^3 + t^2 + 5$ lb. What is the average thrust during this interval?

7. It takes a certain piston 0.1 sec to complete its compression stroke. If the pressure in the chamber during this time is $P = 9000t^2$ psi, what is the average pressure during the compression stroke?

8. The force required to move an object varies with the displacement as $F = (3s+1)^2$. The force is in pounds when s is in feet. (a) Find the average force exerted in moving the object 5 ft. (b) Find the work done on the object over this interval.

9. A car which is moving 25 ft/sec is given an acceleration of $a = 5\sqrt{t}$ ft/sec^2 for a period of 4 sec. What is the average velocity of the car over the indicated interval? How far does the car travel in this time period?

10. The voltage across the plates of a capacitor is $v = 4 - t^{-3/2}$. Find the average current in the capacitor from $t = 1$ to $t = 4$ sec. Assume that the capacitance is $C = 5$.

11. The current in capacitor is $i = 6t - 2t^2$ amperes. (a) What is the *total* charge transferred from $t = 0$ to $t = 4$ sec? (b) How much charge remains on the plates of the capacitor after 4 sec?

12. The voltage across a circuit is given by $v = t - 3$, and the current by $i = t - 1$. The power in the circuit is given by $p = vi$. Find the *total* energy output of the circuit from $t = 0$ to $t = 3$.

13. The velocity of a solenoid actuated rod is $v = t^3 - 12t^2 + 32t$ cm/sec during the time interval $t = 0$ to $t = 8$ sec. (a) Find the *total* distance traveled by the rod in the first 5 sec. (b) How far is the end of the rod from its rest position at the end of 5 sec?

14. Find the average velocity of the rod in Exercise 13 from $t = 0$ to $t = 8$. Explain your answer.

15. The rate at which fuel is being pumped from a storage tank is $dF/dt = 8 - t^3$ gal/sec. Find the *total* fuel transferred from $t = 0$ to $t = 3$ sec.

7.11 APPROXIMATE INTEGRATION

In the previous chapter we showed that every definite integral could be interpreted as the algebraic sum of the areas between the graph of the integrand function and the x-axis. The practice of determining this area by finding an antiderivative $F(x)$ and then computing $F(b) - F(a)$ is based upon the use of the fundamental theorem, but is limited to those cases where it is convenient to find such an antiderivative. In practice, we meet functions which are either: (a) not defined by a "rule" for all values of x but, instead, a table of values is given; or (b) the function is very difficult to antidifferentiate.

To integrate any of this broad set of functions, we resort to our "area under the curve" interpretation. If the area can be divided into parts, the area of each of which is well known, we can often "do" the integration even when we do not know the exact formula for the function.

Example 25. The graph of $f(x)$ is shown in Figure 7.22. Find $\int_1^6 f(x)\ dx$.

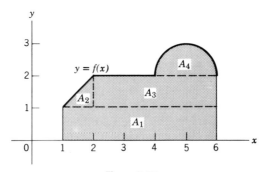

Figure 7.22

Solution.   Here the total area under the curve is given by $A_1 + A_2 + A_3 + A_4$. Each is a well known geometric figure.

$$A_1 = 1 \cdot 5 = 5$$

$$A_2 = \frac{1}{2} \cdot 1 \cdot 1 = \frac{1}{2}$$

$$A_3 = 1 \cdot 4 = 4$$

$$A_4 = \frac{1}{2}\pi(1)^2 = \frac{\pi}{2}$$

$$\int_1^6 f(x)\ dx = 5 + \frac{1}{2} + 4 + \frac{\pi}{2}$$

$$= 9\frac{1}{2} + \frac{\pi}{2} = 9.50 + 1.57 = 11.07$$

In most practical cases the graph of the integrand function will not permit such a quick and precise division as was done in the previous example. In such cases we usually try to approximate the area by using geometric figures *whose area is well known.* For example, if we were to take a rather general situation as shown in Figure 7.23, we could approximate $\int_1^4 f(x)\ dx$ by evaluating the areas of the "outer rectangles" in Figure 7.23a, the "inner rectangles" in Figure 7.23b, or the trapezoids in Figure 7.23c. Notice also that to find the areas of the rectangles or the trapezoid, the functional values need be known only at discrete points on the interval – in this case, at 1, 2, 3, and 4.

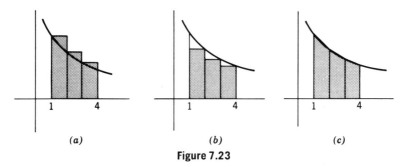

(a)   (b)   (c)

Figure 7.23

The method of approximating $\int_a^b f(x)\ dx$ by using rectangles is exactly the same technique as was shown in Chapter 6. We will now exhibit the technique of approximating with trapezoids in the following example.

Example 26.   Using 3 trapezoids, approximate $\int_2^8 dx/x$.

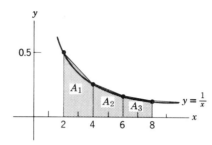

Figure 7.24

Solution.   The graph of the integrand is shown with the approximating trapezoids in Figure 7.24.

Here

$$\int_2^8 \frac{dx}{x} \approx A_1 + A_2 + A_3$$

Recalling that the area of a trapezoid is $(h/2)(b_1 + b_2)$, we have

$$A_1 = \frac{2}{2}\left(\frac{1}{2} + \frac{1}{4}\right) = \frac{3}{4}$$

$$A_2 = \frac{2}{2}\left(\frac{1}{4} + \frac{1}{6}\right) = \frac{5}{12}$$

$$A_3 = \frac{2}{2}\left(\frac{1}{6} + \frac{1}{8}\right) = \frac{7}{24}$$

$$\int_2^8 \frac{dx}{x} \approx \frac{3}{4} + \frac{5}{12} + \frac{7}{24} = \frac{35}{24} = 1.46$$

The exact area is 1.39.

Example 27.   In an experiment to determine the charge transferred past a point, the current was recorded every 0.1 sec. What was the charge transferred in the first half-second if the following current readings were taken?

t(sec)	0	0.1	0.2	0.3	0.4	0.5	0.6	0.7	0.8	0.9	1.0
i(amp)	0	2.3	2.9	2.4	1.9	1.5	1.2	1.0	0.9	0.8	0.6

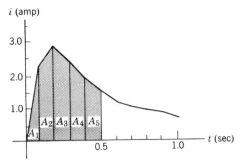

Figure 7.25

Solution.   First we plot the points and draw lines connecting these points. The resulting area is 5 trapezoids (Figure 7.25). The charge transferred during the first half-second is given by

$$q = \int_0^{0.5} idt \approx A_1 + A_2 + A_3 + A_4 + A_5$$

$$= \frac{1}{2}(0 + 2.3)0.1 + \frac{1}{2}(2.3 + 2.9)0.1 + \frac{1}{2}(2.9 + 2.4)0.1$$

$$+ \frac{1}{2}(2.4 + 1.9)0.1 + \frac{1}{2}(1.9 + 1.5)0.1$$

$$= 1.025 \text{ coul}$$

The technique of using trapezoids to estimate areas under curves can be generalized into a formula known as the *trapezoidal rule*. We choose not to state this rule: we encourage you to analyze such problems as we did above instead of trying to memorize a formula.

EXERCISES

Use trapezoids to estimate the values of the given definite integrals. The number of trapezoids to be used is given to the right of the integrals.

1. $\int_0^3 (4x + 3)\, dx$;   1

2. $\int_1^2 z^2 dz$;   4

3. $\int_2^8 \frac{dx}{x^2}$;   3

4. $\int_0^2 \sqrt{4-x^2}\,dx$;   2

5. $\int_1^4 \sqrt{9+y}\,dy$;   3

6. $\int_0^2 (2x-3x^2)\,dx$;   2

7. The velocity of a charged particle is found to be $v = 1/(1+t)$ cm/sec. Use three trapezoids to approximate the displacement of the particle from $t = 0$ to $t = 9$ sec.

8. The angular velocity, ω, of a radar antenna while tracking a satellite is $\omega = 0.1(1+t)^{1/2}$ rad/sec. Use two trapezoids to approximate the angular displacement, θ, from $t = 3$ to $t = 5$ sec.

9. An accelerometer is used to record the vertical acceleration of an airplane due to wind gusts. During a brief period of turbulence the following data was recorded.

t(sec)	0	0.01	0.02	0.03	0.04	0.05	0.06
a(ft/sec^2)	0	0.05	0.12	0.25	0.40	0.45	0.47

Find the average acceleration during this interval.

10. In an experiment to determine the electrical charge transferred to a copper plate emersed in a silver solution, the electrical current is recorded every 0.1 sec. Find the charge transferred if the following table represents the current.

t(sec)	0	0.1	0.2	0.3	0.4	0.5
i(amp)	0	0.1	0.5	0.7	0.8	0.7

11. The rate of fuel consumption, R, of an experimental engine is monitored every 2 minutes by a flowmeter, and the data recorded in the table below. How much fuel is used in the experiment?

t(min)	0	2	4	6	8	10
R(gal/min)	0.2	0.8	1.9	3.0	3.4	3.6

12. The force F delivered by a piston was recorded as a function of the displacement S of the piston from top-dead center. What was the average force delivered by the piston, if the relation between F and s is described by the following table?

S(in.)	0	1	2	3	4	5
F(lb)	140	90	50	30	15	10

Finish for Mon.
Quiz Wednesday

8

Composite Functions

8.1 INTRODUCTION

In the previous chapters we limited our discussion to linear combinations of expressions of the form x^n. In this chapter we shall see that with a few additional rules one can compute the derivative of a much larger class of functions. The fact that the differentiation process can be reduced to attaining a facility with these rules makes it a tool within the grasp of almost everyone.

8.2 THE CHAIN RULE

One of the most important differentiation formulas that you will learn is the so-called *chain rule* because it allows you to decompose a given problem into manageable pieces. The trick involved, as you will see, is to learn exactly how to make this decomposition.

The idea of decomposing a function is suggested to us because many functions are composed of other more elementary functions. Rather than go through the formality of a definition, we will give several examples to show you the general idea of composition of functions. (Remember, that as far as differentiation is concerned, you will be interested in the reverse process of decomposing.)

Example 1. Suppose the velocity of a missile varies with displacement s according to $v = s^{1/2}$ and the displacement in turn is given as a function of elapsed time to be $s = t^3 - 4$. Then the velocity can obviously be written as $v = (t^3 - 4)^{1/2}$. We say that v is a composite function of t through the displacement function s. Alternately we can say that v as a function of time can be decomposed into functions of s and t.

Example 2. Let $y = (x^2 - 2)^3$. Then if we let $u = x^2 - 2$, we will have decomposed y into functions of u and x, that is $y = u^3$ and $u = x^2 - 2$. Notice that in this problem dy/du and du/dx are both found almost by inspection. What we shall attempt to show is how to find dy/dx by knowing dy/du and du/dx.

Now, more generally, if $y = f(u)$ and $u = g(x)$ then we may regard y as a composite function of the variable x. A rather easy succession of steps will at least motivate the formula for the chain rule. We begin by writing

$$\frac{\Delta y}{\Delta x} = \frac{\Delta y}{\Delta x} \cdot \frac{\Delta u}{\Delta u} \qquad \text{if} \qquad \Delta u \neq 0$$

which may be rewritten as

$$\frac{\Delta y}{\Delta x} = \frac{\Delta y}{\Delta u} \cdot \frac{\Delta u}{\Delta x}$$

We are now going to apply the limit process to this equation. Figure 8.1 gives a typical relationship between y and x through u. As you can see, $\Delta u \to 0$ as $\Delta x \to 0$. If we assume that both dy/du and du/dx exist, then we have in the limit

$$\lim_{\Delta x \to 0} \frac{\Delta y}{\Delta x} = \lim_{\Delta u \to 0} \frac{\Delta y}{\Delta u} \cdot \lim_{\Delta x \to 0} \frac{\Delta u}{\Delta x}$$

and, therefore

$$\frac{dy}{dx} = \frac{dy}{du} \cdot \frac{du}{dx}$$

which is the formula ordinarily called the *chain rule*.

We emphasize that what we have done above is no proof, for the all important case $\Delta u = 0$ has not been handled. We shall not carry through

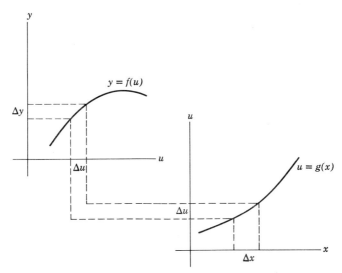

Figure 8.1

this rather delicate case here. We state the chain rule in the following theorem.

THEOREM 8.1   (The Chain Rule)

Let $y = f(u)$ and $u = g(x)$ be differentiable functions, then

$$\frac{d}{dx}[f(u)] = \frac{d}{du}[f(u)] \cdot \frac{d}{dx}[g(x)]$$

(1)

It is interesting to note that *the derivative of a function formed by composition is the product of the derivatives of the composing functions*. To actually apply the chain rule, it will be up to you to "make" the necessary decomposition. Most of the time this is done mentally and is not difficult, but it does take practice.

Example 3.   To find the acceleration of the missile in Example 1, we note that $dv/ds = \frac{1}{2}s^{-1/2}$ and $ds/dt = 3t^2$. Therefore

$$a = \frac{dv}{dt} = \frac{1}{2}\,(s^{-1/2}) \cdot 3t^2$$

It may not be important (or possible) to express the acceleration as a function of t only, but in this case it is easy to see that

$$a = \frac{3}{2}t^2(t^3 - 4)^{-1/2}$$

Example 4.   From Example 2 we note that

$$\frac{dy}{dx} = \frac{dy}{du} \cdot \frac{du}{dx} = 3u^2 2x = 6x(x^2 - 2)^2$$

One last remark about the generality of the chain rule: this rule applies to functions other than algebraic ones; in fact, there is no limitation to the class of functions appropriate for use with the chain rule. For example, $(d/dx)(\sin x^2)$ can be thought of as $(d/dx)(\sin u)$ where $u = x^2$. Thus, $(d/dx)(\sin u) = (d/du)(\sin u) \cdot 2x$. In order to complete the problem, all we need to know is the formula for $(d/du)(\sin u)$ which you will find in Chapter 10.

8.3   THE DERIVATIVE OF u^n

In this section, we apply the chain rule to composite functions of the form $y = [u(x)]^n$ or, as it is usually written, $y = u^n$.

THEOREM 8.2

If u is a differentiable function of x, the derivative of u^n with respect to x is given by

$$\frac{d}{dx}(u^n) = nu^{n-1}\frac{du}{dx} \qquad (2)$$

Proof

The derivative of y with respect to u is

$$\frac{dy}{du} = nu^{n-1}$$

Hence, by the chain rule, the derivative of y with respect to the variable x is

$$\frac{dy}{dx} = \left(\frac{dy}{du}\right)\left(\frac{du}{dx}\right) = nu^{n-1}\frac{du}{dx}$$

Or, replacing y with u^n, we have

$$\frac{d}{dx}(u^n) = nu^{n-1}\frac{du}{dx}$$

The use of Theorem 8.2 to find the derivative of a function raised to a power is demonstrated in the ensuing examples.

Example 5. Find the derivative of y with respect to x of the function $y = (3x+5)^2$.

Solution. In applying Formula 2 to this problem, we let $u = 3x+5$ and $n = 2$, so that

$$\frac{dy}{dx} = 2(3x+5)^1 \cdot \frac{d}{dx}(3x+5) = 2(3x+5)(3) = 6(3x+5)$$

In this case, we can check our answer by expanding the given binomial and then differentiating term by term. Thus

$$y = (3x+5)^2 = 9x^2 + 30x + 25$$

and

$$\frac{dy}{dx} = 18x + 30 = 6(3x+5)$$

This agrees with the result obtained by using Formula 2.

Example 6.   Find the derivative of $y = (x^3 - 4)^5$ with respect to x.

Solution.   Letting $u = x^3 - 4$ and $n = 5$, we have

$$\frac{dy}{dx} = 5(x^3 - 4)^4 \cdot \frac{d}{dx}(x^3 - 4) = 5(x^3 - 4)^4(3x^2) = 15x^2(x^3 - 4)^4$$

Those who feel compelled to do so may verify this result by raising $(x^3 - 4)$ to the fifth power and then differentiating.

Example 7.   Find the derivative of $p = \sqrt[3]{4 - t}$ with respect to t.

Solution.   Writing the cube root as a fractional exponent, the given equation may be written as

$$p = (4 - t)^{1/3}$$

In this form we see that Formula 2 may be applied to find the derivative. Letting $u = 4 - t$ and $n = 1/3$, we have

$$\frac{dp}{dt} = \frac{1}{3}(4 - t)^{-2/3} \cdot \frac{d}{dt}(4 - t) = \frac{1}{3}(4 - t)^{-2/3}(-1) = -\frac{1}{3}(4 - t)^{-2/3}$$

Example 8.   Differentiate $y = 1/(x^2 + x)^3$ with respect to x.

Solution.   This may be written in the form u^n by transferring the expression in the denominator into the numerator with a negative exponent. Thus

$$y = \frac{1}{(x^2 + x)^3} = (x^2 + x)^{-3}$$

Formula 2 may now be used to find the indicated derivative if we let $u = x^2 + x$ and $n = -3$. Hence

$$\frac{dy}{dx} = -3(x^2 + x)^{-4} \cdot \frac{d}{dx}(x^2 + x) = -3(x^2 + x)^{-4}(2x + 1) = -\frac{3(2x + 1)}{(x^2 + x)^4}$$

Example 9.   Find the expression for the instantaneous slope of the graph of $y^2 = x^2 + 4$.

Solution.   The given equation may be rearranged into the form $y = (x^2 + 4)^{1/2}$ by taking the square root of both sides. We then use Formula 2 to find the derivative of the given function. (Remember that the derivative of a function is the expression for the instantaneous slope of the graph of the function.) Letting $u = x^2 + 4$ and $n = \frac{1}{2}$, we have

$$\frac{dy}{dx} = \frac{1}{2}(x^2+4)^{-1/2} \cdot \frac{d}{dx}(x^2+4) = \frac{1}{2}(x^2+4)^{-1/2}(2x) = \frac{x}{\sqrt{x^2+4}}$$

as the desired expression.

Example 10. The theory of relativity predicts that the apparent mass m_a of an object moving at a velocity v is given by

$$m_a = \frac{m_0}{\sqrt{1-\left(\frac{v}{c}\right)^2}}$$

where m_0 is the rest mass and c is the speed of light. Find the expression for the rate of change of mass with respect to the velocity of the object.

Solution.   The expression for the apparent mass can be rewritten as

$$m_a = m_0\left[1-\frac{v^2}{c^2}\right]^{-1/2}$$

This is of the form u^n, where $u = 1 - v^2/c^2$ and $n = -\frac{1}{2}$ and, therefore

$$\frac{dm_a}{dv} = -\frac{m_0}{2}\left[1-\frac{v^2}{c^2}\right]^{-3/2} \cdot \frac{d}{dv}\left[1-\frac{v^2}{c^2}\right]$$

$$= -\frac{m_0}{2}\left[1-\frac{v^2}{c^2}\right]^{-3/2}\left[-\frac{2v}{c^2}\right]$$

$$\frac{dm_a}{dv} = \frac{m_0 v}{c^2}\left[1-\frac{v^2}{c^2}\right]^{-3/2}$$

Example 11. The light intensity, I, from a carbon filament is found to vary with the applied voltage, v, according to $I = \sqrt{\rho\,Rv}$, where ρ is a constant and R is the resistance of the filament. Find the rate of change of light intensity with respect to voltage for a given filament.

Solution.   The resistance R is considered to be fixed since we are concerned with a particular filament. The given equation can be rewritten as

$$I = (\rho\,Rv)^{1/2}$$

To find dI/dv, we let $u = \rho\,Rv$ and $n = \frac{1}{2}$. Hence

$$\frac{dI}{dv} = \frac{1}{2}(\rho Rv)^{-1/2} \cdot \frac{d}{dv}(\rho Rv) = \frac{1}{2}(\rho Rv)^{-1/2}(\rho R) = \frac{\rho R}{2\sqrt{\rho Rv}} = \frac{1}{2}\sqrt{\frac{\rho R}{v}}$$

EXERCISES

Find the derivative of each of the following functions with respect to the independent variable by applying Formula 2.

1. $y = (4-3x)^5$

2. $y = (2x-3)^3$

3. $s = (t^2+4)^4$ $4(t^2+4)^3(2t)$

4. $p = (w^3-1)^9$

5. $y = (5x+1)^{1/5}$ $\frac{1}{5}(5x+1)^{-4/5}(5)$

6. $z = (s^2+4)^{3/2}$

7. $w = \sqrt{4t^2-3t}$ $\frac{1}{2}(4t^2-3t)^{-1/2}(8t-3)$

8. $q = 2\sqrt[3]{1-3t}$

9. $i = (3t+1)^{-2}$ $-2(3t+1)^{-3}(3)$

10. $y = (x^2+3x+4)^{-1}$

11. $y = \frac{2}{(2+5x)^3}$ $2(2+5x)^{-3} = -6(2+5x)^{-4}(5)$

12. $s = \frac{1}{(t^3-t^2)}$

13. $L = \frac{3}{\sqrt{x^2+1}}$ $3(x^2+1)^{-1/2} = -\frac{3}{2}(x^2+1)^{-3/2}(2x)$

14. $y = \frac{1}{\sqrt[5]{(1-x)^3}}$

15. $s = \left(2-\frac{1}{x}\right)^3 = 3\left(2-\frac{1}{x}\right)^2\left(\frac{1}{x^2}\right)$ $x^{-1} = -x^{-1}$

16. $m = (2+3\sqrt{s})^4$

17. $r = \sqrt{1+\pi\sqrt{y}}$ $\frac{1}{2}(1+\pi y^{1/2})^{-1/2}(\pi \cdot y^{-1/2}) ?$

18. $z = (t-1/t^2)^3$

19. $y = \frac{1}{\sqrt[3]{4+3x-2x^2}}$

20. $y = \sqrt[3]{\sqrt{3x-2}}$

21. Find the second derivative of $y = \sqrt[3]{2-3x}$.
22. Find the second derivative of $s = (2t+1)^3$.
23. The distance traveled by a charged particle in a linear accelerator is given by $s = (3t+4)^{3/2}$ cm. Find the expression for the velocity of the particle as a function of time.
24. The velocity of discharge of a horizontal jet of water issuing from an orifice which is h feet below the surface of the water is $v = \sqrt{2gh}$ where $g = 32$ ft/sec². This is Torricelli's theorem. How fast does v change with respect to h?
25. The fundamental frequency f of a vibrating string is $f = (1/2L)\sqrt{T/u}$ where L is length, u is the mass per unit length, and T is the tension in the string. Find the expression for df/dT.
26. The impedance Z of a series RL circuit is given by $Z = \sqrt{R^2+X_L^2}$ where R is the resistance and X_L is the inductive reactance. Find dZ/dX_L.
27. The voltage across the plates of a capacitor varies according to $v = (3t+2)^{-1/3}$. Find the expression for the current in the capacitor.
28. The stagnation pressure at the nose of an object moving with a velocity V through an incompressible medium is given by $P = \frac{1}{2}\rho V^2$ psi, where ρ is the density of the medium. Determine the stagnation pressure, as a function of

time, of an experimental torpedo if the displacement of the torpedo is known to be $s = (2t+4)^{3/2}$ ft.

29. The modulus of rigidity G of a circular shaft is related to the modulus of elasticity E and Poisson's ratio μ by the equation $G = E/2 \, (1+\mu)^{-1}$. Assuming E to be constant, determine the expression for the approximate change of G due to a small change in μ.

30. In Exercise 29, let $E = 30 \times 10^6$ and $\mu = 0.250$. Find the approximate change in G if μ changes to 0.251.

31. The bending moment M of a simply supported beam is a function of the distance x (see Figure 8.2) measured from the left end of the beam. The change in bending moment M with respect to the distance x is called *shear* and is denoted $s = (dM/dx)$. Determine the equation for the shear if the bending moment of a beam is given by $M = (2-3x)^2$. When the units of bending moment are ft-lbs, the unit of shear will be lb.

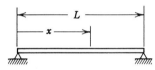

Figure 8.2

32. Evaluate both the bending moment and the shear force 2 ft from the left end of the beam in the above problem.

33. Find the shear equation for a cantilever beam if its bending moment is given by $M = (10-x^2)^3$.

34. In the study of heat transfer, the rate of heat conduction q through a material is given by

$$q = -kA \frac{dT}{dx}$$

where $k =$ a constant called the thermal conductivity
    $A =$ the area normal to the direction of flow
$dT/dx =$ the rate of change of temperature with respect to the thickness of the material

If the temperature of a concrete wall varies with thickness according to $T = (3x+2)^{-2}$, find the expression for the rate of heat conduction.

8.4   THE DERIVATIVE OF A PRODUCT

Let us consider the problem of finding the derivative of a product of two functions of x, such as

$$y = (x^2+3)(2x+1) \tag{3}$$

We observe that the derivative of this product can be found by first expanding it and then by differentiating term by term. Thus

$$y = 2x^3 + x^2 + 6x + 3$$

and

$$\frac{dy}{dx} = 6x^2 + 2x + 6 \tag{4}$$

Now let us try another approach. This time we are going to try to differentiate the product as it is given in Equation 3. Our first inclination might be to assume that the derivative of a product is equal to the product of the derivatives of the individual factors. If we apply this reasoning to the given product, we get

$$\frac{dy}{dx} = (2x)(2) = 4x \tag{5}$$

It is apparent that Equations 4 and 5 do not agree. Since we know that Equation 4 is correct, our original conjecture must be wrong; that is, the derivative of a product is *not* the product of the derivatives. The correct formula for differentiating a product of two functions is given in the following theorem.

THEOREM 8.3

If $f(x)$ and $g(x)$ represent differentiable functions, then

$$\frac{d}{dx}[f(x) \cdot g(x)] = f(x)g'(x) + f'(x)g(x) \tag{6}$$

Proof

Let $h(x) = f(x) \cdot g(x)$, then

$$\begin{aligned}
h(x + \Delta x) - h(x) &= f(x + \Delta x)g(x + \Delta x) - f(x)g(x) \\
&= f(x + \Delta x)g(x + \Delta x) - f(x + \Delta x)g(x) \\
&\quad + f(x + \Delta x)g(x) - f(x)g(x) \\
&= f(x + \Delta x)[g(x + \Delta x) - g(x)] \\
&\quad + g(x)[f(x + \Delta x) - f(x)]
\end{aligned}$$

$$\lim_{\Delta x \to 0} \frac{h(x + \Delta x) - h(x)}{\Delta x} = \lim_{\Delta x \to 0} f(x + \Delta x) \cdot \lim_{\Delta x \to 0} \frac{g(x + \Delta x) - g(x)}{\Delta x}$$

$$+ \lim_{\Delta x \to 0} g(x) \cdot \lim_{\Delta x \to 0} \frac{f(x + \Delta x) - f(x)}{\Delta x}$$

Since f is differentiable and, therefore, continuous

$$\lim_{\Delta x \to 0} f(x + \Delta x) = f(x)$$

Furthermore, $f'(x)$ and $g'(x)$ exist, so that

$$h'(x) = f(x)g'(x) + g(x)f'(x)$$

as was to be proved.

Example 12. Now find the derivative of $y = (x^2+3)(2x+1)$.

Solution. Letting $f(x) = x^2+3$ and $g(x) = 2x+1$, we have

$$\frac{dy}{dx} = (x^2+3) \cdot \frac{d}{dx}(2x+1) + (2x+1) \cdot \frac{d}{dx}(x^2+3)$$

$$= (x^2+3)(2) + (2x+1)(2x) = 6x^2 + 2x + 6 \tag{7}$$

Obviously, Equations 4 and 7 agree.

Example 13. Differentiate $y = 3x^4 \sqrt{2x-3}$.

Solution. This problem can be solved by use of the product formula by letting $f(x) = 3x^4$ and $g(x) = (2x-3)^{1/2}$. We note in passing that the second factor, $(2x-3)^{1/2}$, is a function of x raised to a power and, therefore, must be differentiated using Formula 2 from the last article. The required derivative is then

$$\frac{dy}{dx} = 3x^4 \cdot \frac{d}{dx}[(2x-3)^{1/2}] + (2x-3)^{1/2} \cdot \frac{d}{dx}[3x^4]$$

$$= 3x^4[\tfrac{1}{2}(2x-3)^{-1/2}(2)] + (2x-3)^{1/2}[12x^3]$$

This completes the differentiation process. However, the result can be simplified considerably by factoring $3x^3(2x-3)^{-1/2}$ from each term. Thus

$$\frac{dy}{dx} = 3x^3(2x-3)^{-1/2}[x+4(2x-3)] = \frac{3x^3(9x-12)}{\sqrt{2x-3}}$$

Example 14. Differentiate $y = (5x+6)^3(2x-1)^4$

Solution. Applying Theorem 8.3 to the given product, we have

$$\frac{dy}{dx} = (5x+6)^3 \cdot \frac{d}{dx}[(2x-1)^4] + (2x-1)^4 \cdot \frac{d}{dx}[(5x+6)^3]$$

$$= (5x+6)^3[4(2x-1)^3(2)] + (2x-1)^4[3(5x+6)^2(5)]$$

Simplifying this algebraically by factoring out $(5x+6)^2(2x-1)^3$, we have

$$\frac{dy}{dx} = (5x+6)^2(2x-1)^3[8(5x+6)+15(2x-1)]$$

$$= (5x+6)^2(2x-1)^3(70x+33)$$

8.5 THE DERIVATIVE OF A QUOTIENT

Let $h(x) = f(x)/g(x)$, that is, a quotient of two functions. We may derive a formula for the derivative of $h(x)$ by considering $h(x)$ as a product of $f(x)$ and $[g(x)]^{-1}$. Applying the product rule we have

$$\frac{d}{dx}[h(x)] = \frac{d}{dx}[f(x)[g(x)]^{-1}]$$

$$= f'(x)[g(x)]^{-1} + f(x)\frac{d}{dx}[g(x)]^{-1}$$

$$= \frac{f'(x)}{g(x)} - \frac{f(x)g'(x)}{[g(x)]^2}$$

$$= \frac{g(x)f'(x) - f(x)g'(x)}{[g(x)]^2}$$

This proves the following theorem.

THEOREM 8.4

If $f(x)$ and $g(x)$ are differentiable functions of x, then

$$\frac{d}{dx}\left[\frac{f(x)}{g(x)}\right] = \frac{g(x)\cdot f'(x) - f(x)\cdot g'(x)}{[g(x)]^2} \tag{8}$$

Example 15.   Differentiate $y = x^2/(2-x)^3$.

Solution.   The Application of Theorem 8.4 to the given quotient yields

$$\frac{dy}{dx} = \frac{(2-x)^3[2x] - x^2[3(2-x)^2(-1)]}{[(2-x)^3]^2}$$

$$= \frac{2x(2-x)^3 + 3x^2(2-x)^2}{(2-x)^6}$$

$$= \frac{x(4+x)}{(2-x)^4}$$

Example 16. Differentiate

$$s = \left(\frac{2t+3}{2t-3}\right)^4$$

Solution.   Rewriting the given expression as

$$s = \frac{(2t+3)^4}{(2t-3)^4}$$

we have

$$\frac{ds}{dt} = \frac{(2t-3)^4[4(2t+3)^3(2)] - (2t+3)^4[4(2t-3)^3(2)]}{[(2t-3)^4]^2}$$

$$= \frac{8(2t-3)^4(2t+3)^3 - 8(2t+3)^4(2t-3)^3}{(2t-3)^8}$$

$$= \frac{8(2t-3)^3(2t+3)^3[(2t-3) - (2t+3)]}{(2t-3)^8} = \frac{-48(2t+3)^3}{(2t-3)^5}$$

EXERCISES

Find the derivative in Exercises 1 through 20. *examples*

1. $y = (2x+3)(3x-1)$

$(2x+3)(3)+(3x-1)(2)$

2. $s = (3t^2+2)(5t+4)$

3. $y = \dfrac{3+2x}{3-2x}$ $\dfrac{(3-2x)(2)-(3+2x)(-2)}{(3-2x)^2}$

4. $p = \dfrac{w^2-1}{w^2+1}$

5. $q = t^3(t+1)^2$

6. $f(x) = x^5(x^2-1)^3$

7. $m = \dfrac{3s^2}{s^2-4}$

8. $g(t) = t\sqrt{2+3t}$

9. $y = (3x+2)^3(5x+1)^4$

10. $z = (2-x)^2(2+x)^5$

11. $p(z) = \dfrac{(4z+2)^3}{(3z+1)^2}$

12. $s = \left[\dfrac{t^2+2}{t^2-2}\right]^3$

13. $y = \dfrac{(1-x)^2}{4-x^2}$

14. $B = (q^2+q)^2\sqrt{q^2-q}$

15. $y = \sqrt[5]{\dfrac{m}{m-1}}$

16. $f(t) = \sqrt{\dfrac{1-3t}{1+3t}}$

17. $y = \sqrt[3]{\dfrac{t^2+1}{t^2-1}}$

18. $E = \sqrt{\dfrac{\pi-v}{\pi+v}}$

19. $G = \dfrac{1}{(2x-1)^3(3x+5)^2}$ **20.** $h = \dfrac{4}{3m^4\sqrt{m^2+1}}$

21. Find the second derivative of $y = (x^3-1)^4$.

22. Find the second derivative of $s = \sqrt{3-2t^2}$.

23. The energy in a 100 ohm resistor is given by $E = 100\,tI^2$ where t is time and I is current. Find the power in the resistor at $t = 2$ sec if the current varies according to $I = \sqrt[3]{5t+2}$.

24. In hydraulics the discharge q of water through an open channel is found to be a function of the depth of flow y for a constant specific head H. The defining equation of this relationship is $q = y\sqrt{2g(H-y)}$ where g is the acceleration of gravity. Show that the maximum discharge occurs when $y = \frac{2}{3}H$.

25. The relativistic velocity of an electron is given by

$$v = \frac{cp}{\sqrt{p^2 + m_0^2 c^2}}$$

where m_0 is the rest mass, c is the speed of light, and p is the momentum of the particle. Find the rate of change of v with respect to p.

26. The bending moment of a certain I-beam is given by $M = 3x^2/4(x-2)$. Find the equation for the shear.

27. The voltage output of a laboratory thermocouple varies with temperature according to $v = 3T^2/\sqrt{T^2 - T}$ volts. Find the rate of change of voltage output with respect to temperature.

28. The charge transferred in an ionic solution varied with time as $q = \sqrt{(2t-3)/(2t+3)}$. Find the expression for the electric current in the solution.

8.6 IMPLICIT DIFFERENTIATION

Sometimes we have a relation stated between x and y in which none, one, or more values of y are determined for each value of x; that is, none, one, or more *implicit* functions are defined. We prefer to work with explicit functions but the task of solving for y as an explicit function of x may be prohibitive or unnecessary.

Example.   (a) The relation $x+y+1 = 0$ defines implicitly the explicit function $y = -1-x$.

(b) The relation $x^2+y^2 = 1$ defines implicitly the two explicit functions $y = \sqrt{1-x^2}$ and $y = -\sqrt{1-x^2}$.

(c) The relation $x^2+y^2 = -1$ defines no function since no values of x and y satisfy the relation.

The determination of when a general implicit function defines y as a function of x is beyond the scope of this book, but may be found in more advanced texts under the heading "The Implicit Function Theorem."

We shall not consider examples that cause any substantial theoretical difficulty so that we can assume that a given implicit function defines one or more explicit functions $y(x)$.

It is often desirable to be able to find the derivative of a function directly from the implicit relation. The method by which we accomplish this is called *implicit differentiation*. It is illustrated in the succeeding examples.

Example 17.   Given $x^5 + y^4 = 25$, find dy/dx.

Solution.   While this function can be stated explicitly, we choose to differentiate it as an implicit function. To find the derivative of an implicit function, we take the derivative of both sides with respect to x. Thus

$$\frac{d}{dx}(x^5 + y^4) = \frac{d}{dx}(25)$$

The derivative of a sum is the sum of the derivatives, so

$$\frac{d}{dx}(x^5) + \frac{d}{dx}(y^4) = \frac{d}{dx}(25)$$

Starting with the first term on the left, we get

$$\frac{d}{dx}(x^5) = 5x^4$$

To differentiate the second term, we use Formula 2, since y is an implicit function of x. (This is the key to implicit differentiation.) In Formula 2 we substitute y for u and 4 for n. Thus

$$\frac{d}{dx}(y^4) = 4y^3 \frac{dy}{dx}$$

Differentiating the constant term

$$\frac{d}{dx}(25) = 0$$

These steps combine to form the equation

$$5x^4 + 4y^3 \frac{dy}{dx} = 0$$

which can also be written

$$\frac{dy}{dx} = -\frac{5x^4}{4y^3}$$

Example 18.   Given $y^3 + x^2y^4 + x^3 = 1$, find dy/dx.

Solution.   To find dy/dx, we differentiate each term with respect to x, that is

$$\frac{d}{dx}(y^3) + \frac{d}{dx}(x^2y^4) + \frac{d}{dx}(x^3) = \frac{d}{dx}(1)$$

Note that the second term must be handled as a product of two functions of x since x^2 is certainly a function of x and y^4 is an implied function of x. Using the product rule, the derivative of the second term is

$$\frac{d}{dx}(x^2y^4) = x^2\left[4y^3\frac{dy}{dx}\right] + y^4[2x] = 4x^2y^3\frac{dy}{dx} + 2xy^4$$

Completing the differentiation of the other terms with respect to x, we have

$$3y^2\frac{dy}{dx} + \left[4x^2y^3\frac{dy}{dx} + 2xy^4\right] + 3x^2 = 0$$

Solving for dy/dx

$$\frac{dy}{dx} = -\frac{3x^2 + 2xy^4}{3y^2 + 4x^2y^3}$$

Example 19.   The relationship between displacement (cm) and time (sec) of a charged particle in a cloud chamber is found to be $s^2 - 9t^2 = 16$. Using implicit differentiation, find the velocity of the particle when $t = 1$ sec.

Solution.   Since $v = ds/dt$, we differentiate each term with respect to t, therefore

$$\frac{d}{dt}(s^2) - \frac{d}{dt}(9t^2) = \frac{d}{dt}(16)$$

and

$$2s\frac{ds}{dt} - 18t = 0$$

from which

$$\frac{ds}{dt} = \frac{9t}{s}$$

Note that the velocity is a function of both s and t. Therefore, to find the velocity at $t = 1$, we need the corresponding value of s. Substituting $t = 1$ into the given equation, we get $s = 5$ cm. The velocity of the particle at $t = 1, s = 5$ is then

$$v = \frac{ds}{dt} = \frac{9t}{s} = \frac{9(1)}{5} = 1.8 \text{ cm/sec}$$

EXERCISES

In problems 1 through 14, find dy/dx.

1. $x^2 + y^2 = 25$
2. $4x^2 - 9y^2 = 1$
3. $x^4 + y^3 = 15$
4. $\sqrt{x} + \sqrt{y} = 16$
5. $x + x^2y^2 + y = 1$
6. $x + \sqrt{xy} + y = 25$
7. $x^{2/3} + y^{2/3} = r^{2/3}$

8. $2xy + y^2 = 2x + y$
9. $x^3 + xy + y^3 = 0$
10. $x = x^2y^2 + y$
11. $\frac{1}{x} + \frac{1}{y} = 1$
12. $\frac{1}{x^2} + \frac{1}{y^2} = 1$
13. $\frac{x}{y} - 2 = y$
14. $2x^{2/3} + y^{4/3} = 10$

Find d^2y/dx^2 in Exercises 15 through 18.

15. $x^2 + y^2 = 16$
16. $x^3 + y^3 = 4$
17. $3y^2 + xy = x$
18. $x^3 + 3xy = 9$

19. Find the equation of the straight line tangent to the graph of $x^2 + xy + y^2 = 19$ at the point $(2, 3)$.
20. Find the equation of the straight line normal to the graph of $\sqrt{x} + \sqrt{y} = 6$ at the point $(4, 16)$.

8.7 RELATED RATES

The concept of the derivative as a rate of change was discussed in Chapters 2 and 3 in connection with functions of a single variable. Often the rate of change of a quantity depends upon two or more variables each of which is a function of time. For instance, if a car is moving due north with a velocity v_N, and another car is moving due east with a velocity v_E, the velocity at which the two cars are moving apart is dependent upon both v_N and v_E. Problems in which a relationship exists between the time-rates of change of several variables are called *related* rate problems.

The technique of solving a related rate problem will be briefly summarized and then some examples given to show you how to proceed. The first and most obvious step is that you be able to recognize a related rate problem, after which you should establish a general relationship between those things that are varying and those which are held constant.

Then differentiate the relation implicitly with respect to time, and evaluate the unknown rate at the specific time in question.

Example 20. A 20-ft ladder leans against a vertical wall. If the top of the ladder slides down the wall at a rate of 4 ft/sec, how fast is the bottom of the ladder moving along the ground when it is 16 ft from the wall? (Figure 8.3)

Solution. We recognize this as a rate problem because of the question, "How fast?" We can establish the relationship between the pertinent variables by referring to Figure 8.3. Let y be the height to the top of the ladder, and let x be the corresponding distance from the wall to the bottom of the ladder. The relationship between x, y, and the length of the ladder is given by the Pythagorean theorem to be

$$x^2 + y^2 = 400 \tag{9}$$

As you can see from Figure 8.3, $-dy/dt$ represents the rate at which the top of the ladder is moving downward, and dx/dt represents the rate at which the bottom of the ladder is moving away from the wall. Since x and y are both implied functions of the time t, they may be differentiated with respect to t by employing Formula 2. The relationship between dx/dt and dy/dt is then found by differentiating with respect to t the equation relating x and y. Differentiating Equation 9 with respect to t,

$$\frac{d}{dt}(x^2) + \frac{d}{dt}(y^2) = \frac{d}{dt}(400)$$

or

$$2x\frac{dx}{dt} + 2y\frac{dy}{dt} = 0$$

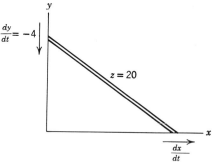

Figure 8.3

Solving for dx/dt

$$\frac{dx}{dt} = -\frac{2y\,dy/dt}{2x} = -\frac{y}{x}\frac{dy}{dt} \tag{10}$$

The value of dx/dt may be found if numerical values of x, y, and dy/dt are available. The values of x and dy/dt are given in the problem. The value of y may be found by substituting $x = 16$ into Equation 9. This yields $y = 12$.

Substituting $x = 16$, $y = 12$, and $dy/dt = -4$ into Equation 10, we get

$$\frac{dx}{dt} = -\frac{12}{16}(-4) = 3 \text{ ft/sec}$$

Example 21.   When a stone is dropped into a pond the ripples form concentric circles of increasing radius. Find the rate at which the area of one of these circles is increasing when the radius is 4 ft, if the radius is increasing at a rate of 0.5 ft/sec.

Solution.   Here we desire the relationship between dA/dt and dr/dt. To obtain this, we note that A is related to r by the formula

$$A = \pi r^2$$

Since A and r are implied functions of time, we may differentiate both sides with respect to t to get

$$\frac{d}{dt}(A) = \frac{d}{dt}(\pi r^2)$$
$$\frac{dA}{dt} = 2\pi r \frac{dr}{dt}$$

Substituting $r = 4$ and $dr/dt = 0.5$ into this equation, we get

$$\frac{dA}{dt} = 2\pi(4)(0.5) = 12.56 \text{ ft}^2/\text{sec}$$

EXERCISES

1. Two cars, A and B, leave a common point at 3 pm. Car A travels east at 80 mph, and car B moves north at 60 mph. How fast are the two cars moving apart at 5 pm?

2. A man walks 200 ft north and then turns west. If he walks 4 ft/sec, at what rate is the distance between the man and the starting point increasing 1 min after he turns west?

3. When a square plate is heated, the side length increases at a rate of 0.05 cm/min. How fast is the area increasing when the side is 10 cm long?

4. A spherical balloon is filled with gas at a rate of 5 ft³/sec. Determine the rate at which the radius of the balloon is increasing when the radius is 2 ft.

5. The force of attraction between two unit charges, separated by a distance s, is $F = 1/s^2$ dyn. Find the rate at which the force is changing when $s = 2$ cm if the two charges are moving closer together at a rate of 3 cm/sec.

6. The power in a resistor is given by $p = Ri^2$ watts. Determine the relationship between dp/dt and di/dt. In a 50 ohm resistor the current changes at a rate of 0.1 amp/sec. At what rate is the power changing when the current is 2 amp?

7. The stagnation pressure at the nose of a ballistic missile is given by $P = \frac{1}{2}\rho V^2$ psi, where ρ is the density of the air and V is the velocity of the missile. At what rate is the stagnation pressure changing when the missile has a velocity of 800 ft/sec and an acceleration of 50 ft/sec²? Assume $\rho = 0.015$.

8. In an experiment the relationship between the tensile strength of a specimen and its temperature was found to be $s = 1000 - T^{1/2}$ psi. If the temperature is increasing at a rate of 10 deg/min, at what rate is the tensile strength changing when $T = 100$ deg.

9. The electrical resistance of a wire varies with temperature according to $R = 50 - 0.3T + 0.001T^2$ ohms. Find the rate at which the resistance is changing when $T = 260$ deg if the temperature is changing at a rate of 5 deg/min.

10. In the ac circuit theory, the reactance X_L of an inductance coil is given by $X_L = 2\pi f L$, where f is the frequency of the alternating current and L is the inductance of the coil. What is the relationship between dX_L/dt and df/dt? At what rate is the reactance of a 10 henry inductance coil changing if the frequency is changing at a rate of 0.5 cps/sec?

11. A meterological balloon is released at a point 5000 ft downrange from a tracking radar. If the balloon rises vertically at a rate of 20 ft/sec, how fast is the distance (slant range) between the balloon and the radar increasing 3 min after the balloon is released?

12. Water is flowing into an 8-ft diameter conical tank at the rate of 20 cu ft/min. Determine the rate of water level rise when the depth of the water is 4 ft if the tank is 12 ft deep.

13. If $y = 5x + x^3$, and x is increasing at the rate of 0.5 units/sec, find how fast the slope of the graph is changing when $x = 2$.

8.8   THE ANTIDERIVATIVE OF $u^n(du/dx)$

Antidifferentiation is the process of reconstructing a function when we know its derivative. Thus, if we are given that

$$\frac{dy}{dx} = (x^2 + 3)^4 \cdot 2x$$

we can see that

$$y = \frac{(x^2 + 3)^5}{5} + C$$

Or, we might note in general that if

$$\frac{dy}{dx} = u^n \frac{du}{dx} \tag{11}$$

then

$$y = \frac{u^{n+1}}{n+1} + C, n \neq -1 \tag{12}$$

Using the notation that we introduced in Chapter 5, we can write the antiderivative of $u^n(du/dx)$ in the form

$$\int \left[u^n \frac{du}{dx} \right] dx \tag{13}$$

This expression is commonly shortened to

$$\int u^n du \tag{14}$$

where it is understood that the differential du is equal to the derivative of u with respect to x times dx; that is, $du = (du/dx)\, dx$. Using Formula 14 in place of Formula 13 it follows from the definition of antidifferentiation that

$$\int u^n du = \frac{u^{n+1}}{n+1} + C, n \neq -1 \tag{15}$$

A word of caution: The beginner sometimes forgets the reason for the "du" in Formula 15. For example, if we ignore the du we might (erroneously) conclude that

$$\int (x^2+3)^4 dx = \frac{(x^2+3)^5}{5} + C$$

This is clearly not the correct answer since

$$\frac{d}{dx}\left[\frac{(x^2+3)^5}{5} + C \right] = (x^2+3)^4 \cdot 2x$$

is not the given integrand function. Where have we gone wrong? The answer to this question lies in the interpretation that we give to Formula 15. Notice that it is written $\int u^n du$ (not $\int u^n dx$). This means that in order to use Formula 15, the differential $du = (du/dx)\, dx$ must be present.

Furthermore, it must be emphasized that unless we have exactly $\int u^n du$, Formula 15 may not be applied.

In the above problem, if we let $u = x^2 + 3$ we see that $du/dx = 2x$ and, therefore, $du = 2x dx$. Since we cannot account for the $2x dx$ in the integrand function, we must conclude that $\int (x^2 + 3)^4 dx$ is not of the form $\int u^n du$ and, therefore, cannot be evaluated using Formula 15. We hasten to point out that this does not mean that $\int (x^2 + 3)^4 dx$ does not exist; it simply means it cannot be found by Formula 15.

The following examples will be useful to you in learning how to use Formula 15.

Example 22.   Find $\int 6x(3x^2 - 1)^{1/2}dx$.

Solution.   Here we let $u = 3x^2 - 1$ and $n = \frac{1}{2}$. Then $du/dx = 6x$ and $du = 6x dx$. The desired antiderivative is then

$$\int \underbrace{(3x^2 - 1)^{1/2}}_{u^n} \underbrace{(6x dx)}_{du} = \underbrace{\frac{(3x^2 - 1)^{3/2}}{3/2}}_{\frac{u^{n+1}}{n+1}} + C = \frac{2}{3}(3x^2 - 1)^{3/2} + C$$

Check this result by differentiation.

Normally we must use our own ingenuity to obtain the form $u^n du$. When attempting to do this, always keep in mind *that only constants may be brought inside or outside the integral sign without changing the problem.*

Example 23.   Evaluate $\int 6\sqrt{3x - 2} \, dx$.

Solution.   First we express again the given antiderivative in the form $\int 6(3x - 2)^{1/2}dx$, and then we let $u = 3x - 2$ and $n = \frac{1}{2}$. Hence, the required differential is $du = 3dx$. By writing 6 as 2×3, we may associate the 3 with dx and take the 2 to the left of the integral sign. After this has been done, Formula 15 may be used to compute the indicated antiderivative.

$$\int 6(3x - 2)^{1/2}dx = 2 \int \overbrace{(3x - 2)^{1/2}}^{u^n}\overbrace{(3dx)}^{du}$$

$$= 2\frac{(3x - 2)^{3/2}}{3/2} + C$$

$$= \frac{4}{3}(3x - 2)^{3/2} + C$$

Example 24. Evaluate $\displaystyle\int \frac{x^2\,dx}{(x^3+1)^2}.$

Solution. In order to find this antiderivative, we first rewrite it in the form $\int x^2(x^3+1)^{-2}dx$. With it in this form, we let $u=x^3+1$ and $n=-2$. We see that, in order to use Formula 15, we need $du=3x^2\,dx$ whereas the given integrand supplies only $x^2\,dx$. *Remember, Formula 15 cannot be applied until the required differential is present.* Since $3x^2\,dx$ and $x^2\,dx$ differ only by a constant factor, we may obtain the required differential by multiplying by 3 and $\frac{1}{3}$ as indicated below.

$$\int x^2(x^3+1)^{-2}dx = \frac{1}{3}\int (x^3+1)^{-2}(3x^2\,dx)$$

$$= \frac{1}{3}\frac{(x^3+1)^{-1}}{-1}+C$$

$$= -\frac{1}{3}(x^3+1)^{-1}+C$$

Example 25. Show that $\int (x^3-3)^2x^5\,dx$ is not of the form $\int u^n\,du$.

Solution. Letting $u=x^3-3$, we see that $du=3x^2\,dx$ whereas $x^5\,dx$ appears in the integrand. *Since a variable cannot be moved across an integral sign, there is no way that $x^5\,dx$ can be rearranged to give $3x^2\,dx$.* We must, therefore, conclude that the given antiderivative is not of the form $\int u^n\,du$.

Example 26. Evaluate $\int (x^3-3)^2x^5\,dx$.

Solution. In the preceding example, we showed that this antiderivative could not be found using Formula 15. However, it can be evaluated by first expanding the integrand. Thus

$$\int (x^3-3)^2x^5\,dx = \int (x^{11}-6x^8+9x^5)\,dx$$

$$= \frac{x^{12}}{12}-\frac{2x^9}{3}+\frac{3x^6}{2}+C$$

Example 27. Evaluate $\int (x^2+1)(x^3+3x)^{1/2}\,dx$.

Solution. Let $u=x^3+3x$, then $du=(3x^2+3)\,dx=3(x^2+1)\,dx$. Multiplying by 3 and $\frac{1}{3}$, we write

$$\int (x^2+1)(x^3+3x)^{1/2}\,dx = \frac{1}{3}\int \overbrace{(x^3+3x)^{1/2}}^{u^n}\overbrace{3(x^2+1)\,dx}^{du}$$

Now applying Formula 15

$$\int (x^2+1)(x^3+3x)^{1/2}dx = \frac{1}{3}\frac{(x^3+3x)^{3/2}}{3/2}+C = \frac{2}{9}(x^3+3x)^{3/2}+C$$

EXERCISES *important*

In Exercises 1 through 16, find the indicated antiderivative.

1. $\int (2x+1)^3dx$ **9.** $\int \dfrac{dx}{(2+4x)^4}$

2. $\int (2+3t)^4dt = u^5\,du$ **10.** $\int \dfrac{3tdt}{(t^2+4)^2}$ $7 = \dfrac{1}{3}\dfrac{(2+3t)^5}{5}\dfrac{(2+3t)^5}{15}+C$
$= \frac{5}{3}(2+3t)+C$

3. $\int (1-5x)^3dx$ $du = 3\,d$ **11.** $\int \dfrac{4r^2dr}{\sqrt{r^3+1}}$

4. $\int (a-bx)^2\,dx$ **12.** $\int \dfrac{(x^2-1)\,dx}{\sqrt[3]{x^3-3x}}$

5. $\int \sqrt{3+4p}\,dp$ $= \frac{1}{4}\cdot\frac{2}{} (3+4p)^2 +c$ **13.** $\int \dfrac{(4+\sqrt{x})^2dx}{\sqrt{x}}$

6. $\int (5x+1)^{1/3}dx$ **14.** $\int \dfrac{\sqrt{x^{1/3}+4}dx}{x^{2/3}}$

7. $\int x(x^2+1)^4dx$ **15.** $\int (x^2+4)^2dx$

8. $\int x^3(4+x^4)^{1/4}dx$ **16.** $\int \sqrt{x}(1+\sqrt{x})^2dx$

In Exercises 17 through 22, evaluate the given definite integral.

17. $\int_0^2 (z+1)\sqrt[3]{z^2+2z}\,dz$ **20.** $\int_0^3 \dfrac{sds}{\sqrt{16+s^2}}$

18. $\int_{-1}^0 (3y+5)^{-2}dy$ **21.** $\int_{-2}^2 \sqrt[3]{4+2x}\,dx$

19. $\int_2^3 \dfrac{ds}{(1-s)^2}$ **22.** $\int_0^4 (\sqrt{x}+1)^2dx$

23. In mechanics the impulse I of a force F is defined as $I = \int_{t_1}^{t_2} F\,dt$, where t is time. Find the impulse of a force which varies with time according to $F = \sqrt{10t+1}$ lb over the time interval $t_1 = 0$ to $t_2 = 0.01$ sec.

24. During an elastic collision of two particles, the force between the particles varies as $F = (1+10t)^3$ dyn. What is the impulse of the force if the duration of the collision is 0.2 sec?

25. The power generated by an internal combustion engine varies with time according to $p = 2t/\sqrt{t^2+9}$ hp. Find the average power generated from $t = 0$ to $t = 4$ sec.

26. The current in a 25 ohm resistor varied according to $i = 0.05t+0.3$ amp. Find the energy dissipated by the resistor from $t = 0$ to $t = 4$ sec, if the power in the circuit is equal to the resistance times the current squared.

27. The velocity of an object is given by $v = 30/\sqrt{6t+4}$ cm/sec. Find the displacement of the object from $t = 0$ to $t = 10$ sec.

28. Find the average velocity over the given interval of the object in Exercise 27.

29. The current supplied to a circuit is given as a function of time so that $i = t\sqrt[3]{t^2 - 1}$ amp. What is the total charge transferred from $t = 0$ to $t = 3$ sec? (Take the direction of flow into consideration.)

30. The bending moment of an elastic beam is given by $M = \int S\,dx$, where S is the shear in the beam. If the shear changes with x according to the equation $S = (3+2x)^{-2}$, determine the equation for the bending moment. Assume $M = 10$ when $x = 0$.

9

Exponential and Logarithmic Functions

9.1 EXPONENTIAL FUNCTIONS

One of the most useful functions for describing physical phenomena is that of a fixed positive number (not equal to 1) to a variable power. Such a function is called an *exponential* function and is denoted

$$y = b^x, \qquad b > 0 \text{ and } b \neq 1.$$

This type of function arises naturally in many physical problems. The following list indicates a few of the physical quantities that can be described by an exponential function.

(1) Electric current
(2) Atmospheric pressure
(3) Decomposition of uranium
(4) Population growth

Far and away the most important "base" used is that for $b = e$, where e is an irrational number approximately equal to 2.718. The number e is defined by

$$e = \lim_{x \to 0} (1+x)^{1/x}$$

The evaluation of this limit is beyond the scope of this book, however, the following table suggests that the value of e is near 2.718. Remember we are not concerned with the value of $(1+x)^{1/x}$ when $x = 0$. This, of course, is not allowed, and is crucial to the understanding of a limit.

x	10	1	0.1	0.01	0.001	0.0001
$(1+x)^{1/x}$	1.271	2.000	2.594	2.705	2.717	2.718

If we take positive values of x smaller than 0.0001, the accuracy of our estimate of e will improve. However, no matter how small we take x, the value of the first four digits will not change. In Chapter 17 we shall see how we can conveniently compute e to as many decimal places as we desire.

When b is greater than 1, exponential functions of the form $y = b^x$ have the following characteristics.

(1) The value of $y = b^x$ is positive for all values of x
(2) When $x = 0$, $y = b^0 = 1$
(3) When $x = 1$, $y = b^1 = b$
(4) As $x \to \infty$, $y \to \infty$
(5) As $x \to -\infty$, $y \to 0$

By making use of these characteristics, we can draw the graph of any exponential function quite rapidly. To obtain the graph of $y = 2^x$ we note: when $x = 0$, $y = 2^0 = 1$; when $x = 1$, $y = 2^1 = 2$; when $x \to \infty$, $y \to \infty$; when $x \to -\infty$, $y \to 0$. This graph is shown in Figure 9.1 along with the graphs of $y = e^x$ and $y = 2^{-x}$ which are obtained in the same manner.

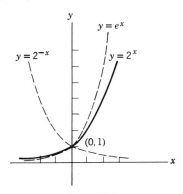

Figure 9.1

Example 1.   Solve the equations $3x + 2y = 6$ and $y = 2^{x-1}$ for x and y.

Solution. To solve these equations simultaneously by algebraic means is impractical. However, the solution can be approximated graphically by plotting the graphs of both curves and estimating their point of intersection.

The linear equation may be plotted by first writing it in slope-intercept form, $y = -\frac{3}{2}x + 3$. The exponential curve may be plotted by considering appropriate values of x; that is, the value of x that makes the exponent equal to 0, and the value of x that makes the exponent equal to 1.

x	$y = 2^{x-1}$
1	1
2	2
$\to \infty$	$\to \infty$
$\to -\infty$	$\to 0$

From Figure 9.2 we estimate that the two lines intersect at $x = 1.2$ and $y = 1.1$. By substituting these values into the linear equation, we see that they are an approximate solution.

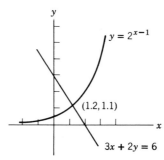

Figure 9.2

EXERCISES

Sketch the graph of each of the following exponential functions.

1. $y = e^x$ **7.** $y = 2e^{2x}$
2. $y = 3^{2x}$ **8.** $y = 3e^{x/2}$
3. $y = e^{-x}$ **9.** $y = e^{x+2}$
4. $y = 2^{-x}$ **10.** $y = e^{1-x}$
5. $y = 2 + e^{3x}$ **11.** $y = 5 - 2e^x$
6. $y = 3 + e^{-x}$ **12.** $y = 10 + e^{3x/2}$

Solve the following systems of equations by graphical means.

13. $y = e^x$ **15.** $y = 3 - 2^x$
      $y + x = 2$ $3y = x - 2$
14. $y = 10^x$ **16.** $y = e^{2-x}$
      $y = 3 - x^2$ $y = x^3$

9.2 LOGARITHMIC FUNCTIONS

You studied logarithms in algebra. A function whose rule is given by a logarithm is called a *logarithmic* function. Indeed, since we know that

$\log_b N = n$ means $b^n = N$, we say that by the function

$$y = \log_b x$$

is meant $x = b^y$. A function defined in terms of another is called the *inverse* of the first one. So in this case we say that *the logarithmic function is the inverse of the exponential function*. We will see again in Chapter 10 how the concept of an inverse function is used in defining new functions.

By virtue of its definition, the logarithmic function $y = \log_b x$, $b > 1$, has the following characteristics.

(1) $\text{Log}_b\, x$ is not defined for $x \leqslant 0$
(2) $\text{Log}_b\, 1 = 0$
(3) $\text{Log}_b\, b = 1$
(4) $\text{Log}_b\, x$ is negative for $x < 1$ and positive for $x > 1$
(5) As $x \to 0$, $\log_b x \to -\infty$
(6) As $x \to \infty$, $\log_b x \to \infty$

The graphs of $y = \log_2 x$, $y = \log_e x$, and $y = \log_{10} x$ are presented in Figure 9.3. Each of these graphs were obtained by considering characteristics listed above.

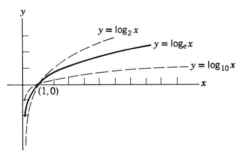

Figure 9.3

Example 2.   Sketch the graph of $y = \log_5 (2x - 1)$.

Solution.   Here we note that the function is not defined for values of $x \leqslant \frac{1}{2}$. Furthermore, when $x = 1$, $y = \log_5 1 = 0$, and when $x = 3$, $y = \log_5 5 = 1$. These facts are summarized in the table and the graph drawn in Figure 9.4.

Although the base of a logarithm may theoretically be any positive number except 1. In practice we seldom use bases other than 10 and e. Therefore, to simplify the writing of logarithmic expressions, we introduce the following notation. Logarithms to the base 10, which are called

x	$y = \log_5(2x-1)$
$x \leqslant \frac{1}{2}$	Not defined
1	0
3	1
$x \to \infty$	$\to \infty$
$x \to \frac{1}{2}$	$\to -\infty$

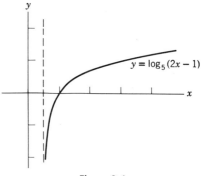

Figure 9.4

common logarithms, will be denoted by

$$\log x$$

The base is understood to be 10 if no base is written. Logarithms to the base e, which are called *natural* logarithms, will be denoted by

$$\ln x$$

This notation will be used throughout the remainder of this book.

In working with logarithmic expressions, it is frequently necessary to rearrange the expression algebraically. The rules for manipulating logarithmic expressions are listed below, where $b > 0$ and $b \neq 1$. The proofs of these rules are usually given when logarithms are treated in algebra and, therefore, will not be given here.

Rule 1

$$\log_b MN = \log_b M + \log_b N$$

Rule 2

$$\log_b \frac{M}{N} = \log_b M - \log_b N$$

Rule 3

$$\log_b M^n = n \log_b M$$

Rule 4

$$\log_b M^{1/n} = (1/n) \log_b M$$

Example 3. Solve $2 \log_3 x + \log_3 (x+1) = \log_3 (x^2+8)$.

Solution. Applying Rule 3, we have

$$\log_3 x^2 + \log_3 (x+1) = \log_3 (x^2+8)$$

Then by Rule 1

$$\log_3 x^2(x+1) = \log_3 (x^2+8)$$

But if two logarithms are equal, their arguments must be equal, so

$$x^2(x+1) = x^2+8$$
$$x^3+x^2 = x^2+8$$
$$x^3 = 8$$
$$x = 2$$

Example 4. Solve $y = e^{2x+1}$ for x in terms of y.

Solution. Using the inverse relationship between the exponential and the logarithmic function, we write the solution as

$$2x+1 = \ln y$$
$$2x = \ln y - 1$$
$$x = \tfrac{1}{2} \ln y - \tfrac{1}{2}$$

Example 5. Solve $\ln y = 3x + \ln x$ for y in terms of x.

Solution. Subtracting $\ln x$ from both sides, we have

$$\ln y - \ln x = 3x$$

By Rule 2

$$\ln \frac{y}{x} = 3x$$

Writing the logarithm in exponential form, we get

$$\frac{y}{x} = e^{3x}$$

or

$$y = xe^{3x}$$

EXERCISES

Sketch the graph of the indicated functions.

1. $y = \ln x$
2. $y = \log_3 x$
3. $y = \ln (x+1)$
4. $y = \log (x-1)$
5. $y = \log 2x$
6. $y = \log x/2$
7. $y = 3 \ln x$
8. $y = 4 \log x$
9. $y = \log x^2$
10. $y = \log x + 2$

In Exercises 11 through 16, find x without using a table.

11. $\ln x = \ln 3 + \ln 5$
12. $\ln x = \ln 4 - \ln 3 + \ln 9$
13. $\ln x + \ln (x+5) = \ln 6$
14. $\ln (x^2 - 4) - \ln (x+2) = \ln 3$
15. $\ln (x+4) + \ln 2 = 2 \ln x$
16. $\log (4x-3)/\log x = 2$

Solve each of the following equations for x in terms of y.

17. $y = \log x$
18. $y = \ln (x+2)$
19. $y = \ln x^2$
20. $y = e^{2x}$
21. $y = 10^x$
22. $\ln y + \ln x = \ln 3$
23. $\ln (x+2) - \ln y = -2$
24. $y = a\, e^{bx}$

9.3 DERIVATIVES OF LOGARITHMIC FUNCTIONS

We continue our discussion by deriving a formula for the derivative of a logarithm. Consider the logarithmic function

$$y = \log_b x$$

The derivative of this function with respect to x can be found by applying the definition of a derivative with $f(x) = \log_b x$. Thus, if $\Delta y = f(x+\Delta x) - f(x)$ we have

$$\Delta y = \log_b (x+\Delta x) - \log_b x$$

or

$$\Delta y = \log_b \frac{x+\Delta x}{x} = \log_b \left(1 + \frac{\Delta x}{x}\right)$$

Using this expression, we obtain

$$\frac{\Delta y}{\Delta x} = \frac{1}{\Delta x} \log_b \left(1 + \frac{\Delta x}{x}\right)$$

The limit of the right-hand member cannot be evaluated as it stands. To evaluate the limit, we first multiply the right member by x/x,

$$\frac{\Delta y}{\Delta x} = \frac{1}{x} \cdot \frac{x}{\Delta x} \log_b \left(1 + \frac{\Delta x}{x}\right)$$

$$= \frac{1}{x} \log_b \left(1 + \frac{\Delta x}{x}\right)^{x/\Delta x}$$

Therefore

$$\frac{dy}{dx} = \lim_{\Delta x \to 0} \frac{1}{x} \log_b \left(1 + \frac{\Delta x}{x}\right)^{x/\Delta x}$$

As $\Delta x \to 0$, $\dfrac{\Delta x}{x} \to 0$ and $\dfrac{x}{\Delta x} \to \infty$, so that $\lim\limits_{\Delta x \to 0} \left(1 + \dfrac{\Delta x}{x}\right)^{x/\Delta x}$

and

$$\frac{dy}{dx} = \frac{1}{x} \log_b e$$

Since $y = \log_b x$, this may be rewritten as,

$$\frac{d}{dx}(\log_b x) = \frac{1}{x} \log_b e \qquad (1)$$

We may extend this result to include logarithmic functions of the form, $\log_b u$, where u is a function of the variable x, by applying the chain rule. Thus

$$\frac{d}{dx}(\log_b u) = \frac{1}{u}\frac{du}{dx} \log_b e \qquad (2)$$

is the formula for the derivative of the logarithm of a function of the variable x.

An important special case of Formula 2 is the case in which base e is used. The right member of (2) becomes $(1/u)(du/dx) \ln e$, where $\ln e = 1$, and so

$$\frac{d}{dx}(\ln u) = \frac{1}{u}\frac{du}{dx} \qquad (3)$$

The simplicity of Formula 3 compared to Formula 2 encourages us to use natural logarithms in calculus.

Example 6.   Differentiate $y = \ln (x^3 + 2)$.

Solution.   Using Formula 3, we let $u = x^3 + 2$ so that $du/dx = 3x^2$.

Thus

$$\frac{dy}{dx} = \frac{1}{x^3 + 2}(3x^2) = \frac{3x^2}{x^3 + 2}$$

Example 7.   Differentiate $s = \log 3t$.

Solution.   Here we use Formula 2 with $b = 10$ and $u = 3t$. Before differentiating we observe that $\log e \approx 0.4343$. This may be verified by looking up the logarithm of 2.718 in a table of common logarithms. The required derivative is then

$$\frac{ds}{dt} = \frac{1}{3t}(3)\log e = \frac{\log e}{t} = \frac{0.4343}{t}$$

Example 8.   Differentiate $y = \ln(3-5x^2)^{1/2}$

Solution.   This may be rewritten as

$$y = \tfrac{1}{2}\ln(3-5x^2)$$

so therefore

$$\frac{dy}{dx} = \frac{1}{2}\left[\frac{1}{3-5x^2}(-10x)\right] = \frac{-5x}{3-5x^2}$$

Example 9.   Differentiate $y = \ln^3 2x$.

Solution.   To clarify this expression we note that $\ln^3 2x$ can be written as $(\ln 2x)^3$. In this form we see that we have a function of x raised to a power. The derivative of a function of x raised to a power is given by

$$\frac{d}{dx}(v^n) = nv^{n-1}\frac{dv}{dx}$$

Letting $n = 3$ and $v = \ln 2x$, we have

$$\frac{dy}{dx} = 3(\ln 2x)^2 \cdot \frac{d}{dx}(\ln 2x)$$

Now, using Formula 3 to evaluate $d/dx\,(\ln 2x)$, we have

$$\frac{dy}{dx} = 3(\ln 2x)^2\left[\frac{1}{2x}\cdot 2\right] = \frac{3}{x}\ln^2 2x$$

Example 10.   Differentiate $p = \ln\dfrac{t}{t^2-1}$.

Solution.   The most expedient method of finding the derivative is to first simplify the expression algebraically. Thus, the given equation may be written

$$p = \ln t - \ln(t^2-1)$$

and

$$\frac{dp}{dt} = \frac{1}{t}(1) - \frac{1}{t^2-1}(2t) = \frac{1}{t} - \frac{2t}{t^2-1} = -\frac{t^2+1}{t(t^2-1)}$$

Example 11. If the temperature T is held constant, the chemical potential μ of a perfect gas varies with pressure P according to

$$\mu = RT \ln P + M.$$

Find the rate at which the chemical potential changes with respect to pressure, if R, T, and M are constants.

Solution.   Using Formula 3 we have

$$\frac{d\mu}{dP} = RT \left[\frac{1}{P} \right] = \frac{RT}{P}$$

Example 12. The tensile strength of a new plastic to be used in the impeller of a centrifugal pump varies with temperature according to $S = 300 \ln (T + 10) - 2T + 5000$. At what temperature does the plastic have its maximum tensile strength?

Solution.   The maximum value of a function is found by setting the first derivative equal to zero and solving for the variable. The derivative of tensile strength with respect to temperature is

$$\frac{dS}{dT} = \frac{300}{T + 10} - 2$$

Setting this equal to zero

$$\frac{300}{T + 10} - 2 = 0$$
$$300 - 2T - 20 = 0$$
$$T = 140 \deg$$

To check this value we will use the second derivative test.

$$\frac{d^2S}{dT^2} = \frac{-300}{(T + 10)^2}$$

Since this is negative for $T = 140$, we conclude that the tensile strength is a maximum at this temperature.

EXERCISES

Differentiate each of the expressions in Exercises 1 through 20.

1. $y = \log 2x$
2. $y = \log \frac{1}{2}x$
3. $s = \ln (2t - 1)$
4. $p = \ln 2(r + 1)$

5. $z = \ln p^2$

6. $y = \log (x^2 + 1)$

7. $q = 3 \ln (x - x^2)$

8. $r = 3 + \ln (1 - s^2)$

9. $f(x) = \ln x^3 (4 - x)^2$

10. $B = \ln I(I^2 + 4)^3$

11. $s = \ln \sqrt{t^3 - 1}$

12. $w = \log (t^2 + 3t + 1)^{2/3}$

13. $g(r) = \ln \dfrac{r^2 + 1}{(r^2 - 1)^3}$

14. $i = \ln \sqrt{\dfrac{2 - v}{2 + v}}$

15. $Q = \ln^4 (3T + 2)$

16. $m = \sqrt{\ln (p^2 - 1)}$

17. $C = \dfrac{R^2}{\ln (R^2 - 2)}$

18. $y = (3x + 2)^3 \ln (8x^2 - 3)$

19. $E = V^2 \ln (3V + 2)$

20. $G(x) = 3x + \dfrac{x}{\ln (x - 1)}$

21. A coaxial cable consists of an inner cylindrical conductor of radius r inside of a thin-walled conducting tube of radius R. If the two conductors are oppositely charged, the potential difference between the conductors is $V = 100 \ln (R/r)$. Find the rate of change of V with respect to r if R is held constant.

22. In Exercise 21 find the rate of change of V with respect to R if r is held constant.

23. The Chezy discharge coefficient for water flowing in a wide channel is $C = 42 \ln (R/p)$, where R is the hydraulic radius and p is the roughness of the channel. Find dC/dp assuming R to be constant.

24. The displacement of a car varies with time according to $s = \ln (t^3 + 1)$ ft. Find the expression for the velocity of the car.

25. Find the expression for the acceleration of the car in Exercise 24.

26. The current in a 5 henry inductor varies as $i = \ln (2t - 1)$ amp. Find the expression for induced voltage in the coil. What is the voltage when $t = 3$ sec?

27. The tensile strength S of a viscoelastic material, used for damping mechanical vibrations, is a function of temperature T. Find the rate of change of tensile strength with respect to temperature if $S = \ln [3/(1 + 3T)]$ psi.

28. The Link Trucking Company has a fleet of trucks which are serviced by company mechanics. The cost of maintenance is given by $C = \ln (3m + 2)$ dollars, where m is the interval in miles between repairs. Find the rate of change of maintenance cost with respect to the repair interval.

29. The energy dissipated by a resistor during an experiment is given by $E = \ln (t + 1) - \frac{1}{2}t + 10$ joules. At what time is the energy output a maximum?

30. What is the maximum energy output of the resistor in Exercise 29?

9.4   LOGARITHMIC DIFFERENTIATION

By combining the results of the last article with that of implicit differentiation, we may expand the set of functions which we can differentiate. To illustrate this, consider differentiating the function $y = x^x$. This does not fit the formula for a power function, because the exponent also contains the variable; that is, this is a variable raised to a variable power. The derivative of functions of this type may be found by a technique known as *logarithmic* differentiation. The basic steps in the process are below.

(1) Take the natural logarithm of both sides of the given relation.
(2) Simplify the logarithmic expressions.
(3) Differentiate the resulting equation implicitly with respect to x.
(4) Solve for the required derivative.

Example 13.   Find the derivative of $y = x^x$.

Solution.   (1) $\ln y = \ln x^x$
            (2) $\ln y = x \ln x$

Before differentiating, note that y is a function of x and, therefore, d/dx $(\ln y) = (1/y)(dy/dx)$

$$(3)\ \frac{1}{y}\frac{dy}{dx} = x\frac{1}{x} + \ln x(1) = 1 + \ln x$$

To solve for dy/dx, we multiply both sides of this equation by y.

$$(4)\ \frac{dy}{dx} = (1 + \ln x)y$$

But, from the given equation $y = x^x$, so

$$\frac{dy}{dx} = (1 + \ln x)x^x$$

Example 14.   Differentiate $y = (x^2 - 3)^4/(3x + 1)^2$ using logarithmic differentiation.

Solution.   (1) $\ln y = \ln \dfrac{(x^2 - 3)^4}{(3x + 1)^2}$

$$(2)\ \ln y = \ln (x^2 - 3)^4 - \ln (3x + 1)^2$$
$$= 4 \ln (x^2 - 3) - 2 \ln (3x + 1)$$

$$(3)\ \frac{1}{y}\frac{dy}{dx} = \frac{8x}{x^2 - 3} - \frac{6}{3x + 1} = \frac{2(9x^2 + 4x + 9)}{(x^2 - 3)(3x + 1)}$$

$$(4)\ \frac{dy}{dx} = \frac{2(9x^2 + 4x + 9)}{(x^2 - 3)(3x + 1)} \cdot y = \frac{2(9x^2 + 4x + 9)}{(x^2 - 3)(3x + 1)} \cdot \frac{(x^2 - 3)^4}{(3x + 1)^2}$$

$$= \frac{2(9x^2 + 4x + 9)(x^2 - 3)^3}{(3x + 1)^3}$$

Example 15.   The amplitude of a damped oscillation varies with time according to $A = e^{-t}$ cm. Find the expression for the rate of change of amplitude with respect to time.

Solution.   To find the required rate we must use logarithmic differentiation

$$(1)\ \ln A = \ln e^{-t}$$

$$(2)\ \ln A = -t \ln e = -t$$

$$(3)\ \frac{1}{A}\frac{dA}{dt} = -1$$

$$(4)\ \frac{dA}{dt} = -A = -e^{-t}\ \text{cm/sec}$$

EXERCISES

Find the derivative of the indicated function by using logarithmic differentiation

1. $y = e^{2x}$

2. $y = x^{3x}$

3. $y = (2x-3)^{1/3}(2x+3)^{2/3}$

4. $s = (t^2+4)^3(3t+4)^5$

5. $v = t^2 e^{2t}$

6. $y = \dfrac{x}{e^x}$

7. $i = 5e^{-3t}$

8. $s = 10e^{2/t}$

9. $y = \sqrt{\dfrac{x^2+1}{x^2-1}}$

10. $z = \sqrt[3]{\dfrac{1-2s}{1+2s}}$

11. $B = (4L^2-5)^{\sqrt{L^2+1}}$

12. $y = (1-x^3)^{(1-x^3)}$

13. $y = \dfrac{(2x+1)(3x+2)}{4x+3}$

14. $C = T^3(T^2+1)^4(T^2-1)^4$

9.5   DERIVATIVES OF EXPONENTIAL FUNCTIONS

Consider an exponential function of the form

$$y = b^u$$

where b is a constant and u is a function of the variable x. The formula for the derivative of functions of this type can readily be obtained by applying logarithmic differentiation to the given function. Taking the natural logarithm of both sides

$$\ln y = \ln b^u = u \ln b$$

Differentiating this equation with respect to x

$$\frac{1}{y}\frac{dy}{dx} = \frac{du}{dx}\ln b$$

Solving for dy/dx

$$\frac{dy}{dx} = y \cdot \frac{du}{dx}\ln b$$

Replacing y with b^u, we have

$$\frac{d}{dx}(b^u) = b^u\frac{du}{dx}\ln b \qquad (4)$$

$215.3.$ $\mathcal{BRAN}$

When $b = e$, we have the important special case

$\mathcal{BAKER}$

$$\frac{d}{dx}(e^u) = e^u\frac{du}{dx} \qquad (5)$$

Example 16.   Differentiate $y = 10^{2x}$.

Solution.   If $y = 10^{2x}$

$$\frac{dy}{dx} = 10^{2x}(2)\ln 10$$

Example 17.   Find the derivative of $p = t^2 e^{t^3}$.

Solution.   The product rule applies here. Hence

$$\frac{dp}{dt} = t^2 \cdot \frac{d}{dt}(e^{t^3}) + e^{t^3} \cdot \frac{d}{dt}(t^2)$$

$$= t^2[e^{t^3}(3t^2)] + e^{t^3}[2t]$$

$$\frac{dp}{dt} = 3t^4 e^{t^3} + 2t e^{t^3} = t e^{t^3}(3t^3 + 2)$$

Example 18.   The current in a 10 henry inductance coil increases according to $i = 2 - e^{-5t}$ amp. Find the expression for the induced voltage.

Solution.   The voltage induced in a coil is

$$v = -L\frac{di}{dt}$$

Therefore

$$v = -10\frac{d}{dt}(2 - e^{-5t}) = -10[-e^{-5t}(-5)] = -50e^{-5t} \text{ volts}$$

Example 19. When the brake is applied to a rotating flywheel, the angular speed of the flywheel decreases according to the equation $\omega = 5e^{-t^2}$ rad/sec. At what rate is ω decreasing when $t = 2$ sec.

Solution.   The derivative is

$$\frac{d\omega}{dt} = 5e^{-t^2}(-2t) = -10te^{-t^2}$$

Substituting $t = 2$

$$\frac{d\omega}{dt} = -20e^{-4} = -20\,(0.01832) = -0.3664\,\frac{\text{rad/sec}}{\text{sec}}$$

The value of e^{-4} was obtained from Table 4 in the Appendix.

Example 20. When a heated object is immersed in a cooling medium, the temperature of the object at any time after immersion is approximated by the formula $T = T_0 + Ce^{kt}$, where T_0 is the temperature of the cooling medium, C is the temperature gradient between the object and the cooling medium before immersion, and k is a constant of proportionality. If $T_0 = 10$, $C = 80$, and $k = -0.5$, find (a) the temperature of the object 2 sec after immersion, and (b) the rate at which the temperature is changed at this time.

Solution.   (a) The equation for T is obtained by replacing T_0, C, and k with their numerical values. Thus

$$T = 10 + 80\,e^{-0.5t}$$

is the desired equation. Substituting $t = 2$

$$T = 10 + 80\,e^{-1} = 10 + 80\,(0.3679) = 39.4 \text{ deg}$$

(b) The rate of change of temperature is

$$\frac{dT}{dt} = 80\,e^{-0.5t}\,(-0.5) = -40\,e^{-0.5t}$$

Letting $t = 2$

$$\frac{dT}{dt} = -40e^{-1} = -40(0.3679) = -14.7 \text{ deg/sec}$$

EXERCISES

In Exercises 1 through 18 find the first derivative of the indicated function.

1. $s = e^{2t}$

2. $m = 10^{2r-1}$

3. $y = 4e^{x^2}$

4. $i = 5e^{3t/2}$

5. $v = \dfrac{4}{e^{3x}}$

6. $J = \dfrac{1}{10^K}$

7. $y = e^{\sqrt{x}}$

8. $z = 3e^{(1-x)}$

9. $B = T^3 e^T$

10. $G(x) = \dfrac{x}{e^{2x}}$

11. $w = \dfrac{e^v}{v^2}$

12. $y = \dfrac{1}{xe^x}$

13. $\theta = \dfrac{e^r + 1}{e^r - 1}$

14. $f(t) = \dfrac{1}{2}(e^t - e^{-t})$

15. $I = e^{\ln R}$

16. $q = \ln(2 + e^{2x})$

17. $P = e^t \ln t$

18. $x = e^y \sqrt{1 + \ln y}$

In Exercises 19 through 24 find the second derivative of the indicated function.

19. $y = e^{-x}$

20. $y = e^{3x}$

21. $p = e^{t^2}$

22. $Q = e^{\sqrt{s}}$

23. $y = t e^{t^2}$

24. $p(f) = f^2 e^{-f}$

25. The growth of current in an inductance coil connected in series to a resistor is given by $i = (V/R)(1 - e^{-Rt/L})$ where R, L, and V are constants and t is time. Find di/dt.

26. The saturation current density J at the surface of a cathode is a function of the temperature T of the cathode. The relationship between saturation current density and cathode temperature was found by Dushman to be $J = AT^2 e^{-K/T}$ where A and K are constants. Find the rate of change of J with respect to T.

27. The shear s of a simple beam is equal to the derivative of the bending moment M with respect to the distance x, measured from one end of the beam. Find the shear equation of a beam whose bending moment is $M = xe^x$.

28. When a cable hangs between two poles it forms a curve known in mathematics as a *catenary*. The equation of a catenary is $y = (H/2)(e^{x/H} + e^{-x/H})$, where H is the height of the cable at the center point and x is the horizontal distance from the center point, as in Figure 9.5. Find the slope of the cable at $x = 20$ ft if $H = 100$ ft.

29. The atmosphere pressure varies with altitude according to $p = p_0 e^{-kz}$ where p_0 and k are constants and z is the altitude. Find the rate of change of p with respect to z.

30. The reliability function $R(t)$ of a collection of objects which are subject to failure is defined as the derivative of the failure function $F(t)$ where t is a

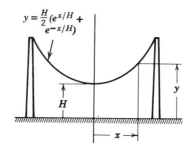

Figure 9.5

unit of time. If $F(t) = 1 - e^{-3t}$ is the failure function of Shurfire spark plugs, what is the reliability function for the spark plugs?

31. The force F required to accelerate an object of weight W is given by Newton's second law to be $F = (W/g)a$, where g is the acceleration of gravity. If W is given in lb and a and g in ft/sec², then F will be in lb. Find the expression for the force exerted on an object by a resisting medium if the velocity of the object is given by $v = 10(1 - e^{-0.1t})$ ft/sec. What is the magnitude of the force when $t = 20$ sec if the object weighs 96 lb?

32.   In a critically damped vibrating system, the displacement of the system from its equilibrium position is given by $x = (A + Bt)e^{-pt}$ where A, B and p are constants and t represents time. Determine the expression for the velocity of the system dx/dt.

9.6   ANTIDERIVATIVES OF RECIPROCAL FUNCTIONS

In the previous chapter, we introduced the formula

$$\int u^n \, du = \frac{u^{n+1}}{n+1} + C$$

At that time we indicated that the formula was invalid for $n = -1$. We will now show how to evaluate antiderivatives of the form

$$\int u^{-1} \, du = \int \frac{du}{u}$$

The formula for the derivative of $\ln u$ with respect to x is

$$\frac{d}{dx}(\ln u) = \frac{1}{u}\frac{du}{dx}$$

Therefore

$$\int \left[\frac{1}{u}\frac{du}{dx}\right] dx = \ln u + C \tag{6}$$

or using the differential notation $du = (du/dx)\,dx$ we can write (6) as

$$\int \frac{du}{u} = \ln u + C \tag{7}$$

In Formula 7, as in all other antidifferentiation formulas, the symbols have a very precise meaning. Thus, (7) applies only to antiderivatives in which the integrand function appears in the form $1/u$, and is accompanied by the differential of u. Of course, we may alter the form of the antiderivative by multiplying and dividing by constants.

Example 21.   Find $\displaystyle\int \frac{dx}{3x+2}$.

Solution.   In this problem, let $u = 3x + 2$, then the differential needed in order to use Formula 7 is $du = 3dx$. Since the desired differential can be obtained by multiplying dx by 3 and the integral by $1/3$, we may evaluate this antiderivative by (7). Thus

$$\int \frac{dx}{3x+2} = \frac{1}{3}\int \frac{3dx}{3x+2} = \frac{1}{3}\ln (3x+2) + C$$

Example 22.   Show that $\displaystyle\int \frac{xdx}{x^3+1}$ is not of the form $\int du/u$.

Solution.   Letting $u = x^3 + 1$, we see that the required differential is $du = 3x^2\,dx$. Since the available differential, xdx, cannot be altered into $3x^2dx$, the given integral is not of the form $\int du/u$. We reiterate that a variable cannot be moved across an integral sign.

Example 23.   During an exothermic chemical reaction the rate at which heat is generated is given by $dH/dt = t/(t^2+4)$ Btu/sec. Find the amount of heat generated from $t = 0$ to $t = 2$ sec.

Solution.   The amount of heat generated during this interval of time is given by the definite integral

$$H = \int_0^2 \left(\frac{dH}{dt}\right) dt = \int_0^2 \frac{t}{t^2+4}\, dt$$

Multiplying the integrand function by 2 and the integral by $\frac{1}{2}$

$$H = \frac{1}{2}\int_0^2 \frac{2t\,dt}{t^2+4} = \frac{1}{2}[\ln (t^2+4)]_0^2$$

$$= \frac{1}{2}[\ln 8 - \ln 4] = \frac{1}{2}\ln 2 = 0.347 \text{ Btu}$$

Example 24.   Find the area under the curve $y = 1/x$, from $x = 1$ to $x = 3$. (See Figure 9.6.)

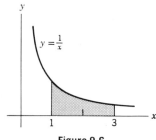

Figure 9.6

Solution.   The indicated area is given by $\int_1^3 y\,dx$, where $y = 1/x$. Hence

$$A = \int_1^3 \frac{dx}{x} = [\ln x]_1^3 = \ln 3 - \ln 1 = \ln 3 = 1.0986 \text{ sq units}$$

EXERCISES

In Exercises 1–20, find the indicated antiderivative.

1. $\displaystyle\int \frac{2\,dx}{x}$

2. $\displaystyle\int \frac{dx}{3x}$

3. $\displaystyle\int \frac{dy}{1-2y}$

4. $\displaystyle\int \frac{ds}{3s+4}$

5. $\displaystyle\int \frac{t\,dt}{t^2+4}$

6. $\displaystyle\int \frac{z\,dz}{1-z^2}$

7. $\displaystyle\int \frac{e^x\,dx}{e^x+1}$

8. $\displaystyle\int \frac{(2x+1)\,dx}{x^2+x}$

9. $\displaystyle\int \frac{x\,dx}{(x^2-3)^2}$

10. $\displaystyle\int \frac{B^{1/2}\,dB}{1+B^{3/2}}$

11. $\displaystyle\int \frac{\ln^2 x\,dx}{x}$

12. $\displaystyle\int \frac{(x^2-1)\,dx}{x^3-3x+5}$

13. $\displaystyle\int \frac{dx}{x\ln x}$

14. $\displaystyle\int \frac{x\,dx}{x^2+2}$

15. $\displaystyle\int \frac{e^x-e^{-x}}{e^x+e^{-x}}\,dx$

16. $\displaystyle\int \frac{dy}{\sqrt{y}(1+\sqrt{y})}$

17. $\displaystyle\int \frac{dI}{I(2+3\ln I)}$

18. $\displaystyle\int \frac{da}{a(1+\ln a^2)}$

19. $\displaystyle\int \frac{dz}{z(1+\ln z)^2}$

20. $\displaystyle\int \frac{(x^2+2x+3)\,dx}{x+2}$

21. When a rocket of initial mass m_0 is fired, the mass is constantly changing because the fuel is being burned. Through the principle of impulse and momentum, it can be shown that the velocity of the rocket at any time is given by $v = \int (qu/(m_0 - qt) - g)\, dt$, where q is the rate of fuel consumption, u is the exhaust of the burned fuel, and g is the acceleration of gravity. Determine the expression for v if the rocket starts from rest.

22. In the study of fluid resistance, the rate of change of the velocity of the fluid with respect to the distance from the boundry layer is $dv/dy = 2.5\, s/y$, where s is the shear velocity of the given fluid. Assuming s to be constant, find the velocity of the fluid as a function of the distance y. Assume that $v = 10$ when $y = 1$.

23. From $t = 0$ to $t = 0.5$ sec, the force acting on an object is $F = 1/(1 + 2t)$ lb. Compute the impulse of the force during this time interval, if $I = \int_{t_1}^{t_2} F\, dt$.

24. The current in a capacitor is found to vary as $i = t/(t^2 + 1)$ amp, from $t = 0$ to $t = 2$ sec. Find the charge transferred to the capacitor during this time period.

25. The rate at which oxygen is consumed by a guinea pig after being given an experimental drug is found to be $dV/dt = 2 + 1/(t + 1)$ cm³/sec. How much oxygen does a guinea pig consume in the first 5 sec after an injection?

26. The work done in moving an object is $W = \int_{s_1}^{s_2} F\, ds$, where F is the force and s is the displacement of the object. Find the work done in moving an object 10 ft if the force varies with displacement according to $F = 1/(2x + 3)$.

27. The velocity of a charged particle in a magnetic field is found to be $v = 10/(12t + 1)$ cm/sec. How far does the particle move from $t = 0$ to $t = 1$ sec?

28. Find the expression for the acceleration of the particle in Exercise 27. What is the velocity and the acceleration of the particle when $t = 1$ sec?

29. Find a function whose derivative is $f'(x) = \dfrac{1}{2x}$.

30. Find a function whose derivative is $g'(t) = t/(t^2 + 4)$.

9.7   ANTIDERIVATIVES OF EXPONENTIAL FUNCTIONS

Since the derivative of b^u with respect to x is

$$\frac{d}{dx}(b^u) = b^u \frac{du}{dx} \ln b$$

it follows that

$$\int b^u\, du = \frac{b^u}{\ln b} + C \tag{8}$$

or, more importantly, that

$$\int e^u\, du = e^u + C \tag{9}$$

Example 25.   Evaluate $\int e^{3x}\, dx$.

Solution.   Here, we note that before Formula 9 can be applied, the integrand must contain the differential of the exponent. Letting $u = 3x$, we see that $du = 3dx$ is the required differential. Therefore, multiplying dx by 3 and the integral by $\frac{1}{3}$, we have

$$\int e^{3x} \, dx = \frac{1}{3} \int e^{3x}(3dx) = \frac{1}{3} e^{3x} + C$$

Example 26.   Evaluate $\displaystyle\int \frac{e^x \, dx}{\sqrt{e^x + 2}}$.

Solution.   First, we rewrite the given integrand as

$$\int (e^x + 2)^{-1/2} e^x \, dx$$

Now, let $u = e^x + 2$. The differential of this is $du = e^x \, dx$ and, therefore, the antiderivative is of the form $\int u^n \, du$. Accordingly, we have

$$\int (e^x + 2)^{-1/2} e^x \, dx = \frac{(e^x + 2)^{1/2}}{1/2} + C = 2\sqrt{e^x + 2} + C$$

Example 27.   During an experiment, the velocity of an alpha particle varied according to $v = e^{t/2}$ cm/sec. Find the distance traveled by the alpha particle during the time interval $t = 1$ to $t = 3$ sec.

Solution.   The displacement is the integral of the velocity equation. Therefore, to find the displacement from $t = 1$ to $t = 3$, we evaluate the definite integral

$$s = \int_1^3 e^{t/2} \, dt$$

The required differential is $\frac{1}{2}dt$. Therefore

$$s = 2 \int_1^3 e^{t/2}\left(\frac{1}{2} \, dt\right) = 2\,[e^{t/2}]_1^3 = 2\,[e^{1.5} - e^{0.5}]$$

$$= 2\,[4.48 - 1.65] = 5.66 \text{ cm}$$

EXERCISES

In Exercises 1 through 20, find the indicated antiderivative.

1. $\displaystyle\int 2e^x \, dx$ **2.** $\displaystyle\int 10^{2y} \, dy$

3. $\int 2^{-x} \, dx$

12. $\int e^{3x}(e^{3x}+2)^4 \, dx$

4. $\int \dfrac{dz}{e^z}$

13. $\int \dfrac{\sqrt{1-e^{-x}} \, dx}{e^x}$

5. $\int (e^{2x}+e^{5x}) \, dx$

14. $\int \dfrac{e^{1/x} \, dx}{x^2}$

6. $\int \left(1+\dfrac{1}{e^x}\right) dx$

15. $\int (e^x+e^{-x})^2 \, dx$

7. $\int se^{s^2} \, ds$

16. $\int \left(2+\dfrac{1}{e^x}\right)^2 dx$

8. $\int (x-1)e^{(x^2-2x)} \, dx$

17. $\int \sqrt[3]{e^s} \, ds$

9. $\int x^2 (e^{x^3}+x) \, dx$

18. $\int \sqrt{e^{2x}} \, dx$

10. $\int (1+xe^{x^2}) \, dx$

19. $\int \dfrac{(1+e^{3x}) \, dx}{e^x}$

11. $\int \dfrac{e^{\sqrt{y+1}} \, dy}{\sqrt{y+1}}$

20. $\int (e^{2x}-e^{-x})e^{3x} \, dx$

21. The transient current in a capacitor is given by $i = (V/R)e^{-t/RC}$ amp. Find the expression for the voltage across the plates of the capacitor.

22. Find the average force exerted on a piston during the time interval $1 \leqslant t \leqslant 2$ sec if $F = 15 - e^{-t}$ lb is the relation between force and time.

23. If the velocity of an object is given by $v = 4e^{2t}$ mph from $t = 0$ to $t = 2$ hr, what is the displacement of the object during this interval?

24. Find the area under the curve $y = e^{x/2}$ from $x = 0$ to $x = 4$.

25. Find the moment about the x-axis of the area given in Exercise 24.

26. The current in a 1000 ohm resistor is $i = e^{5t}$ amp. Find the energy dissipated by the resistor from $t = 0$ to $t = 0.5$ sec.

27. The population N of virus in a test tube increases at a rate of $dN/dt = 2e^{2t}$. Find the population when $t = 1.5$ if the initial population is 100.

28. Experiments show that when a casting is placed in an oven the temperature of the casting increases at a rate $dT/dt = 5e^{0.05t}$ deg/min. Find the temperature of the casting after it has been in the oven for 5 min if its initial temperature is 70 deg.

29. How long does it take to heat the casting in the above problem to 250 deg?

30. Find the volume generated by revolving segments of the curve $y = e^{x/2}$, from $x = 0$ to $x = 2$, about the x-axis.

31. The Laplace transform of a square wave is $L\{F(t)\} = \int_0^\pi e^{-st} \, dt$, where π is the half period of the wave and s is a constant. Evaluate $L\{F(t)\}$.

10

Trigonometric Functions

10.1   INTRODUCTION

Trigonometry has as its goal the study of two specific functions, their quotients and reciprocals. Thus, we define $\sin \theta$ and $\cos \theta$ in terms of the points on a unit circle and then make the following definitions

$$\tan \theta = \frac{\sin \theta}{\cos \theta}; \qquad \cot \theta = \frac{\cos \theta}{\sin \theta}; \qquad \sec \theta = \frac{1}{\cos \theta}; \qquad \csc \theta = \frac{1}{\sin \theta}$$

As you know, the six trigonometric functions are related by many relations which have come to be known as identities, meaning that they are true for all values of the argument for which the functions are defined. Besides the defining equations mentioned above, there are three other identities which you should know well since so many others are derivable from them.

$$\sin^2 A + \cos^2 A = 1 \tag{1}$$
$$\sin (A + B) = \sin A \cos B + \cos A \sin B \tag{2}$$
$$\cos (A + B) = \cos A \cos B - \sin A \sin B \tag{3}$$

We will not bother proving these three important relations but you should know how to derive others from them.

Example 1.   Show that $\tan^2 t + 1 = \sec^2 t$.

Solution.   Using (1), and dividing by $\cos^2 t$, we obtain

$$\frac{\sin^2 t}{\cos^2 t} + 1 = \frac{1}{\cos^2 t}$$

Recognizing that $\sin t / \cos t = \tan t$, and $1/\cos t = \sec t$, we have the desired result.

Example 2.   Find some formulas for $\cos 2x$.

Solution.   Using (3) we obtain

$$\cos 2x = \cos (x + x) = \cos^2 x - \sin^2 x$$

224

Then using (1) in the form $\cos^2 x = 1 - \sin^2 x$, we have

$$\cos 2x = 1 - 2 \sin^2 x$$

If instead we use $\sin^2 x = 1 - \cos^2 x$, then

$$\cos 2x = 2 \cos^2 x - 1$$

The values of the six trigonometric functions are presented in tables for our use. Even though tables are available, students are ordinarily expected to know, without reference to such tables, the values of the six trigonometric functions for $\theta = 0°$, $30°$, $45°$, $60°$, and $90°$. These are tabulated below.

θ	$\sin \theta$	$\cos \theta$	$\tan \theta$	$\cot \theta$	$\sec \theta$	$\csc \theta$
$0°$	0	1	0	Undefined	1	Undefined
$30°$	$1/2$	$\sqrt{3}/2$	$\sqrt{3}/3$	$\sqrt{3}$	$2/\sqrt{3}$	2
$45°$	$\sqrt{2}/2$	$\sqrt{2}/2$	1	1	$\sqrt{2}$	$\sqrt{2}$
$60°$	$\sqrt{3}/2$	$1/2$	$\sqrt{3}$	$\sqrt{3}/3$	2	$2/\sqrt{3}$
$90°$	1	0	Undefined	0	Undefined	1

Most of the trigonometric formulas that we encounter in calculus require that the argument be stated in radians, where *one radian is defined as the measure of a central angle which intercepts an arc equal in length to the radius*. (See Figure 10.1.)

Since the circumference of a circle is $2\pi r$, where r is the radius, it is clear that one complete revolution is measured as 2π radians or $360°$. Therefore

$$2\pi \text{ radians} = 360 \text{ degrees}$$

from which we have

$$1 \text{ radian} = \frac{180}{\pi} \text{ degrees}$$

and

$$1 \text{ degree} = \frac{\pi}{180} \text{ radians}$$

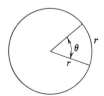

Figure 10.1

Example 3.   Convert 30° to radians.

Solution.   We multiply the degrees by $\pi/180$ to obtain radians. Thus

$$30° = 30\left(\frac{\pi}{180}\right) \text{radians} = \frac{\pi}{6} \text{radians}$$

Example 4.   Convert $\pi/4$ radians to degrees.

Solution.   Multiplying the number of radians by $180/\pi$, we have

$$\frac{\pi}{4} \text{radians} = \frac{\pi}{4}\left(\frac{180}{\pi}\right) \text{degrees} = 45°$$

As a matter of practice, one should know the degree measure of angles whose measure in radians is an easy multiple of π. The table below gives you an idea of which ones you should know without referring to a book.

Radians	$\frac{\pi}{6}$	$\frac{\pi}{4}$	$\frac{\pi}{3}$	$\frac{\pi}{2}$	$\frac{2\pi}{3}$	$\frac{3\pi}{4}$	$\frac{5\pi}{6}$	π	$\frac{3\pi}{2}$	2π
Degrees	30	45	60	90	120	135	150	180	270	360

10.2   GRAPHS OF TRIGONOMETRIC FUNCTIONS

Functions that repeat themselves at regular intervals are called *periodic* functions. The interval required for the function to make one complete cycle is called the *period* of the function. The trigonometric functions are familiar examples of periodic functions. The functions $y = \sin x$, $y = \cos x$, $y = \sec x$, and $y = \csc x$ all have periods of 2π radians, while $y = \tan x$ and $y = \cot x$ have periods of π radians.

The trigonometric functions appear frequently in all phases of engineering and therefore you should be familiar with the graphs of the six basic trigonometric functions. These are presented in Figure 10.2 for your reference.

In practice, we must deal with the basic trigonometric functions altered in three possible ways: (1) by multiplying the function by a constant, (2) by multiplying the argument by a constant, and (3) by adding a constant to the argument. We will use $y = \sin x$ to demonstrate each of these cases.

(1) Multiplication of the Function by a Constant.   If we multiply the function $y = \sin x$ by a positive constant A, we write $y = A \sin x$. Since $\sin x$ is bounded by $-1 \leqslant \sin x \leqslant 1$, $A \sin x$ is bounded by

$$-A \leqslant A \sin x \leqslant A.$$

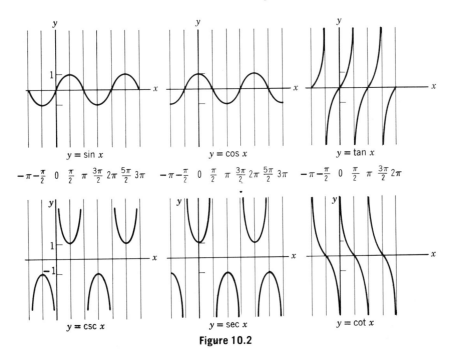

Figure 10.2

If A is greater than one, the amplitude of the basic wave is increased and, if A is less than one, the amplitude is decreased. Figure 10.3 shows $y = A \sin x$ for $A = 1, A = \frac{1}{2}$, and $A = 2$.

(2) Multiplication of the Argument by a Constant. Assuming that

$$y = \sin x$$

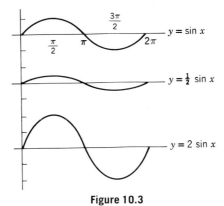

Figure 10.3

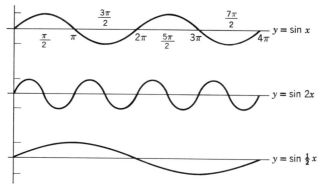

Figure 10.4

is the basic function, the function $y = \sin Bx$ represents a sine function in which the argument has been multiplied by the constant B. The argument of $\sin Bx$ is now Bx and since the sine function repeats itself for every increase in the argument of 2π we have that one period of $\sin Bx$ is contained in the interval $0 \leqslant Bx \leqslant 2\pi$, that is, $0 \leqslant x \leqslant 2\pi/B$. Therefore, multiplying the argument by a constant has the effect of altering the period of the basic function. The period of $y = \sin Bx$ is then given by $2\pi/B$. Notice that the period will be decreased if B is greater than one and increased if it is less than one. This is illustrated in Figure 10.4 for $B = 1$, $B = 2$, and $B = \frac{1}{2}$.

(3) Addition of a Constant to the Argument.   The effect of adding a constant to the argument is best handled by noticing that $y = \sin(Bx + C)$ is zero when $x = -(C/B)$, that is, the graph of $\sin(Bx + C)$ is exactly the same as $\sin Bx$, but displaced to the left by the amount $x = -(C/B)$. The

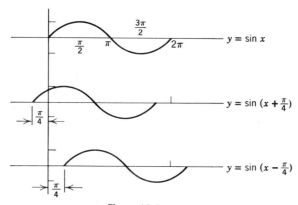

Figure 10.5

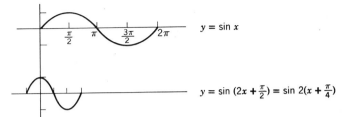

Figure 10.6

amount that the graph of the function is moved to the left (or right) is called the *phase shift*. When $B = 1$, we get $y = \sin(x+C)$, which will be displaced C units to the left if C is positive, and C units to the right if C is negative. The amplitude and period of the wave are unaffected by the addition of a constant term to the argument. The values $C = 0$, $C = \pi/4$, and $C = -\pi/4$ are illustrated in Figure 10.5.

The relationship between $y = \sin x$ and $y = \sin(Bx+C)$ for $B \neq 1$ is illustrated in Figure 10.6. In this case $B = 2$ and $C = \pi/2$. The phase shift is then $\pi/4$ units to the left.

Example 5.   Sketch the graph of $y = 3\cos\left[\tfrac{1}{2}x + (\pi/4)\right]$.

Solution.   The amplitude of the graph is 3, since the function is multiplied by 3. The period of $\cos x$ is 2π, so the period of the given function is $2\pi/\tfrac{1}{2} = 4\pi$. The phase shift is $\pi/2$. Combining these facts we have the graph shown in Figure 10.7.

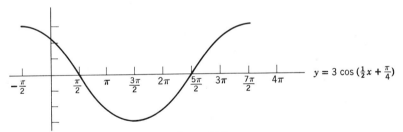

Figure 10.7

EXERCISES

Convert degrees to radians.

1. $45°$	**4.** $135°$
2. $60°$	**5.** $120°$
3. $90°$	**6.** $270°$

Convert radians to degrees.

7. $\pi/12$ rad
8. π rad
9. $5\pi/6$ rad
10. $4\pi/3$ rad
11. $5\pi/4$ rad
12. $5\pi/2$ rad

Sketch the graphs of the given functions.

13. $y = 3 \sin x$
14. $y = \frac{1}{2} \sin x$
15. $y = \tan x/2$
16. $y = \cos (x + \pi/3)$
17. $y = \tan (x + \pi/2)$
18. $y = 2 \sin \frac{1}{3}x$
19. $y = \cot (\pi/4 - x)$
20. $y = \cos (2x + \pi)$
21. $y = 2 \sec (x - \pi/2)$
22. $y = \sin 2(x + \pi/6)$
23. $y = 2 \cos \frac{1}{2} (x - \pi)$
24. $y = \tan (2x + \pi/3)$

*check out -
look at*

10.3   A FUNDAMENTAL INEQUALITY

The inequality

$$\sin \theta < \theta < \tan \theta$$

is fundamental to trigonometric analysis since it relates the arc length to two of the trigonometric functions. To demonstrate the validity of this inequality we construct a unit circle as shown in Figure 10.8.

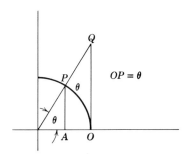

Figure 10.8

From the figure it is obvious that

$$AP < OP < OQ$$

Or, since the circle is unit, we have

$$\sin \theta < \theta < \tan \theta$$

Example 6.   Show that $\lim\limits_{x \to 0} \dfrac{\sin x}{x} = 1.$

Solution.   By the fundamental inequality we have

$$\sin x < x < \tan x$$

Dividing through by $\sin x$ (we are assuming x is positive) we obtain

$$1 < \frac{x}{\sin x} < \frac{1}{\cos x}$$

or

$$1 > \frac{\sin x}{x} > \cos x$$

Since in the limit as x approaches 0, the ratio $\sin x/x$ is trapped between two quantities that are approaching 1, the result is obvious.

10.4   DERIVATIVES OF THE TRIGONOMETRIC FUNCTIONS

To find the derivative of $f(x) = \sin x$, let us first evaluate the derivatives of $\sin x$ and $\cos x$ at $x = 0$. We can guess these values by inspection of their graphs. Thus, from Figure 10.2 we would probably guess (correctly) that the slope of the graph of $y = \cos x$ at $x = 0$ is zero; that is

$$\frac{d}{dx}(\cos x)\Big|_{x=0} = \lim_{\Delta x \to 0} \frac{\cos \Delta x - 1}{\Delta x} = 0$$

You might be less sure of yourself about the derivative of $\sin x$ at $x = 0$; although you might guess it is close to 1. We can verify this value by noting that the derivative of $\sin x$ at $x = 0$ is given by

$$\frac{d}{dx}(\sin x)\Big|_{x=0} = \lim_{\Delta x \to 0} \frac{\sin \Delta x}{\Delta x}$$

Then using the results of Example 6, we have the desired verification; that is

$$\frac{d}{dx}(\sin x)\Big|_{x=0} = \lim_{\Delta x \to 0} \frac{\sin \Delta x}{\Delta x} = 1$$

Now we shall see that just by knowing the values of $(d/dx)(\sin x)$ and $(d/dx)(\cos x)$ at one point (that is at $x = 0$), we can find the values of the derivatives of these functions everywhere.

Derivative of sin u
The derivative of $\sin x$ is by definition

$$\frac{d}{dx}(\sin x) = \lim_{\Delta x \to 0} \frac{\sin (x + \Delta x) - \sin x}{\Delta x}$$

Using the formula for sin $(A+B)$, this can be rewritten in the form

$$\frac{d}{dx}(\sin x) = \lim_{\Delta x \to 0} \frac{\sin x \cos \Delta x + \cos x \sin \Delta x - \sin x}{\Delta x}$$

$$= \lim_{\Delta x \to 0} \frac{\cos x \sin \Delta x + \sin x(\cos \Delta x - 1)}{\Delta x}$$

$$= \cos x \lim_{\Delta x \to 0} \frac{\sin \Delta x}{\Delta x} + \sin x \lim_{\Delta x \to 0} \frac{\cos \Delta x - 1}{\Delta x}$$

$$= \cos x \cdot \frac{d}{dx}(\sin x)\Big|_{x=0} + \sin x \cdot \frac{d}{dx}(\cos x)\Big|_{x=0}$$

$$= \cos x$$

If u represents a function of x, the chain rule yields the more general formula

$$\frac{d}{dx}(\sin u) = \cos u \frac{du}{dx} \tag{4}$$

Example 7.   Differentiate $y = \sin 3x^2$.

Solution.   This is the above formula with $u = 3x^2$. Therefore

$$y' = \cos 3x^2(6x) = 6x \cos 3x^2$$

Derivative of cos u
The formula for the derivative of cos u can be found by making use of Formula 4 and trigonometric identities $\cos u = \sin (\pi/2 - u)$, and $\sin u = \cos (\pi/2 - u)$. To find the derivative of cos u, we note that

$$\frac{d}{dx}(\cos u) = \frac{d}{dx}[\sin (\pi/2 - u)]$$

By (4), we obtain

$$\frac{d}{dx}(\cos u) = \cos (\pi/2 - u) \cdot \frac{d}{dx}(\pi/2 - u) = \cos (\pi/2 - u)\left(-\frac{du}{dx}\right)$$

or

$$\frac{d}{dx}(\cos u) = -\sin u \frac{du}{dx} \tag{5}$$

Example 8.   Differentiate $y = \cos (x^2 + 1)$.

Solution.   Using Formula 5 with $u = x^2 + 1$,

$$\frac{dy}{dx} = -\sin(x^2+1) \cdot \frac{d}{dx}(x^2+1) = -2x \sin(x^2+1)$$

Example 9.   Differentiate $y = \sin^3 2x$.

Solution.   The given equation may be regarded as a function of x raised to the third power; that is, $y = v^3$ where $v = \sin 2x$. Differentiating accordingly,

$$\frac{dy}{dx} = 3v^2 \frac{dv}{dx}$$

Replacing v with $\sin 2x$,

$$\frac{dy}{dx} = 3 \sin^2 2x \cdot \frac{d}{dx}(\sin 2x)$$

By Formula 4

$$\frac{d}{dx}(\sin 2x) = \cos 2x \cdot \frac{d}{dx}(2x) = 2 \cos 2x$$

and, therefore,

$$\frac{dy}{dx} = 3 \sin^2 2x(2 \cos 2x) = 6 \sin^2 2x \cos 2x$$

The derivative formulas for the remaining four trigonometric functions are obtained by using Formulas 4 and 5, and the basic trigonometric identities.

Derivative of tan u

Writing $\tan u = \sin u / \cos u$, we have

$$\frac{d}{dx}(\tan u) = \frac{d}{dx}\left(\frac{\sin u}{\cos u}\right)$$

Differentiating the right side by using the formula for the derivative of a quotient,

$$\frac{d}{dx}(\tan u) = \frac{\cos u \left[\cos u \,(du/dx)\right] - \sin u \left[-\sin u \,(du/dx)\right]}{\cos^2 u}$$

$$= \frac{(\cos^2 u + \sin^2 u)\,(du/dx)}{\cos^2 u}$$

or

$$\frac{d}{dx}(\tan u) = \sec^2 u \frac{du}{dx} \tag{6}$$

Derivative of cot u

The formula for the derivative of the cot u is found by a similar procedure.

$$\frac{d}{dx}(\cot u) = -\csc^2 u \frac{du}{dx} \tag{7}$$

Derivative of sec u

Here we let sec $u = 1/\cos u$. Then

$$\frac{d}{dx}(\sec u) = \frac{d}{dx}\left(\frac{1}{\cos u}\right)$$

$$= \frac{0 - [-\sin u \, du/dx]}{\cos^2 u} = \frac{1}{\cos u} \cdot \frac{\sin u}{\cos u} \frac{du}{dx}$$

$$\frac{d}{dx}(\sec u) = \sec u \tan u \frac{du}{dx} \tag{8}$$

Derivative of csc u

Using a similar procedure to that used above, we have

$$\frac{d}{dx}(\csc u) = -\csc u \cot u \frac{du}{dx} \tag{9}$$

Example 10.   Differentiate $s = \sec 3t$.

Solution.   By Formula 8

$$\frac{ds}{dt} = \sec 3t \tan 3t \cdot \frac{d}{dt}(3t) = 3 \sec 3t \tan 3t$$

Example 11.   After a tuning fork is sounded, the amplitude of the vibration decreases until it becomes zero. Assuming that the displacement of the fork from the rest position decreases according to

$$s = t^{-1} \sin \tfrac{1}{2}t,$$

determine the expression for the velocity of the prong at any time t after sounding.

Solution.   The velocity of the end point of the vibrating prong is $v = ds/dt$. To differentiate the given expression we must use the form for the derivative of the product of two functions of t, since t^{-1} and $\sin \tfrac{1}{2}t$ are both functions of t. Hence

$$\frac{ds}{dt} = t^{-1} \cdot \frac{d}{dt}\left(\sin\frac{1}{2}t\right) + \sin\frac{1}{2}t \cdot \frac{d}{dt}(t^{-1})$$

$$= t^{-1}\left[\frac{1}{2}\cos\frac{1}{2}t\right] + \sin\frac{1}{2}t\left[-t^{-2}\right]$$

$$\frac{ds}{dt} = \frac{1}{2}t^{-1}\cos\frac{1}{2}t - t^{-2}\sin\frac{1}{2}t$$

Example 12. The magnetic induction B of a charged particle moving through a uniform magnetic field at an angle θ is given by

$$B = (F/qv)\csc\theta,$$

where F, q, and v are constants. Find the equation for the rate of change of B with respect to θ.

Solution. To obtain the derivative of the given equation with respect to θ, we use Formula 9. Hence

$$\frac{dB}{d\theta} = \frac{F}{qv} \cdot \frac{d}{d\theta}[\csc\theta] = -\frac{F}{qv}\csc\theta\cot\theta$$

EXERCISES

In Exercises 1 through 24, find the first derivative of the given function.

1. $v = \sin\frac{\pi}{2}t$

2. $y = \tan 3x$

3. $z = 2\cos 3t$

4. $i = \sec \pi t$

5. $p = \frac{1}{2}\cot(t+1)$

6. $s = 2\csc(3r+4)$

7. $P = \sin T^3$

8. $f(y) = 3\tan\sqrt{2y+1}$

9. $s = \cos e^t$

10. $y = x^2 \mid \sec x^2$

11. $w = \cos^3 t$

12. $C = \sec^2 r + \tan r$

13. $Z = \sqrt{\sin 2X}$

14. $h(x) = \dfrac{1}{\sqrt{\cos \pi x}}$

15. $i = e^t \sin t$

16. $q = \cos 2t \sin 2t$

17. $y = \ln\sqrt{x}\cot 2x$

18. $y = x^3 \tan^4 x$

19. $p = \ln\sin 5s$

20. $L = \ln\sec z^2$

21. $G(t) = e^{\cos 2t}$

22. $I = V^2 e^{\sin V}$

23. $y = \sin(\cos x - \cot x)$

24. $R = \dfrac{e^{\sin\theta}}{\sin\theta}$

25. Find the second derivative of $y = \sin x^2$.

26. Find the second derivative of $y = \tan 2x$.

27. Derive Formula 7.

28. Derive Formula 9.

29. If a force F is parallel to the plane of motion, the magnitude of the force necessary to cause impending motion up an incline is $F = W \cos \theta$, where W is the weight of the object and θ is the angle of inclination of the plane. Find the rate of change of F with respect to θ.

30. The horizontal displacement of a simple oscillator is given by $x = A \cos 2\pi ft$, where A and f are constants and t is time. Find the expression for the horizontal velocity of the oscillation.

31. The current in an RL series circuit is $i = I_m \sin (\omega t - \phi)$, where I_m is the peak current, ϕ is the phase angle, ω is the angular frequency, and t is the elapsed time. Find the expression for the induced voltage in the inductance coil.

32. The torque T acting on a loop of wire, carrying a current i in a magnetic field of flux density B, is given by $T = iBA \cos \alpha$, where A is the area of the loop and α is the angle between the plane of the loop and the magnetic field. Find $dT/d\alpha$.

33. The total bending B of electromagnetic waves received from outside the atmosphere is described by $B = N \cot E$ when the elevation angle E is greater than 5 deg. N is a constant called the surface refractivity. What is the rate of change of total bending with respect to elevation angle?

34. The azimuth error Δ_a of the angle tracker of a radar will increase as the target rises in elevation E according to $\Delta_a = \Delta_t \sec E$, where Δ_t is a constant bias error. Find $d\Delta_a/dE$ for the angle tracker.

35. The expression for the shearing stress on a plane located by the angle ϕ and caused by a system of stresses is $S = -\frac{1}{2}(\sigma_x - \sigma_y) \sin 2\phi + T \cos 2\phi$. The quantities σ_x, σ_y, and T are initially known constants. Find $dS/d\phi$.

36. The equation of motion of a damped vibration is $s = e^{-at} \cos bt$. Show that the velocity is given by $v = -e^{-at}(b \sin bt + a \cos bt)$.

10.5   INVERSE TRIGONOMETRIC FUNCTIONS

Recall that any function is in reality a pairing of numbers x and y that we describe by the rule $y = f(x)$. A function $y = f(x)$ is said to be *one-to-one* if there is only one value of x corresponding to each value of y. If there is more than one value of x corresponding to some y, the function is said to be *many-to-one*. You can determine whether a given function is 1–1 or not by observing that any horizontal line drawn through the graph of a 1–1 function will intersect it in, at most, one place; otherwise, the function is many-to-one. In Figure 10.9, the first graph represents a 1–1 function and the second graph does not.

Pairings of numbers that form 1–1 functions have the attractive property that their inverse pairings also form 1–1 functions. Many-to-one functions do not have this property. We encountered an example of inverse functions in the previous chapter in connection with the exponential and logarithmic functions.

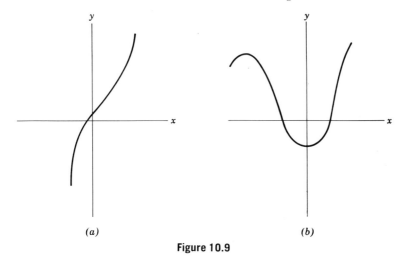

(a)   *(b)*

Figure 10.9

We see from an analysis of the six trigonometric functions that none of them are 1–1 and, therefore, none of the six has an inverse function. For a variety of reasons, this is an impractical situation that we rectify simply by limiting the allowable values of x in each of the six functions. The allowable values for each function are called the *principal values* of the function. The following table shows the principal values for the sine, cosine, and tangent. There are principal values defined for the other three functions, but they are of little practical value and will not be considered.

Function	$\sin x$	$\cos x$	$\tan x$
Principal values	$-\dfrac{\pi}{2} \leqslant x \leqslant \dfrac{\pi}{2}$	$0 \leqslant x \leqslant \pi$	$-\dfrac{\pi}{2} < x < \dfrac{\pi}{2}$

We shall not attempt to justify the choice of these intervals for the principal values but we observe that each of the three functions is 1–1 if restricted to its principal values. Thus, an inverse function exists for each one.

The inverse function for $x = \sin y$, where y is limited to the principal values of the sine function, is denoted by $y = sin^{-1} x$, or $y = arc\ sin\ x$. The notation $y = \sin^{-1} x$ (read: "inverse sine of x") is particularly deceptive, since it suggests taking the reciprocal of $\sin x$. Therefore, the notation $y = arc\ sin\ x$ will be used in this book.

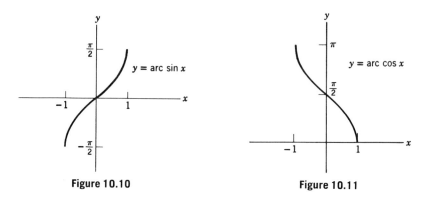

Figure 10.10   Figure 10.11

The graph of arc sin x is shown in Figure 10.10. It is just the graph of the sine function wrapped around the y-axis but limited to

$$-\pi/2 \leqslant y \leqslant \pi/2.$$

(Just from the graph of $y =$ arc sin x, what can you say about the derivative of arc sin x?)

The defining equation for $y =$ arc cos x is $x =$ cos y, with y limited to values between 0 and π. The graph of $y =$ arc cos x then follows immediately and is shown above. (See Figure 10.11.)

By this time you have probably guessed that $y =$ arc tan x defines a function determined by the equation $x =$ tan y for $-\pi/2 < y < \pi/2$. The graph is shown in Figure 10.12.

We shall not discuss the inverse cotangent, inverse secant, and inverse cosecant, because the need for them in applications is extremely limited. They may, therefore, be omitted without loss of continuity.

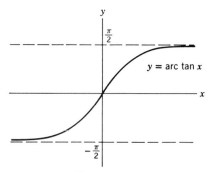

Figure 10.12

EXERCISES

Evaluate the function at the indicated point.

1. $y = \arc \sin x$; $x = \sqrt{2}/2$ 4. $p = \arc \cos x$; $x = -1$
2. $\theta = \arc \tan t$; $t = -1$ 5. $y = \arc \sin x$; $x = \sqrt{3}/2$
3. $s = \arc \cos t$; $t = \frac{1}{2}$ 6. $B = \arc \sin C$; $C = 0$

10.6 DERIVATIVES OF INVERSE TRIGONOMETRIC FUNCTIONS

The graphs of the three inverse trigonometric functions are a source of information about their derivatives. Notice that we would not expect the derivative of arc sin x to exist at $x = -1$ and $x = 1$. Furthermore, we would expect its sign to be positive throughout the interval. Similarly, the derivative of the arc cos x is not defined at $x = -1$ and $x = 1$, and is negative over the interval. The derivative of the arc tan x seems to be positive and bounded, reaching its maximum at $x = 0$ and becoming arbitrarily close to zero for large x. All this information is obtained without a formula!

Derivative of arc sin u

The formula for the derivative of the inverse sine function follows immediately from its definition and the use of implicit differentiation. Thus, by

$$y = \arc \sin u \qquad (10)$$

we mean

$$u = \sin y \qquad (11)$$

If we now differentiate this with respect to x, we obtain

$$\frac{du}{dx} = \cos y \, \frac{dy}{dx}$$

or solving for dy/dx

$$\frac{dy}{dx} = \frac{1}{\cos y} \frac{du}{dx}$$

Using the trigonometric identity $\cos y = \sqrt{1 - \sin^2 y}$, we write

$$\frac{dy}{dx} = \frac{1}{\sqrt{1 - \sin^2 y}} \frac{du}{dx}$$

But, from Formula 11, $\sin y = u$. Thus we may express the required derivative as

$$\frac{d}{dx} (\arc \sin u) = \frac{1}{\sqrt{1 - u^2}} \frac{du}{dx} \qquad (12)$$

Derivative of arc cos u

The derivation of this formula is similar to that used for arc sin u. It is left to the student to prove

$$\frac{d}{dx}(\text{arc cos } u) = -\frac{1}{\sqrt{1-u^2}}\frac{du}{dx} \tag{13}$$

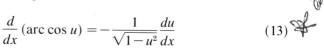

Derivative of arc tan u

To find the derivative of arc tan u, let

$$y = \text{arc tan } u$$

Then

$$u = \tan y$$

and

$$\frac{du}{dx} = \sec^2 y \frac{dy}{dx}$$

Solving for dy/dx

$$\frac{dy}{dx} = \frac{1}{\sec^2 y}\frac{du}{dx}$$

Here, we use $\sec^2 y = 1 + \tan^2 y$, and $\tan y = u$ to obtain the formula

$$\frac{d}{dx}(\text{arc tan } u) = \frac{1}{1+u^2}\frac{du}{dx} \tag{14}$$

Example 13.   Find the derivative of $y = \text{arc sin } 3x^2$.

Solution.   Here we let $u = 3x^2$. Then applying Formula 12, we have

$$\frac{dy}{dx} = \frac{1}{\sqrt{1-(3x^2)^2}} \cdot \frac{d}{dx}(3x^2) = \frac{6x}{\sqrt{1-9x^4}}$$

Example 14.   Find the slope of $y = 4x$ arc tan $2x$ at $x = \frac{1}{2}$.

Solution.   The given function is the product of two functions of x; namely, $4x$ and arc tan $2x$. We must therefore apply the product rule in differentiating the given function. Hence

$$\frac{dy}{dx} = 4x\left[\frac{1}{1+(2x)^2} \cdot \frac{d}{dx}(2x)\right] + \text{arc tan } 2x \cdot \frac{d}{dx}(4x)$$

$$= \frac{8x}{1+4x^2} + 4 \text{ arc tan } 2x$$

Letting $x = \frac{1}{2}$, we have

$$\frac{dy}{dx} = \frac{4}{2} + 4 \text{ arc tan } 1 = 2 + 4\left(\frac{\pi}{4}\right) = 5.14$$

Example 15. A winch on the bed of a truck is used to move a large casting. If the winch winds in the cable at the rate of 2 ft/sec, and if the truck is 5 ft high, at what rate is the angle θ changing when there is 10 ft of cable out? (See Figure 10.13.)

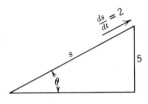

Figure 10.13

Solution.   Referring to Figure 10.13, angle θ may be written

$$\theta = \text{arc sin } \frac{5}{s}$$

The rate at which θ is changing is then found by differentiating this function with respect to t.

$$\frac{d\theta}{dt} = \frac{1}{\sqrt{1 - (25/s^2)}} \cdot \frac{d}{dt}\left(\frac{5}{s}\right) = \frac{1}{\sqrt{(s^2 - 25)/s^2}}\left(-\frac{5}{s^2}\right)\frac{ds}{dt}$$

$$= \frac{-5 \, ds/dt}{s\sqrt{s^2 - 25}}$$

The cable is being wound in at a rate $ds/dt = -2$ ft/sec. Therefore, when $s = 10$ ft,

$$\frac{d\theta}{dt} = \frac{-5(-2)}{10\sqrt{75}} = \frac{1}{\sqrt{75}} = 0.115 \text{ rad/sec}$$

EXERCISES

Differentiate each of the given functions.

1. $\theta = \text{arc sin } 3t$
2. $y = \text{arc sin } x^3$
3. $z = \text{arc tan } \frac{1}{2}y$
4. $B = \text{arc cos } q^2$
5. $s = \text{arc sin } \sqrt{t}$
6. $T = \text{arc tan } 1/z$

7. $y = \arctan x/3$

8. $h(x) = \arccos \sqrt{1 - x^2}$

9. $p = \arctan e^t$

10. $f(x) = x^2 \arcsin 3x$

11. $m = \sqrt{x} \arcsin \sqrt{x}$

12. $i = \dfrac{\arccos 3t}{t}$

13. $q = \dfrac{\arctan 2t}{e^t}$

14. $y = \sqrt{\arctan 2x}$

15. $H = \dfrac{1}{\sqrt{\arccos \sqrt{\phi}}}$

16. $y = x \arcsin x - \sqrt{1 - x^2}$

17. $y = x \arctan x - \ln \sqrt{1 + x^2}$

18. Derive Formula 13.

19. Find the slope of the graph of $y = x \arcsin x$ at $x = \frac{1}{2}$.

20. Find the slope of the graph of $y = e^x \arccos x$ at $x = 0$.

21. When an ac generator is applied to a series RLC circuit, the voltage and current are out of phase by some angle ϕ. The magnitude of ϕ is described by $\phi = \arctan X/R$ where X is the reactance of the circuit and R is the resistance. Find $d\phi/dR$.

22. In Exercise 21, find $d\phi/dX$.

23. When a thermocouple is placed in a gas stream, the angle by which the peak thermocouple temperature lags the peak temperature of the gas stream is $\theta = \arctan T/\omega$, where T is a constant and ω is the angular frequency of the temperature variation. Find $d\theta/d\omega$.

24. A balloon rises vertically from a point which is 1000 ft from an observer. At what rate is the angle of elevation changing when the balloon is 200 ft high if the balloon is rising at a rate of 15 ft/sec?

25. An airplane flying at an altitude of 10,000 ft and a velocity of 500 ft/sec passes directly over a radar station. At what rate is the angle of elevation of the radar changing 20 sec later.

10.7   ANTIDERIVATIVES OF TRIGONOMETRIC FUNCTIONS

Antidifferentiation is the inverse operation to differentiation; therefore, every new differentiation formula yields a new antidifferentiation formula. The following formulas are a direct consequence of differentiation Formulas 4 through 9.

$$\int \sin u \, du = - \cos u + C \tag{15}$$

$$\int \cos u \, du = \sin u + C \tag{16}$$

$$\int \sec^2 u \, du = \tan u + C \tag{17}$$

$$\int \csc^2 u \, du = - \cot u + C \tag{18}$$

$$\int \sec u \tan u \, du = \sec u + C \tag{19}$$

$$\int \csc u \cot u \, du = - \csc u + C \tag{20}$$

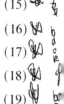

The technique for antidifferentiating the circular functions can be eased if we recall from Chapter 8 that the du in the formula may be treated formally as the differential of u. As we pointed out, a formula may be applied only when the given antiderivative fits the formula exactly. Using this kind of formal analysis, we note that $\int \sin x^3 \, dx$ may not be evaluated by Formula 15, because the differential of $u = x^3$ (which is $du = 3x^2 \, dx$) does not appear in the integrand.

Example 16. Find $\int x \cos 3x^2 \, dx$.

Solution.   If we let $u = 3x^2$ in this Exercise, the required differential is seen to be $du = 6x \, dx$. Consequently, before we can apply Formula 16, we must multiply the given differential by 6 and the integral by $\frac{1}{6}$. Performing this operation, we get

$$\int \cos 3x \, dx = \frac{1}{6} \int \cos 3x^2 (6x \, dx) = \frac{1}{6} \sin 3x^2 + C$$

Example 17. Find $\int \csc \frac{x}{2} \cot \frac{x}{2} \, dx$.

Solution.   Letting $u = x/2$, we see that $du = (1/2) \, dx$. Therefore, we must multiply the given integrand by 1/2, and the integral by 2 before we can use Formula 20. Hence

$$\int \csc \frac{x}{2} \cot \frac{x}{2} \, dx = 2 \int \csc \frac{x}{2} \cot \frac{x}{2} \left(\frac{1}{2} \, dx \right) = -2 \csc \frac{x}{2} + C$$

Example 18. Find $\int \sin^3 \theta \cos \theta \, d\theta$.

Solution.   This antiderivative is not one of the standard forms given in Formulas 15 through 20. However, it can be evaluated as $\int v^n \, dv$, if we let $v = \sin \theta$ and $n = 3$. The necessary differential is then $dv = \cos \theta d\theta$. Since each of these components is present in the given integrand, we conclude that

$$\int \sin^3 \theta \cos \theta \, d\theta = \frac{\sin^4 \theta}{4} + C$$

To conclude this article, we present the antiderivatives of $\tan u$, $\cot u$, $\sec u$, and $\csc u$.

Antiderivative of tan u
The formula for the antiderivative of $\tan u$ can be found by using the trigonometric identity $\tan u = \sin u/\cos u$. Thus

$$\int \tan u \, du = \int \frac{\sin u \, du}{\cos u} = -\int \frac{-\sin u \, du}{\cos u}$$

Since the differential of cos u is $-\sin u\, du$, we have

$$\int \tan u\, du = -\ln (\cos u) + C$$
$$= \ln (\cos u)^{-1} + C$$
$$\int \tan u\, du = \ln \sec u + C \tag{21}$$

Antiderivative of cot u

By a similar procedure the formula for antiderivative of cot u is

$$\int \cot u = \ln \sin u + C \tag{22}$$

Antiderivative of sec u

The formula for the antiderivative of sec u is

$$\int \sec u\, du = \ln (\sec u + \tan u) + C \tag{23}$$

We will verify this by showing that the differential of the right side of Formula 23 is sec $u\, du$.

$$d\,[\ln (\sec u + \tan u)] = \frac{1}{\sec u + \tan u} \cdot (\sec u \tan u + \sec^2 u)\, du$$

$$= \frac{\sec u(\tan u + \sec u)\, du}{\sec u + \tan u}$$

$$= \sec u\, du$$

Antiderivative of csc u

The formula for the antiderivative of csc u is

$$\int \csc u\, du = \ln (\csc u - \cot u) + C \tag{24}$$

This may be verified in the same manner as Formula 23.

Example 19.   Evaluate $\int x \sec x^2\, dx$.

Solution.   Here we let $u = x^2$, then $du = 2x\, dx$. By associating the x with the dx, and then multiplying this product by 2, we have the required differential. Hence

$$\int x \sec x^2\, dx = \frac{1}{2}\int \sec x^2(2x\, dx) = \frac{1}{2}\ln (\sec x^2 + \tan x^2) + C$$

Example 20.   The angle of twist θ of the free end of a cylindrical bar of length L due to a torque T is given by

$$\theta = \frac{1}{GJ}\int_0^L T\, dx$$

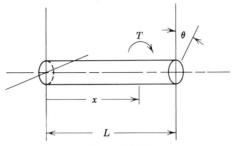

Figure 10.14

where G and J are constants. Find the equation for the angle of twist if the torque applied to the bar varies with distance from the fixed end according to $T = \cos(\pi x/2L)$. (See Figure 10.14.)

Solution. Substituting the expression for torque into the given integral, we have

$$\theta = \frac{1}{GJ} \int_0^L \cos \frac{\pi x}{2L} \, dx$$

If $u = \pi x/2L$, then $du = (\pi/2L)\,dx$. Multiplying dx by $\pi/2L$, and the integral by $2L/\pi$, we may write

$$\theta = \frac{2L}{\pi GJ} \int_0^L \cos \frac{\pi x}{2L} \left(\frac{\pi}{2L} \, dx \right) = \frac{2L}{\pi GJ} \left[\sin \frac{\pi x}{2L} \right]_0^L$$

$$= \frac{2L}{\pi GJ} \left[\sin \frac{\pi}{2} - \sin 0 \right] = \frac{2L}{\pi GJ}$$

Example 21. The current in an ac circuit is given by $i = I_m \sin \omega t$, where I_m is the maximum amplitude of the current, and ωt is the angular displacement of the generator.* Show that the average current over $\frac{1}{2}$ cycle is $I_{avg} = 0.637\, I_m$. (See Figure 10.15.)

Solution. Since the period of $\sin \omega t$ is $2\pi/\omega$, then $\frac{1}{2}$ cycle is the interval $0 \leq \omega t \leq \pi$. Recalling the formula for the average value of a function, we have

$$I_{avg} = \frac{1}{\pi - 0} \int_0^\pi I_m \sin \omega t \, d(\omega t) = \frac{I_m}{\pi} \left[-\cos \omega t \right]_0^\pi = \frac{I_m}{\pi} \left[-(-1) - (-1) \right]$$

$$= \frac{2}{\pi} I_m = 0.637 I_m$$

*Assuming that the generator has angular velocity of ω rad/sec, the angular displacement can be written $\theta = \omega t$. In ac circuits it is common practice to write the angular displacement as ωt to indicate the dependence of the sinusoidal function upon time.

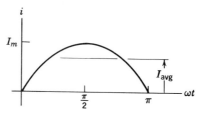

Figure 10.15

EXERCISES

Find each of the following antiderivatives.

1. $\int \sin 3x \, dx$

2. $\int \sec^2 \frac{1}{2} x \, dx$

3. $\int \tan 2t \, dt$

4. $\int \sec 5x \, dx$

5. $\int \cot \omega t \, d(\omega t)$

6. $\int x \cos x^2 \, dx$

7. $\int 6x \sec x^2 \tan x^2 \, dx$

8. $\int (x-1) \cos (x^2 - 2x + 1) \, dx$

9. $\int \frac{\sin e^{-x}}{e^x} \, dx$

10. $\int \frac{\tan \sqrt{\theta} \, d\theta}{\sqrt{\theta}}$

11. $\int \csc \frac{x}{2} \, dx$

12. $\int \csc^2 3x \, dx$

13. $\int \csc \omega t \cot \omega t \, dt$

14. $\int \frac{25}{x^2} \cos \frac{1}{x} \, dx$

15. $\int \sin^5 \theta \cos \theta \, d\theta$

16. $\int \frac{\sin x \, dx}{\sqrt{\cos x}}$

17. $\int \sec^3 \phi \tan \phi \, d\phi$

18. $\int \csc^4 x \cot x \, dx$

19. $\int e^{\sin 2\phi} \cos 2\phi \, d\phi$

20. $\int \frac{\sec^2 y \, dy}{1 + \tan y}$

21. $\int \frac{\sec^2 x \, dx}{(1 + \tan x)^2}$

22. $\int \frac{dz}{\sin^2 z}$

23. $\int (1 + \sec \omega t)^2 \, d(\omega t)$

24. $\int \frac{\cos 2\omega t + \sin 2\omega t}{\cos 2\omega t} \, d(\omega t)$

25. The x-coordinate of the centroid of a circular arc of radius r which subtends an angle $\phi = \beta$ at the center is given by $\bar{x} = 2/\beta \int_0^{\beta/2} r \cos \phi \, d\phi$. Find $\bar{x}$ for a circular arc with $r = 10$ in. and $\beta = \pi/2$ rad.

26. Find the average height of the graph of $y = \tan x$, from $x = 0$ to $x = \pi/3$.

27. The work done, dW, in moving the block through a distance ds is $dW = F \cos \theta \, ds$, where F is the applied force and θ is the angle the force makes with the plane. Find the work done in moving the block 10 ft, if $\theta = 30°$ and $F = s^2$ lb.

28. The work done, dW, in rotating a magnet in a magnetic field through an angle $d\phi$ is $dW = MH \cos \phi \, d\phi$, where M and H are constants. Find the work done in rotating the magnet from $\phi = 0$ to $\phi = 90°$.

29. The total radiation through a hemispherical surface is given by

$$R = 2\pi h \int_0^{\pi/2} \cos \theta \sin \theta \, d\theta,$$

where h is a constant and θ is the angle of radiation. Show that the total radiation is $R = \pi h$.

30. The amount of energy, E, which is scattered by electrons in a cm³ of matter is

$$E = \frac{\pi I n e^4}{m^2 c^4} \int_0^\pi (1 + \cos^2 \theta) \sin \theta \, d\theta$$

where θ is the angle of scatter. Show that

$$E = \frac{8\pi n e^4 I}{3 m^2 c^4}.$$

31. When the bar shown below is rotated about point A, work is done on the spring. The work done in rotating an 8 ft bar to point C is given by $W = 960 \int_0^{\pi/4} (3 \cos \alpha - 2) \sin \alpha \, d\alpha$. Evaluate W. (See Figure 10.16)

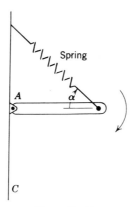

Figure 10.16

10.8 ADDITIONAL TRIGONOMETRIC INTEGRANDS

In many instances, antiderivatives with trigonometric integrands occur that cannot be evaluated immediately by one of the formulas presented in the preceding article. In some of these instances, however, the integrand can be reduced to a workable form by substitution of an appropriate trigonometric identity.

$\int$ **sin² u du and $\int$ cos² u du**   The antidifferentiation of sin² u du is readily accomplished by the substitution of the trigonometric identity $\sin^2 u = \frac{1}{2}(1 - \cos 2u)$. Similarly, the antiderivative of cos² u du can be evaluated by first making the substitution $\cos^2 u = \frac{1}{2}(1 + \cos 2u)$.

Example 22.   The instantaneous value of an alternating current varies continuously from a maximum in one direction to a maximum in the opposite, and so on. In practice it is more convenient to describe alternating currents by their effective values. The effective value of a varying current is defined as that steady current which develops the same quantity of heat. We would like to show that for a sinusoidal current of the form $i = I_m \sin \omega t$, the effective current I_{eff} is equal to $1/\sqrt{2}$ times the maximum current I_m.

Solution.   The power in a resistance R is given by $p = i^2 R$ and, therefore, the average power over one cycle of the alternating current

$$i = I_m \sin \omega t$$

is

$$P = \frac{1}{2\pi - 0} \int_0^{2\pi} (I_m \sin \omega t)^2 \, R \, d(\omega t) = \frac{I_m^2 R}{2\pi} \int_0^{2\pi} \sin^2 \omega t \, d(\omega t)$$

$$= \frac{I_m^2 R}{2\pi} \int_0^{2\pi} \frac{1}{2}(1 - \cos 2\omega t) \, d(\omega t) = \frac{I_m^2 R}{4\pi}\left[\omega t - \frac{1}{2}\sin 2\omega t\right]_0^{2\pi}$$

$$P = \frac{I_m^2 R}{4\pi}[2\pi] = \frac{I_m^2 R}{2}$$

Now if we let I_{eff} represent the steady state current that produces the same power, we can write

$$I_{eff}^2 R = \frac{1}{2}I_m^2 R$$

or

$$I_{eff} = \frac{I_m}{\sqrt{2}}$$

Hence for sinusoidal currents, the effective current is equal to $1/\sqrt{2}$ times the maximum current.

Example 18 of the preceding article describes the technique by which $\int \sin^3 \theta \cos \theta \, d\theta$ can be found. Using the fact that $\cos \theta \, d\theta$ is the exact differential of $\sin \theta$, it is possible to reduce the integrand to the form $v^n \, dv$. This same technique can be used to evaluate trigonometric integrands that involve more complex combinations of powers of trigonometric functions. In each of the following problems the solution is effected by rearranging the integrand into the form $v^n \, dv$.

$\int \sin^m u \cos^n u \, du$ **(where either m or n is odd and positive).**   The method to be described applies to antiderivatives of the indicated form as long as one of the exponents is odd and positive. If m is odd and positive, we may rearrange the integrand into the form $v^n \, dv$ by first factoring out $-\sin u$ and writing

$$\int \sin^m u \cos^n u \, du = -\int \sin^{m-1} u \cos^n u \, (-\sin u \, du)$$

where the factor $-\sin u \, du$ is the differential of $\cos u$. The antidifferentiation may then be effected by using the identity $\sin^2 u = 1 - \cos^2 u$ to transform the product $\sin^{m-1} u \cos^n u$ into cosine terms only. This transformation is possible because $m - 1$ is even and positive. Likewise, if n is odd and positive, the antiderivative may be found by factoring out $\cos u$ and then using $\cos^2 u = 1 - \sin^2 u$.

Example 23.   Evaluate $\int \sqrt{\sin x} \cos^5 x \, dx$.

Solution.   The exponent on the cosine term is odd and positive so we factor out $\cos x$. Thus

$$\int \sqrt{\sin x} \cos^5 x \, dx = \int \sin^{1/2} x \cos^4 x \, (\cos x \, dx)$$

Writing $\cos^4 x$ as $(\cos^2 x)^2$, we have

$$\int \sqrt{\sin x} \cos^5 x \, dx = \int \sin^{1/2} x \, (\cos^2 x)^2 (\cos x \, dx)$$

$$= \int \sin^{1/2} x \, (1 - \sin^2 x)^2 (\cos x \, dx)$$

$$= \int (\sin^{1/2} x - 2 \sin^{5/2} x + \sin^{9/2} x) \, (\cos x \, dx)$$

Since $\cos x \, dx$ is the differential of $\sin x$,

$$\int \sqrt{\sin x} \cos^5 x \, dx = \frac{2}{3} \sin^{3/2} x - \frac{4}{7} \sin^{7/2} x + \frac{2}{11} \sin^{11/2} x + C$$

Example 24.   Find $\int \sin^3 2\theta \, d\theta$.

Solution.   Factoring out $-\sin 2\theta$, we have

$$\int \sin^3 2\theta \, d\theta = -\int \sin^2 2\theta \, (-\sin 2\theta \, d\theta)$$
$$= -\int (1-\cos^2 2\theta)(-\sin 2\theta \, d\theta)$$

Is $-\sin 2\theta \, d\theta$ the differential of $\cos 2\theta$? The answer is no. The differential of $\cos 2\theta$ is $-2 \sin 2\theta \, d\theta$. Therefore, we must multiply the differential by 2 and the integral by $\frac{1}{2}$ before proceeding. Hence

$$\int \sin^3 2\theta \, d\theta = -\frac{1}{2}\int (1-\cos^2 2\theta)(-2\sin 2\theta \, d\theta)$$

$$= \frac{1}{2}\int 2\sin 2\theta \, d\theta + \frac{1}{2}\int \cos^2 2\theta \, (-2\sin 2\theta \, d\theta)$$

$$= -\frac{1}{2}\cos 2\theta + \frac{1}{6}\cos^3 2\theta + C$$

$\int \tan^m u \sec^n u \, du$ **and** $\int \cot^m u \csc^n u \, du$ **(where n is even and positive).**   In the first antiderivative we recognize that $\sec^2 u \, du$ is the differential of $\tan u$. Since n is even and positive, $\sec^2 u$ can be factored out and any remaining powers of $\sec^2 u$ can be transformed by the identity $\sec^2 u = 1 + \tan^2 u$. In the second antiderivative, we proceed similarly by factoring out $-\csc^2 u$ and then using the identity $\csc^2 u = 1 + \cot^2 u$.

Example 25.   Find $\displaystyle\int \frac{\csc^4 \theta}{\cot^6 \theta} \, d\theta$.

Solution.   Factoring out $-\csc^2 \theta$, and rewriting the cotangent term in the numerator

$$\int \frac{\csc^4 \theta}{\cot^6 \theta} \, d\theta = -\int \cot^{-6} \theta \csc^2 \theta (-\csc^2 \theta \, d\theta)$$

$$= -\int \cot^{-6} \theta (1+\cot^2 \theta)(-\csc^2 \theta \, d\theta)$$

$$= -\int (\cot^{-6} \theta + \cot^{-4} \theta)(-\csc^2 \theta \, d\theta)$$

$$= \frac{1}{5}\cot^{-5} \theta + \frac{1}{3}\cot^{-3} \theta + C$$

EXERCISES

Find each of the following antiderivatives.

1. $\displaystyle\int \cos^2 \theta \, d\theta$ **2.** $\displaystyle\int \tan^2 x \, dx$

3. $\int \sin^2 3\phi \, d\phi$

10. $\int \dfrac{\sin^3 x \, dx}{\cos x}$

4. $\int \sin^3 x \cos^2 x \, dx$

11. $\int \csc t \, (\sin^3 t + \cos t) \, dt$

5. $\int \sin^4 2t \cos^3 2t \, dt$

12. $\int \left[\dfrac{\sec \phi}{\tan \phi} \right]^4 d\phi$

6. $\int \sin^5 x \sqrt{\cos x} \, dx$

13. $\int \dfrac{\cos^2 x}{\tan x} \, dx$

7. $\int \tan \theta \sec^4 \theta \, d\theta$

14. $\int \sec^4 t \, dt$

8. $\int \dfrac{\csc^4 y \, dy}{\cot^2 y}$

15. $\int (1 + \csc^2 \theta)^2 \, d\theta$

9. $\int \cos^3 \omega t \, d(\omega t)$

16. The moment of inertia I_x of a circular area of radius R is

$$I_x = (R^4/4) \int_0^{2\pi} \sin^2\theta \, d\theta.$$

Determine the expression for I_x.

17. The elemental moment of inertia of a sine curve is $dI = (1/12) \sin^3 x \, dx$. Determine the value of I for a half cycle of the sine curve.

18. The velocity of a particle varies as $v = \sin^2 t \cos^3 t$ ft/sec. Find the expression for the displacement of the particle if $s = 0$ when $t = 0$.

19. The power in a resistor is $p = \tan^2 3t$ watts. Find the energy dissipated by the resistor from $t = 0$ to $t = (\pi/12)$ sec.

10.9 ANTIDERIVATIVES THAT YIELD INVERSE TRIGONOMETRIC FUNCTIONS

Since the differential of arc sin u/a is

$$d\left(\text{arc sin } \frac{u}{a} \right) = \frac{1}{\sqrt{1 - (u/a)^2}} \cdot \frac{1}{a} \, du = \frac{du}{\sqrt{a^2 - u^2}}$$

we conclude that

$$\int \frac{du}{\sqrt{a^2 - u^2}} = \text{arc sin } \frac{u}{a} + C \qquad\qquad (25)$$

By similar means we can show that

$$\int \frac{du}{a^2 + u^2} = \frac{1}{a} \text{ arc tan } \frac{u}{a} + C \qquad\qquad (26)$$

The use of these formulas is explained in the following examples.

Example 26.   Find $\displaystyle\int \frac{ds}{\sqrt{4-9s^2}}$.

Solution.   This may be evaluated by Formula 25, if we let $a=2$ and $u=3s$. Then

$$\int \frac{ds}{\sqrt{4-9s^2}} = \int \frac{ds}{\sqrt{2^2-(3s)^2}}$$

The differential of u is $du=3ds$, so we multiply the integrand by 3 and the integral by $1/3$. Hence

$$\int \frac{ds}{\sqrt{4-9s^2}} = \frac{1}{3}\int \frac{3ds}{\sqrt{2^2-(3s)^2}} = \frac{1}{3}\arcsin\frac{3s}{2}+C$$

Example 27.   Find $\displaystyle\int \frac{dx}{\sqrt{1+4x-x^2}}$.

Solution.   At first, the expression under the radical does not appear to be of the form a^2-u^2. However, the given quadratic expression can be reduced to this form by completing the square. This yields

$$1+4x-x^2 = 1-(x^2-4x) = 1+4-(x^2-4x+4) = (\sqrt{5})^2-(x-2)^2$$

The required antiderivative may then be found by Formula 25 with $a=\sqrt{5}$, $u=x-2$, and $du=dx$. Hence

$$\int \frac{dx}{\sqrt{1+4x-x^2}} = \int \frac{dx}{\sqrt{(\sqrt{5})^2-(x-2)^2}} = \arcsin\frac{x-2}{\sqrt{5}}+C$$

Example 28.   Find $\displaystyle\int \frac{e^x\,dx}{1+e^{2x}}$.

Solution.   By writing the denominator as $1+(e^x)^2$, we recognize that $a=1$, $u=e^x$ and $du=e^x\,dx$. Therefore

$$\int \frac{e^x\,dx}{1+e^{2x}} = \int \frac{e^x\,dx}{1+(e^x)^2} = \arctan e^x+C$$

EXERCISES

Find each of the following antiderivatives.

1. $\displaystyle\int \frac{dy}{4+y^2}$

2. $\displaystyle\int \frac{dx}{\sqrt{1-4x^2}}$

3. $\displaystyle\int \frac{dv}{\sqrt{3-9v^2}}$

4. $\displaystyle\int \frac{dz}{3z^2+5}$

5. $\displaystyle\int \frac{dx}{2+3x^2}$

6. $\displaystyle\int \frac{y\,dy}{4+y^4}$

7. $\displaystyle\int \frac{dx}{x^2+4x+8}$

8. $\displaystyle\int \frac{3dx}{x^2+2x+5}$

9. $\displaystyle\int \frac{z\,dz}{3z^2+5}$

10. $\displaystyle\int \frac{z\,dz}{(3z^2+5)^2}$

11. $\displaystyle\int \frac{dx}{\sqrt{4x-x^2}}$

12. $\displaystyle\int \frac{dx}{\sqrt{5+4x-x^2}}$

13. $\displaystyle\int \frac{dy}{4y^2+8y+13}$

14. $\displaystyle\int \frac{\cos\theta\,d\theta}{\sqrt{4-\sin^2\theta}}$

15. $\displaystyle\int \frac{(x+2)\,dx}{x^2+2x+3}$

16. $\displaystyle\int \frac{e^{2x}\,dx}{1+e^{4x}}$

17. $\displaystyle\int \frac{dy}{5(y^2+3)}$

18. $\displaystyle\int \frac{2dp}{\sqrt{1-4p}}$

19. $\displaystyle\int \frac{(x+2)\,dx}{x^2+1}$

20. $\displaystyle\int \frac{(3-x)\,dx}{9+4x^2}$

21. The acceleration of an object in a resisting medium is $a = 1/(1+9t^2)$ m/sec². Find an expression for the velocity of the object. Assume $v = 100$ when $t = 0$.

22. Find the average height of the curve $y = 1/\sqrt{1-x^2}$ over the interval from $x = 0$ to $x = \frac{1}{2}$.

23. The current in a circuit is $i = \dfrac{1}{4(t^2+1)}$ amp. What is the average current in the circuit from $t = 0$ to $t = 1$ sec?

24. If $dy/dx = 4y^2+3$, determine an expression for x as a function of y.

25. A vertical column begins to buckle at critical or Euler load. The slope of the column at this load is given by $(dy/dx)^2 = -k^2y^2+c^2$. Find the expression for x, if k and c are constants.

11

Methods of Antidifferentiation and Integration

11.1 INTRODUCTION

In the previous chapters we have shown how to antidifferentiate some of the so-called *standard forms*. The only inventiveness required was in recognizing that the integrand fit one of these standard forms. In this chapter no new standard formulas are derived, but some of the more important techniques of effecting the antidifferentiation of a function are presented.

11.2 SUBSTITUTION

Sometimes we may reduce a given integrand to one of the standard forms by a change in variable or substitution. There are many substitutions that might be used, and there are no incorrect ones. However, we are usually not too pleased with a given substitution unless it leads to one of the standard forms.

Since it would be impossible to show all the various substitutions, we shall show the technique in only one particular case—integrands of the form $x^m(ax+b)^{p/q}$. If you learn the technique for this case, you should be able to do almost any kind of substitution. The change of variable to use here is

$$z = (ax+b)^{1/q}$$

Then

$$z^q = ax+b$$

and

$$x = \frac{z^q - b}{a}$$

which means that

$$dx = \frac{q}{a} z^{q-1} \, dz$$

Example 1. Evaluate $\displaystyle\int \frac{x \, dx}{\sqrt{x-3}}$.

Solution.   We make the substitution $z = (x-3)^{1/2}$ into the integrand to transform it into the variable z. Note that

$$x = z^2 + 3$$

and

$$dx = 2z\,dz$$

Thus

$$\int \frac{x\,dx}{\sqrt{x-3}} = \int \frac{(z^2+3)(2z\,dz)}{z}$$

$$= 2\int (z^2+3)\,dz$$

$$= 2\left(\frac{z^3}{3} + 3z\right) + C$$

$$= \frac{2z^3}{3} + 6z + C$$

Reversing the original substitution. we have as the solution

$$\int \frac{x\,dx}{\sqrt{x-3}} = \frac{2(x-3)^{3/2}}{3} + 6(x-3)^{1/2} + C$$

When evaluating definite integrals by the method of substitution, it is convenient to change the limits of integration to agree with the substituted variable. By rewriting the limits of integration in terms of the new variable, the definite integral can be evaluated without reversing the original substitution.

Example 2.   Evaluate $\int_1^6 x\sqrt{x+3}\,dx$.

Solution.   Let $z = (x+3)^{1/2}$; then $x = z^2 - 3$, and $dx = 2z\,dz$. Also, when $x = 1$, $z = 2$ and when $x = 6$, $z = 3$. The given integral may then be written

$$\int_1^6 x\sqrt{x+3}\,dx = \int_2^3 (z^2 - 3)z(2z\,dz)$$

$$= 2\int_2^3 (z^4 - 3z^2)\,dz$$

$$= 2\left[\frac{z^5}{5} - z^3\right]_2^3$$

$$= 2\left[\left(\frac{243}{5} - 27\right) - \left(\frac{32}{5} - 8\right)\right]$$

$$= \frac{232}{5}$$

EXERCISES

Find each of the following antiderivatives by using an appropriate substitution.

1. $\int x\sqrt{x+2}\,dx$

2. $\int x\sqrt{2x+3}\,dx$

3. $\int (t+2)\sqrt{t-3}\,dt$

4. $\int (x-1)\sqrt{2x-1}\,dx$

5. $\int \dfrac{3s^2\,ds}{\sqrt{1-s}}$

6. $\int \dfrac{x\,dx}{(x-1)^{1/3}}$

7. $\int \dfrac{dy}{(4y+13)\sqrt{y-1}}$

8. $\int \dfrac{(3x-1)\,dx}{\sqrt{2x+1}}$

9. $\int \dfrac{dx}{x+x^{1/2}}$

10. $\int \dfrac{dy}{(1+y)^{1/2}}$

11. $\int x\sqrt[3]{x-1}\,dx$

12. $\int \dfrac{dt}{t+t^{2/3}}$

13. $\int \dfrac{3t\,dt}{(2t+1)^{3/4}}$

14. $\int \dfrac{x\,dx}{(3x+4)^{2/3}}$

15. $\int x(2x+1)^{2/3}\,dx$

16. $\int \dfrac{dw}{\sqrt[3]{w}+w}$

17. The power being generated by a system is given by $p = (2t-1)\sqrt{t+3}$ watts. Find the work done by the system from $t = 1$ to $t = 6$ sec.

18. The current in a resistor varies as $i = t/(t+1)^{2/3}$. Find the amount of charge q transferred through the resistor from $t = 0$ to $t = 7$ sec.

19. The acceleration of a shaper head during its return stroke varies according to $a = t/(t^{1/2}+t^{3/2})$ ft/sec^2. Find the equation for the velocity of the head if $v = 0$ when $t = 0$.

20. The force required to motivate a particular mechanism varies with displacement according to $F = 3s/\sqrt{s-1}$ lb. Find the work done in moving from $s = 2$ to $s = 5$ ft.

11.3 PARTIAL FRACTIONS

A *rational* function is one that can be expressed as the ratio of two polynomial functions. If the degree of the polynomial in the numerator is less than the degree of the polynomial in the denominator, it is called a *proper* rational function; otherwise, the function is said to be *improper*. Any improper rational function can be reduced to a form consisting of the sum of a polynomial and a proper rational function.

For example

$$\frac{6x^4 + 7x^3 + 6x^2 + 32x - 7}{3x^2 + 5x - 2} = 2x^2 - x + 5 + \frac{5x + 3}{3x^2 + 5x - 2}$$

A proper rational function can frequently be written or decomposed into a sum of functions which turn out to be one of the standard forms. This sum is often called the *partial fractions decomposition* of the rational function. It is literally the reverse of what we learn in elementary algebra, that is, there we learn that

$$\frac{2}{x - 3} - \frac{1}{x + 1} = \frac{x + 5}{x^2 - 2x - 3}$$

Now we will show how to "recover" the elements of the sum making up

$$\frac{x + 5}{x^2 - 2x - 3}$$

Parenthetically, we note that if we had to evaluate

$$\int \frac{x + 5}{x^2 - 2x - 3} \, dx$$

we could write

$$\int \frac{x + 5}{x^2 - 2x - 3} \, dx = \int \frac{2}{x - 3} \, dx - \int \frac{1}{x + 1} \, dx$$

$$= 2 \ln (x - 3) - \ln (x + 1) + C$$

$$= \ln \frac{(x - 3)^2}{x + 1} + C$$

Our problem then is to describe the process by which a proper fraction can be decomposed into its partial fractions. In higher algebra courses, we prove that a proper fraction can always be expressed as the sum of partial fractions if the denominator of the proper fraction can be expressed as the product of factors of the form $ax + b$ and $ax^2 + bx + c$. Further, the form of the partial fraction decomposition is completely determined by the form of the factors making up the denominator. The following rules govern the form of the partial fraction decomposition of a given rational function. In each case it is the factors of the denominator of the rational function that are being considered.

Rule 1. If the denominator has an unrepeated linear factor of the form $(ax+b)$, a fraction of the form

$$\frac{A}{ax+b}$$

must be included in the partial fraction decomposition.

Rule 2. If the denominator has a linear factor $(ax+b)$ which is repeated n times, a sum of fractions of the form

$$\frac{A}{ax+b}+\frac{A_2}{(ax+b)^2}+\cdots+\frac{A_n}{(ax+b)^n}$$

must be included in the partial fraction decomposition.

Rule 3. If the denominator has an unrepeated quadratic factor (ax^2+bx+c), a fraction of the form

$$\frac{Ax+B}{ax^2+bx+c}$$

must be included in the partial fraction decomposition.

The following examples show how to utilize the knowledge of the form of the partial fraction decomposition to find the actual one. The examples are stated within the context of performing antidifferentiations of rational functions since that is our only application at this time.

Example 3. Evaluate

$$\int \frac{(x-11)\,dx}{2x^2+5x-3}$$

Solution.   The factors of the denominator are $x+3$ and $2x-1$. Hence, the given fraction can be expressed as the sum of the partial fractions

$$\frac{A}{x+3} \quad \text{and} \quad \frac{B}{2x-1}$$

where A and B are constants to be determined. Thus, we write

$$\frac{x-11}{(x+3)(2x-1)}=\frac{A}{x+3}+\frac{B}{2x-1}$$

To solve for A and B, we proceed by clearing the fractions from the equation to get

$$x-11=A(2x-1)+B(x+3)$$

Expanding

$$x - 11 = 2Ax - A + Bx + 3B$$

Collecting like terms in x

$$x - 11 = (2A + B)x + (3B - A)$$

For this equality to hold, the coefficients of corresponding powers of x must be equal, so that

$$2A + B = 1$$

and

$$3B - A = -11$$

Solving these equations simultaneously, we get $A = 2$ and $B = -3$. Hence

$$\frac{x - 11}{(x+3)(2x-1)} = \frac{2}{x+3} - \frac{3}{2x-1}$$

Substituting into the given antiderivative, we have

$$\int \frac{(x-11)\,dx}{(x+3)(2x-1)} = \int \frac{2\,dx}{x+3} - \int \frac{3\,dx}{2x-1}$$

$$= 2 \ln (x+3) - \frac{3}{2} \ln (2x-1) + C$$

$$= \ln \frac{(x+3)^2}{(2x-1)^{3/2}} + C$$

Example 4. Evaluate $\displaystyle\int \frac{x^2\,dx}{(x+1)^3}$.

Solution. We assume that the partial fractions may be written

$$\frac{x^2}{(x+1)^3} = \frac{A}{x+1} + \frac{B}{(x+1)^2} + \frac{C}{(x+1)^3}$$

Clearing the fractions

$$x^2 = A(x+1)^2 + B(x+1) + C$$

Expanding

$$x^2 = Ax^2 + 2Ax + A + Bx + B + C$$

Collecting like terms in x

$$x^2 = Ax^2 + (2A + B)x + (A + B + C)$$

Equating the like powers of x, we get

$$A = 1$$
$$2A + B = 0$$
$$A + B + C = 0$$

These equations yield the values $A = 1$, $B = -2$, and $C = 1$. Making these substitutions, we have

$$\frac{x^2}{(x+1)^3} = \frac{1}{x+1} - \frac{2}{(x+1)^2} + \frac{1}{(x+1)^3}$$

The given antiderivative may then be written as

$$\int \frac{x^2\, dx}{(x+1)^3} = \int \frac{dx}{x+1} - \int \frac{2dx}{(x+1)^2} + \int \frac{dx}{(x+1)^3}$$

$$= \ln(x+1) + \frac{2}{x+1} - \frac{1}{2(x+1)^2} + C$$

Example 5.   Evaluate

$$\int \frac{(2x+8)\, dx}{x^3 + 4x^2 + 4x}$$

Solution.   The factors of the denominator are x and $(x+2)^2$ and, therefore, the partial fractions are

$$\frac{2x+8}{x(x+2)^2} = \frac{A}{x} + \frac{B}{x+2} + \frac{C}{(x+2)^2}$$

Clearing the fractions yields

$$2x + 8 = A(x+2)^2 + Bx(x+2) + Cx$$

By the method used in the preceding examples, we obtain the simultaneous equations

$$A + B = 0$$
$$4A + 2B + C = 2$$
$$4A = 8$$

The solution of these equations yields $A = 2$, $B = -2$, and $C = -2$. The integral is then

$$\int \frac{(2x+8)\,dx}{x(x+2)^2} = \int \frac{2dx}{x} - \int \frac{2dx}{x+2} - \int \frac{2dx}{(x+2)^2}$$

$$= 2\ln x - 2\ln(x+2) + \frac{2}{x+1} + C$$

$$= \ln\left(\frac{x}{x+2}\right)^2 + \frac{2}{x+1} + C$$

Example 6. Evaluate

$$\int \frac{(9x+14)\,dx}{(x-2)(x^2+4)}$$

Solution. The denominator is composed of a linear factor $(x-2)$ and a quadratic factor (x^2+4). The required partial fractions are therefore

$$\frac{9x+14}{(x-2)(x^2+4)} = \frac{A}{x-2} + \frac{Bx+C}{x^2+4}$$

Clearing the fractions

$$9x+14 = A(x^2+4) + (Bx+C)(x-2)$$

Expanding the right side and collecting like terms

$$9x+14 = (A+B)x^2 + (C-2B)x + (4A-2C)$$

Equating the coefficients of corresponding powers of x, we find

$$A+B = 0$$
$$C-2B = 9$$
$$4A-2C = 14$$

from which $A = 4$, $B = -4$, and $C = 1$. Substituting these values in the partial fractions, we obtain

$$\int \frac{(9x+14)\,dx}{(x-2)(x^2+4)} = \int \frac{4dx}{x-2} + \int \frac{(-4x+1)\,dx}{x^2+4}$$

The first antiderivative on the right-hand side is easy to find, but the second is not. To find its value we proceed in the following manner.

$$\int \frac{(-4x+1)\,dx}{x^2+4} = -\int \frac{4x\,dx}{x^2+4} + \int \frac{dx}{x^2+4}$$

$$= -2\ln(x^2+4) + \frac{1}{2}\arctan\frac{x}{2}$$

The desired antiderivative is then

$$\int \frac{(9x+14)\,dx}{(x-2)(x^2+4)} = 4\ln(x-2) - 2\ln(x^2+4)$$

$$+ \frac{1}{2}\arctan\frac{x}{2} + C$$

$$= \ln\frac{(x-2)^4}{(x^2+4)^2} + \frac{1}{2}\arctan\frac{x}{2} + C$$

Denominators that contain repeated quadratic factors of the form $(x^2+bx+c)^m$ may also be resolved into partial fractions. However, we choose to delete denominators of this type from our discussion. This presents no serious limitations, since repeated quadratic factors occur much less frequently than the other three types.

EXERCISES

In Exercises 1 through 21, find the indicated antiderivatives.

1. $\int \frac{(x+1)\,dx}{x^2+4x-5}$

2. $\int \frac{(x+2)\,dx}{x^2-x-6}$

3. $\int \frac{dx}{x^2+x}$

4. $\int \frac{2dx}{x^3-4x}$

5. $\int \frac{(2z^2+5z+5)\,dz}{z(z+5)(z+2)}$

6. $\int \frac{(x+12)\,dx}{x^3+x^2-6x}$

7. $\int \frac{(x+3)\,dx}{x^2+6x+5}$

8. $\int \frac{dz}{(2z+3)^3}$

9. $\int \frac{(y^2+3y-6)\,dy}{y(y-1)^2}$

10. $\int \frac{(x+1)\,dx}{x^3-x^2}$

11. $\int \frac{ds}{s^3+2s^2+s}$

12. $\int \frac{(t^2+11)\,dt}{(t-5)(t+1)^2}$

13. $\int \dfrac{(w^2 - 3w + 6)\,dw}{(w^2 - 1)(3 - 2w)}$

18. $\int \dfrac{(2x^2 - x + 5)\,dx}{(x + 1)(x^2 + 4)}$

14. $\int \dfrac{dr}{(r + 4)^3}$

19. $\int \dfrac{(x^2 + 1)\,dx}{(2x + 3)(x^2 + 2x + 4)}$

15. $\int \dfrac{(2x - 1)\,dx}{(x - 1)^2}$

20. $\int \dfrac{(5 + 3s - s^2)\,ds}{(s^2 + 2s - 3)(s^2 + 5)}$

16. $\int \dfrac{2(x^2 + 1)\,dx}{x^3 + 2x}$

21. $\int \dfrac{(x^3 + x^2 - x + 1)\,dx}{(x - 1)^2(x^2 + 1)}$

17. $\int \dfrac{(t^2 + t + 2)\,dt}{t^3 + 2t}$

In Exercises 22 to 25, evaluate the indicated definite integrals

22. $\displaystyle\int_1^2 \dfrac{(2x + 3)\,dx}{x(x + 3)}$

24. $\displaystyle\int_0^2 \dfrac{dx}{x^2 + 4x + 20}$

23. $\displaystyle\int_0^1 \dfrac{(3y + 8)\,dy}{y^2 + 5y + 6}$

25. $\displaystyle\int_0^1 \dfrac{(x + 3)\,dx}{(x + 1)^2}$

11.4 INTEGRATION BY PARTS

A very powerful tool for evaluating antiderivatives comes from the formula for the derivative of a product of two functions. From Section 8.5 we have

$$\frac{d}{dx}(uv) = u\frac{dv}{dx} + v\frac{du}{dx}$$

which may be rewritten in differential form as

$$d(uv) = u\,dv + v\,du$$

Rearranging

$$u\,dv = d(uv) - v\,du \tag{1}$$

and antidifferentiating each term, we obtain

$$\int u\,dv = uv - \int v\,du \tag{2}$$

This result is referred to as the *formula for integration by parts*.* In this formula, the product $u\,dv$ indicates that the integrand is to be separated

---

*Formula 2 has traditionally been called the formula for integration by parts. We shall follow tradition and use this label, with the understanding that it is actually an antidifferentiation formula.

into two factors or "parts": a function u and a differential dv.* Once the "parts" have been chosen, Formula 2 clearly indicates that the anti-differentiation of $u\,dv$ depends upon the antidifferentiation of dv and $v\,du$. This may appear to be no improvement but, as we shall see, there are many cases in which $v\,du$ is a recognizable form when $u\,dv$ is not.

The major problem in using Formula 2 is in choosing the "parts" u and dv. No general rules govern the choice and the ability to know how to select the proper parts comes only with practice.

Example 7.   Evaluate $\int 2x \cos x \, dx$.

Solution.   This may be evaluated by letting

$$u = 2x \qquad \text{and} \qquad dv = \cos x \, dx$$

Then by differentiating u and antidifferentiating dv, we get

$$du = 2dx \qquad \text{and} \qquad v = \sin x\dagger$$

Substituting these values into the parts formula, we get

$$\int \underset{\downarrow}{u} \quad \underset{\downarrow}{dv} \quad = \quad \underset{\downarrow}{u} \quad \underset{\downarrow}{v} \quad - \int \underset{\downarrow}{v} \quad \underset{\downarrow}{du}$$

$$\int (2x)(\cos x \, dx) = (2x)(\sin x) - \int (\sin x)(2dx)$$

$$\int 2x \cos x \, dx = 2x \sin x + 2 \cos x + C$$

Sometimes the parts formula must be applied more than once, as the next example shows.

Example 8.   Evaluate $\int 3x^2 e^{6x} \, dx$.

Solution.   Choose the parts

$$u = 3x^2 \qquad \text{and} \qquad dv = e^{6x}dx$$

then

$$du = 6x \, dx \qquad \text{and} \qquad v = \frac{1}{6} e^{6x}$$

*The differential dv does not represent the differential of the function u; if it did, the integral could be evaluated immediately.

†It is not necessary to add an arbitrary constant at this point. All that is required is that a constant be added after the final antidifferentiation.

Substituting into Formula 2

$$\int 3x^2 e^{6x}\, dx = \frac{1}{2}x^2 e^{6x} - \int xe^{6x}\, dx$$

Since $\int xe^{6x}\, dx$ is not a standard form, it is necessary to apply Formula 2 again, this time choosing the parts

$$u = x \qquad \text{and} \qquad dv = e^{6x}\, dx$$

so that

$$du = dx \qquad \text{and} \qquad v = \frac{1}{6}e^{6x}$$

Thus

$$\int 3x^2 e^{6x}\, dx = \frac{1}{2}x^2 e^{6x} - \left[\frac{1}{6}xe^{6x} - \frac{1}{6}\int e^{6x}\, dx\right]$$

$$= \frac{1}{2}x^2 e^{6x} - \frac{1}{6}xe^{6x} + \frac{1}{36}e^{6x} + C$$

$$= \frac{1}{2}e^{6x}\left(x^2 - \frac{1}{3}x + \frac{1}{18}\right) + C$$

Example 9.   Evaluate $\int x(3x+2)^{-1/3}dx$.

Solution.   The method of substitution would work here but we can use the parts formula, also. Letting

$$u = x \qquad \text{and} \qquad dv = (3x+2)^{-1/3}\, dx$$

we have

$$du = dx \qquad \text{and} \qquad v = \frac{1}{2}(3x+2)^{2/3}$$

and

$$\int x(3x+2)^{-1/3}\, dx = \frac{1}{2}x(3x+2)^{2/3} - \frac{1}{2}\int (3x+2)^{2/3}\, dx$$

$$= \frac{1}{2}x(3x+2)^{2/3} - \frac{1}{10}(3x+2)^{5/3} + C$$

Example 10.   Evaluate $\int_2^4 t \ln t\, dt$.

Solution.   Letting

$$u = \ln t \qquad \text{and} \qquad dv = t\,dt$$

we have

$$du = \frac{dt}{t} \qquad \text{and} \qquad v = \frac{t^2}{2}$$

Substituting into Formula 2

$$\int_2^4 t \ln t\,dt = \left[\frac{t^2}{2}\ln t\right]_2^4 - \int_2^4 \left(\frac{t^2}{2}\right)\frac{dt}{t}$$

$$= \left[\frac{t^2}{2}\ln t\right]_2^4 - \frac{1}{2}\int_2^4 t\,dt$$

$$= \left[\frac{t^2}{2}\ln t - \frac{t^2}{4}\right]_2^4$$

$$= (8\ln 4 - 4) - (2\ln 2 - 1)$$
$$= 8\ln 4 - 2\ln 2 - 3$$
$$= 8(1.3863) - 2(0.6931) - 3$$
$$= 6.7042$$

Example 11.   Evaluate $\int e^{4x}\sin 2x\,dx$.

Solution.   This antiderivative can be found by applying the parts formula twice and then solving the resulting equation for the unknown which in this instance is the $\int e^{4x}\sin 2x\,dx$. We proceed by letting

$$u = e^{4x} \qquad \text{and} \qquad dv = \sin 2x\,dx$$

then

$$du = 4e^{4x}\,dx \qquad \text{and} \qquad v = -\frac{1}{2}\cos 2x$$

$$\int e^{4x}\sin 2x\,dx = -\frac{1}{2}e^{4x}\cos 2x + 2\int e^{4x}\cos 2x\,dx$$

Repeating the above process with

$$u = e^{4x} \qquad \text{and} \qquad dv = \cos 2x\,dx$$

we have

$$du = 4e^{4x}\,dx \qquad \text{and} \qquad v = \frac{1}{2}\sin 2x$$

so

$$\int e^{4x} \sin 2x \, dx = -\frac{1}{2} e^{4x} \cos 2x + 2 \left[\frac{1}{2} e^{4x} \sin 2x - 2 \int e^{4x} \sin 2x \, dx \right]$$

$$= -\frac{1}{2} e^{4x} \cos 2x + e^{4x} \sin 2x - 4 \int e^{4x} \sin 2x \, dx$$

We see that the antiderivative on the right is the same as the given anti-derivative. Thus

$$5 \int e^{4x} \sin 2x \, dx = -\frac{1}{2} e^{4x} \cos 2x + e^{4x} \sin 2x$$

or

$$\int e^{4x} \sin 2x \, dx = \frac{1}{10} e^{4x} (2 \sin 2x - \cos 2x) + C$$

EXERCISES

In Exercises 1 through 20, find the indicated antiderivatives.

1. $\int x \sin x \, dx$

2. $\int x \cos 3x \, dx$

3. $\int t e^{-t} \, dt$

4. $\int y \, e^y \, dy$

5. $\int \theta^2 \sin \theta \, d\theta$

6. $\int 3x^2 \cos x \, dx$

7. $\int z^3 \ln z \, dz$

8. $\int x^2 \sin x^3 \, dx$

9. $\int \arctan y \, dy$

10. $\int \arcsin t \, dt$

11. $\int \arccos 3x \, dx$

12. $\int x\sqrt{x+1} \, dx$

13. $\int t(3t^2 - 2)^{-4} \, dt$

14. $\int x \sec^2 x^2 \, dx$

15. $\int \dfrac{x \, dx}{(2x - 5)^{1/4}}$

16. $\int y \tan y \sec^2 y \, dy$

17. $\int \phi^3 \, e^{\phi^2} \, d\phi$

18. $\int \cos^2 \theta \sin \theta \, d\theta$

19. $\int e^{-x} \cos 3x \, dx$

20. $\int e^{x/2} \sin \dfrac{x}{2} \, dx$

In Exercises 21 through 24, evaluate the given definite integral.

21. $\int_1^2 \ln t \, dt$

22. $\int_0^{\pi/2} \theta \sin \theta \, d\theta$

23. $\int_0^{1/2} \arccos r \, dr$

24. $\int_0^{1/2} y \, e^{2y} \, dy$

25. Find the function whose derivative is $dy/dx = x^{1/2} \ln x$, if $y = 0$ when $x = 1$.
26. Find the area bounded by the curve $y = (x+2)e^x$, from $x = 0$ to $x = 1$.
27. The velocity of a particle is given by $v = $ arc tan t cm/sec. Find the displacement of the particle from $t = 0$ to $t = 1$ sec.
28. Find the centroid of the area bounded by the curve $y = e^x$ and the ordinates $x = 0$ and $x = 2$.
29. The resistance of a resistor varies as $r = t + 10$ ohms over the interval $t = \cdot$ 0 to $t = 1$. Find the energy dissipated by the resistor during this time interval if the current varies according to $i = e^{-2t}$ amp.

11.5 IMPROPER INTEGRALS

All of the definite integrals that we have encountered so far have had finite limits of integration and continuous integrand functions. Both of these properties are required for the integral to be defined as *proper*. If either is missing, the integral is, strictly speaking, not defined and is referred to as being *improper*. We now wish to show how to give meaning to improper integrals.

The first kind of "improperness" is that where one or both of the limits of integration become infinite. We write

$$\int_a^\infty f(x) \, dx$$

to describe the situation and by it we mean the limit of the proper integral $\int_a^k f(x) \, dx$ as k increases without bound; that is

$$\int_a^\infty f(x) \, dx = \lim_{k \to \infty} \int_a^k f(x) \, dx \tag{3}$$

Similarly

$$\int_{-\infty}^a f(x) \, dx = \lim_{k \to -\infty} \int_k^a f(x) \, dx \tag{4}$$

If the limit on the right-hand side exists, the improper integral is said to *converge*; otherwise, we say the integral is *divergent*.

Example 12. The graph of the function $y = x^{-2}$ is shown in Figure 11.1. Show that a value can be assigned to the area under the curve as x increases without bound from the point $x = 1$.

Solution. The desired area is given by

$$\int_1^\infty x^{-2} \, dx$$

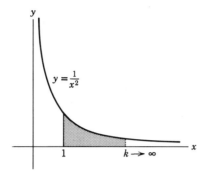

Figure 11.1

But this integral is evaluated as

$$\int_1^\infty x^{-2}\,dx = \lim_{k\to\infty}\int_1^k x^{-2}\,dx = \lim_{k\to\infty}\left[-x^{-1}\right]_1^k = \lim_{k\to\infty}\left[-\frac{1}{k}+1\right] = 1$$

Therefore, the improper integral converges and has the value 1.

Example 13. Show that $\int_1^\infty \dfrac{dx}{x}$ does not exist.

Solution. By definition

$$\int_1^\infty \frac{dx}{x} = \lim_{k\to\infty}\int_1^k \frac{dx}{x} = \lim_{k\to\infty}\left[\ln x\right]_1^k = \lim_{k\to\infty}\left[\ln k - \ln 1\right] = \infty$$

Since $\ln k \to \infty$ as $k \to \infty$, the integral diverges.

Example 14. The force between two charged particles is given by Coulomb's law to be $F = Qq/4\pi ks^2$, where Q and q are the respective charges, k is a constant, and s is the distance separating the charges. Find the formula for the work done in bringing a charge q of one coulomb from a great distance to a point d meters from a charge of Q coulombs.

Solution. Since $q = 1$, the force between Q and q can be written $F = Q/4\pi ks^2$. The work done by a force is $W = \int_{s_1}^{s_2} F\,ds$. Thus

$$W = \int_d^\infty \frac{Q}{4\pi ks^2}\,ds = \frac{Q}{4\pi k}\int_d^\infty \frac{ds}{s^2} = \lim_{m\to\infty}\frac{Q}{4\pi k}\int_d^m \frac{ds}{s^2} = \lim_{m\to\infty}\left[-\frac{Q}{4\pi ks}\right]_d^m$$

$$= \lim_{m\to\infty}\left[-\frac{Q}{4\pi km} + \frac{Q}{4\pi kd}\right] = \frac{Q}{4\pi kd}$$

In physics, the work done in bringing a charge of one coulomb to a point in an electric field is called the *electric potential* at that point. The formula derived above is then the formula for the electric potential d meters from a charge of Q coulombs.

In some instances, definite integrals arise which have finite limits of integration, but are improper because the integrand function becomes infinite for some value of x within the interval of integration. Improper integrals of this kind are also defined as the limit of a proper integral. Thus, if the function $f(x)$ has an infinite discontinuity at the upper limit of integration $x = b$, we define the value of the integral by

$$\int_a^b f(x)\ dx = \lim_{q \to b} \int_a^q f(x)\ dx \tag{5}$$

where q approaches b from within the interval of integration.

Likewise, if the function $f(x)$ has an infinite discontinuity at the lower limit of integration $x = a$, we define the value of the integral from a to b by

$$\int_a^b f(x)\ dx = \lim_{q \to a} \int_q^b f(x)\ dx \tag{6}$$

where q approaches a from within the interval of integration.

As in the case of improper integrals of the first kind, the integral on the left converges if the limit on the right exists and diverges if it does not.

Example 15.   Evaluate $\displaystyle\int_0^2 \frac{dx}{\sqrt{x}}$. (See Figure 11.2.)

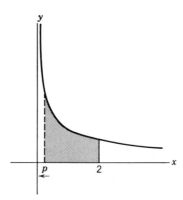

Figure 11.2

Solution. Since the integrand has an infinite discontinuity at $x = 0$, we choose a value $x = p$ within the interval of integration for which the integrand is defined and then take the limit of the integral as $p \to 0$. Hence, the integral can be written

$$\int_0^2 \frac{dx}{\sqrt{x}} = \lim_{p \to 0} \int_p^2 x^{-1/2}\, dx = \lim_{p \to 0} [2x^{1/2}]_p^2 = \lim_{p \to 0} [2\sqrt{2} - 2\sqrt{p}] = 2\sqrt{2}$$

Geometrically, this is interpreted to mean that the area under the curve $y = 1/\sqrt{x}$ approaches $2\sqrt{2}$ as a limit when p approaches arbitrarily close to zero.

Example 16. Evaluate $\int_0^3 \frac{dx}{(x-3)^2}$.

Solution. The integrand becomes discontinuous at $x = 3$; therefore

$$\int_0^3 \frac{dx}{(x-3)^2} = \lim_{q \to 3} \int_0^q (x-3)^{-2}\, dx = \lim_{q \to 3} \left[-\frac{1}{x-3} \right]_0^q$$

$$= \lim_{q \to 3} \left[-\frac{1}{q-3} - \frac{1}{3} \right] = \infty$$

Thus, the given integral diverges.

If the function $f(x)$ has an infinite discontinuity at some value $x = c$, which lies between $x = a$ and $x = b$, then the integral of $f(x)$ from $x = a$ to $x = b$ can be expressed as the sum of two improper integrals,

$$\int_a^b f(x)\, dx = \int_a^c f(x)\, dx + \int_c^b f(x)\, dx$$

where c is the point of infinite discontinuity. Each of the integrals on the right can then be handled as in the previous examples, since the first integral on the right is discontinuous at its upper limit and the second is discontinuous at its lower limit. If either of the two integrals on the right diverges, the integral on the left diverges.

Example 17. Evaluate $\int_{-1}^3 dx/x^2$.

Solution. The integrand is discontinuous at $x = 0$ which is an interior point of the interval of integration. The given integral may then be written as

$$\int_{-1}^3 \frac{dx}{x^2} = \int_{-1}^0 \frac{dx}{x^2} + \int_0^3 \frac{dx}{x^2}$$

Writing the improper integrals as the limit of proper integrals, we have

$$\int_{-1}^{3} \frac{dx}{x^2} = \lim_{p \to 0} \int_{-1}^{p} x^{-2}\, dx + \lim_{q \to 0} \int_{q}^{3} x^{-2}\, dx$$

$$= \lim_{p \to 0} \left[-\frac{1}{x} \right]_{-1}^{p} + \lim_{q \to 0} \left[-\frac{1}{x} \right]_{q}^{3}$$

$$= \lim_{p \to 0} \left[-\frac{1}{p} - 1 \right] + \lim_{q \to 0} \left[-\frac{1}{3} + \frac{1}{q} \right] = \infty$$

You should now be on the alert for improper integrals; unless you are, you may treat an improper integral as proper and be led into a serious error. For example, if we evaluate the improper integral of the last example as if it were proper, we would have

$$\int_{-1}^{3} \frac{dx}{x^2} = \left[-\frac{1}{x} \right]_{-1}^{3} = \left[-\frac{1}{3} - 1 \right] = -\frac{4}{3}$$

which is meaningless for two reasons.

(a) The integrand is always positive and yet a negative answer was obtained.
(b) It contradicts the result of the previous example.

EXERCISES

Evaluate each of the following integrals if they exist.

1. $\displaystyle \int_{1}^{\infty} \frac{dx}{x^3}$

2. $\displaystyle \int_{-\infty}^{-1} \frac{dz}{z^2}$

3. $\displaystyle \int_{0}^{4} \frac{dx}{\sqrt{x}}$

4. $\displaystyle \int_{-1}^{0} \frac{dx}{x^{1/3}}$

5. $\displaystyle \int_{0}^{\infty} \frac{dy}{y+1}$

6. $\displaystyle \int_{0}^{1} \frac{t\, dt}{t^2 - 1}$

7. $\displaystyle \int_{0}^{2} \frac{dx}{x^{3/2}}$

8. $\displaystyle \int_{0}^{\infty} e^{-t}\, dt$

9. $\displaystyle \int_{0}^{\pi/2} \frac{\sin \theta\, d\theta}{1 - \cos \theta}$

10. $\displaystyle \int_{2}^{\infty} \frac{dx}{(x+1)^3}$

11. $\displaystyle\int_0^\infty \frac{dy}{\sqrt{2y+1}}$

15. $\displaystyle\int_{-1}^2 \frac{dx}{\sqrt[3]{x}}$

12. $\displaystyle\int_0^1 \frac{dx}{\sqrt{1-x^2}}$

16. $\displaystyle\int_1^3 \frac{dz}{(z-2)^3}$

13. $\displaystyle\int_0^\infty \frac{dt}{1+t^2}$

17. $\displaystyle\int_0^2 \frac{dx}{\sqrt[3]{x-1}}$

14. $\displaystyle\int_0^2 \frac{e^x\,dx}{1-e^x}$

18. $\displaystyle\int_{-1}^0 \frac{dt}{\sqrt{t+2}}$

12

The Conic Sections

12.1 INTRODUCTION

Curves that can be formed by cutting a right circular cone with a plane are called conic sections. Greek mathematicians in 300 BC, who were the first to study such curves, discovered that four distinct curves can be obtained by cutting a right circular cone with a plane: the circle, the ellipse, the hyperbola, and the parabola. The properties of the conics discovered by the Greeks include those that we shall use as definitions in this chapter.

If two right circular cones are placed vertex-to-vertex with a common axis, the resulting figure is referred to as a cone with two *nappes*. To see how the four conics can be generated from a cone with two nappes, we refer to Figure 12.1.

(a) A *circle* is obtained when the cutting plane is perpendicular to the axis, providing it does not pass through the vertex.

(b) An *ellipse* is obtained when the cutting plane is inclined so as to cut entirely through one nappe of the cone without cutting the other nappe.

(c) A *parabola* is obtained when the cutting plane is parallel to one element in the side of the cone.

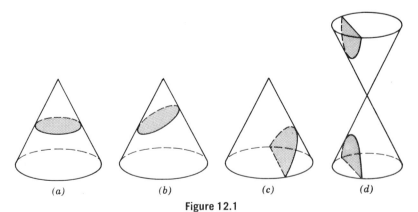

(a) (b) (c) (d)

Figure 12.1

(d) A *hyperbola* is obtained when the cutting plane is inclined so that it cuts through both nappes.

Recalling that analytic geometry is the study of the relationship between algebra and geometry, we wish to establish the algebraic representation of the four conic sections.

12.2 THE PARABOLA

In this article we shall define and discuss the parabola.

Definition
A parabola is the set of all points in a plane equidistant from a fixed point and a fixed line.

The fixed point is called the *focus* and the fixed line, the *directrix* of the parabola. The line through the focus and perpendicular to the directrix is called the *axis* of the parabola. By definition, the midpoint between the focus and the directrix is a point on the parabola and is known as the *vertex*.

In Figure 12.2 we have drawn a parabola with vertex located at the origin, focus at $F(a, 0)$, and directrix perpendicular to the x-axis at $D(-a, 0)$. It should be observed that a is the distance from the vertex to the focus and is sometimes referred to as the *focal distance*.

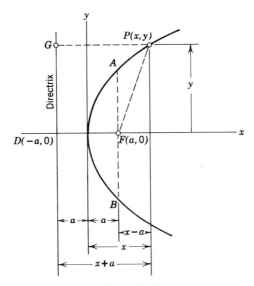

Figure 12.2

To find the algebraic equation of this parabola, consider a point $P(x, y)$ on the parabola. Then by definition, we have

$$GP = FP \tag{1}$$

But from Figure 12.2 we see that

$$GP = x + a$$

and

$$FP = \sqrt{(x-a)^2 + y^2}$$

Substituting these expressions in Equation 1, we get

$$x + a = \sqrt{(x-a)^2 + y^2}$$

or, squaring both sides

$$(x+a)^2 = (x-a)^2 + y^2$$

Expanding and collecting like terms, the equation of the parabola becomes

$$y^2 = 4ax \tag{2}$$

Equation 2 is referred to as the *standard form* of the equation of a parabola with vertex at the origin and focus on the x-axis. It is clearly symmetric about its axis. The focus is at $(a, 0)$ if the coefficient $4a$ is positive. In this case the parabola opens to the right. If the coefficient $4a$ is negative, the focus is at $(-a, 0)$ and the parabola opens to the left.

The chord AB through the focus and perpendicular to the axis is called the *right chord*. The length of the right chord is found by letting $x = a$ in Equation 2. Making this substitution, we find

$$y^2 = 4a^2 \qquad \text{and} \qquad y = \pm 2a$$

The length of the right chord is, therefore, equal numerically to $4a$. This fact is useful because it helps to define the shape of the parabola by giving us an idea of the "opening" of the parabola.

The standard form of a parabola with vertex at the origin and focus on the y-axis is

$$x^2 = 4ay \tag{3}$$

The derivation of this Equation parallels that of Equation 2 and is left for the student to attempt. The parabola represented by Equation 3 is symmetric about the y-axis. It opens upward if the coefficient $4a$ is positive, and downward if it is negative.

The equation of a parabola is characterized by one variable being linear and the other variable squared. *The direction of the axis of the parabola is indicated by the linear variable.*

A unique property of the parabola is that it will reflect any rays emitted from the focus such that they travel parallel to the axis of the parabola. This feature makes the parabola a particularly desirable shape for reflectors in spotlights, reflecting telescopes, and for radar antennas.

A rough sketch of the parabola can be drawn if the location of the vertex and the extremities of the right chord are known. As we will see in the following examples, all of this information can be obtained from the standard form of the equation.

Example 1.   Discuss and sketch the graph of the equation $x^2 = 8y$.

Solution.   This equation has the form (3) with $4a = 8$, or $a = 2$. The y-axis is the axis of the parabola, and the focus is at $(0, 2)$ and the directrix is the line $y = -2$. The end-points of the right chord are then $(-4, 2)$ and $(4, 2)$. The parabola is sketched in Figure 12.3.

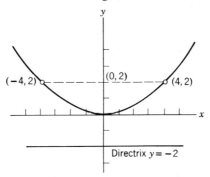

Figure 12.3

Example 2.   Find the equation of the parabola with focus at $(-1, 0)$ and directrix $x = 1$, and sketch the curve.

Solution.   The focus lies on the x-axis to the left of the directrix so the parabola opens to the left. The desired equation is then form (2) with $a = -1$; that is

$$y^2 = -4x$$

To sketch the parabola we note that the length of the right chord is $AB = 4$. Its extremities are therefore $(-1, 2)$ and $(-1, -2)$. The curve is shown in Figure 12.4.

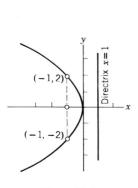

Figure 12.4

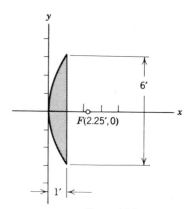

Figure 12.5

Example 3.   A parabolic reflector is to be built with a focal distance of 2.25 ft. What is the diameter of the reflector if it is to be 1-ft deep at its axis? (See Figure 12.5.)

Solution.   To solve this problem we need the equation of the parabola used to generate the reflector. Referring to Figure 12.5, we choose to locate the vertex at the origin and the focus on the x-axis. (We could have just as well placed the focus on the y-axis.) Then using Equation 2 with $a = 2.25$, we have the equation

$$y^2 = 4(2.25)\, x = 9x$$

The diameter of the reflector can now be found by substituting $x = 1$ into this equation. Thus

$$y^2 = 9 \quad \text{and} \quad y = \pm 3$$

which means that the diameter of the reflector is 6 ft.

EXERCISES

In Exercises 1 to 10, find the coordinates of the focus, the end-points of the right chord and the equation of the directrix of each of the parabolas. Sketch the graph of each parabola.

1. $y^2 = -8x$
2. $x^2 = 12y$
3. $2x^2 = 12y$
4. $x^2 = -24y$
5. $y^2 + 16x = 0$

6. $y + 2x^2 = 0$
7. $y^2 = 3x$
8. $3x^2 = 4y$
9. $y^2 = -2x$
10. $y^2 = 10x$

In Exercises 11 to 18, find the equation of the parabolas having the given properties. Sketch each curve.

11. Focus at $(0, 2)$, directrix $y = -2$
12. Focus at $(0, -\frac{1}{2})$, directrix $y = \frac{1}{2}$
13. Focus at $(\frac{3}{2}, 0)$, directrix $x = -\frac{3}{2}$
14. Focus at $(-10, 0)$, directrix $x = 10$
15. End-points of right chord $(2, -1)$ and $(-2, -1)$, vertex at $(0, 0)$
16. End-points of right chord $(3, 6)$ and $(3, -6)$, vertex at $(0, 0)$
17. Vertex at $(0, 0)$, vertical axis, one point of the curve $(2, 4)$
18. Vertex at $(0, 0)$, horizontal axis, one point of the curve $(2, 4)$

In Exercises 19 to 22, solve the given system of equations graphically.

19. $2x + 4y = 0$
$x^2 - 4y = 0$

21. $y^2 = 12x$
$y = 5 \sin\left(x + \dfrac{\pi}{4}\right)$

20. $y = e^x$
$y^2 = -3x$

22. $y = \cos\dfrac{x}{2}$
$x^2 = 8y$

23. The supporting cable of a suspension bridge hangs in the shape of a parabola. Find the equation of a cable hanging from two 400 ft high supports which are 1000-ft apart, if the lowest point of the cable is 250-ft below the top of the supports. Choose the origin in the most convenient location.

24. A parabolic antenna is to be constructed by revolving the parabola $y^2 = 24x$. Sketch the cross-section of the antenna if the diameter of the circular front is to be 12 ft. Locate the focus.

12.3 THE ELLIPSE

An *ellipse* can be constructed from a loop of string in the following way. Place two pins at F and F' as shown in Figure 12.6, and place the loop of string over them. Pull the string taut with the point of a pencil and then move the pencil keeping the string taut. The figure generated is an ellipse. We observe from this construction that the sum of the distances from the two fixed points to the point P is always the same since the loop of string is kept taut. This property characterizes the ellipse.

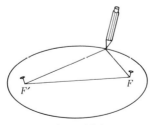

Figure 12.6

Definition

An ellipse is the set of all points in a plane the sum of whose distances from two fixed points in the plane is constant.

The fixed points are called the *foci* of the ellipse. The midpoint of a line through the foci is called the *center* of the ellipse. We will use the center of the ellipse to locate the ellipse in the plane in the same way we used the vertex to locate the parabola.

The equation of an ellipse is obtained by considering an ellipse with foci located on the x-axis such that the origin is midway between them, as in Figure 12.7.

Letting the foci be the points $F(c, 0)$ and $F'(-c, 0)$ and the sum of the distances from a point $P(x, y)$ of the ellipse to the foci be $2a$, where $a > c$, we have

$$PF + PF' = 2a$$

From Figure 12.7 it is clear that

$$PF = \sqrt{(x-c)^2 + y^2}$$

and

$$PF' = \sqrt{(x+c)^2 + y^2}$$

so that

$$\sqrt{(x-c)^2 + y^2} + \sqrt{(x+c)^2 + y^2} = 2a$$

Transposing the first radical and squaring

$$(x+c)^2 + y^2 = 4a^2 - 4a\sqrt{(x-c)^2 + y^2} + (x-c)^2 + y^2$$

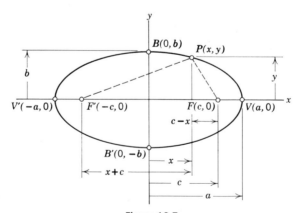

Figure 12.7

Expanding and collecting like terms,

$$a\sqrt{(x-c)^2+y^2} = a^2 - cx$$

Squaring again and simplifying

$$(a^2 - c^2)x^2 + a^2y^2 = a^2(a^2 - c^2)$$

Substituting $b^2 = a^2 - c^2$, this equation becomes

$$b^2x^2 + a^2y^2 = a^2b^2$$

Finally, dividing through by the nonzero quantity a^2b^2 we can write the equation of an ellipse with its center at the origin and its foci on the x-axis as

$$\frac{x^2}{a^2} + \frac{y^2}{b^2} = 1 \tag{4}$$

Equation 4 is clearly symmetric about both axes and the origin, a fact that will be useful in sketching ellipses. By letting $y = 0$ in Equation 4, we see that the x-intercepts of the ellipse are $(a, 0)$ and $(-a, 0)$. The segment of the line through the foci from $(a, 0)$ to $(-a, 0)$ is called the *major axis* of the ellipse. Clearly, the length of the major axis is $2a$, which is also the value chosen for the sum of the distances PF and PF'. The length a is then known as the *semimajor axis* and the end-points of the major axis as the *vertices* of the ellipse. The y-intercepts of the ellipse are found to be $(0, b)$ and $(0, -b)$ by letting $x = 0$ in Equation 4. The segment of the line perpendicular to the major axis from $(0, b)$ to $(0, -b)$ is called the *minor axis*. Since the length of the minor axis is $2b$, the length of the *semiminor axis* is b. The graph of an ellipse can readily be sketched once the semimajor and semiminor axes are known. In this book, the semimajor axis will always be represented by the letter a, and the semiminor axis by the letter b.

The foci of an ellipse can be located by solving the equation $b^2 = a^2 - c^2$ for the focal distance c. Thus, the desired equation is

$$c = \sqrt{a^2 - b^2}$$

A similar derivation will show that

$$\frac{x^2}{b^2} + \frac{y^2}{a^2} = 1 \tag{5}$$

is the equation of an ellipse with its center at the origin and its foci on the *y*-axis. Note that the letter *a* is used to represent the semimajor axis.

Example 4.   Find the equation of an ellipse centered at the origin with foci on the *x*-axis, if the major axis is 10 and the minor axis is 4. (See Figure 12.8.)

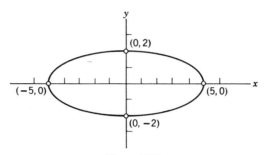

Figure 12.8

Solution.   In this case the major axis is on the *x*-axis. The semimajor axis is $a = 5$, and the semiminor axis is $b = 2$. Substituting these values into Equation 4, we have

$$\frac{x^2}{5^2} + \frac{y^2}{2^2} = 1$$

or

$$\frac{x^2}{25} + \frac{y^2}{4} = 1$$

as the required equation.

Example 5.   Find the equation of the ellipse with vertices at $(0, 5)$ and $(0, -5)$ and foci at $(0, 4)$ and $(0, -4)$. (See Figure 12.9.)

Solution.   From the given information we are able to conclude that the foci are on the *y*-axis, the center of the ellipse is at the origin, and the semimajor axis is $a = 5$. To find the semiminor axis, we use the relation $b^2 = a^2 - c^2$. Thus, $b = \sqrt{25 - 16} = \sqrt{9} = 3$. Substituting $a = 5$ and $b = 3$ into Equation 5, we have

$$\frac{x^2}{9} + \frac{y^2}{25} = 1$$

Example 6.   Sketch the graph of the ellipse $16x^2 + 4y^2 = 64$.

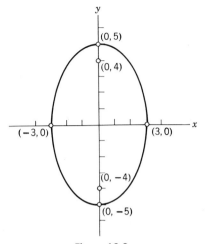

Figure 12.9

Solution.   Dividing the given equation by 64, it can be written in the form

$$\frac{x^2}{4} + \frac{y^2}{16} = 1$$

The major axis lies along the y-axis, since the denominator of the y term in the equation is larger than the denominator of the x term. Consequently, the semimajor axis is $a = 4$ and the semiminor axis is $b = 2$. The vertices are then $(0, 4)$ and $(0, -4)$, and the end-points of the minor axis are $(2, 0)$ and $(-2, 0)$. The ellipse is sketched in Figure 12.10.

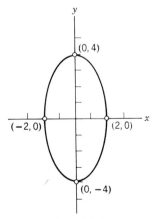

Figure 12.10

One of the first scientific applications of the ellipse was in astronomy. The astronomer Kepler (*c.* 1600) discovered that the planets moved in elliptical orbits about the sun with the sun at one focus. Artificial satellites also move in elliptical orbits about the earth. Another application is in the design of machines where elliptic gears are used to obtain a slow powerful movement with a quick return. A third application of the ellipse is found in electricity where the magnetic field of a single phase induction motor is elliptical under normal operating conditions.

12.4 THE CIRCLE

We shall now define and discuss the circle.

Definition:
A circle is the set of all points in a plane equidistant from a fixed point in the plane.

The fixed point is the *center*, and the given distance is the *radius* of the circle. The circle is probably the most familiar of the conic sections as well as the most important. After all, where would we be without the wheel?

We can obtain the equation of a circle centered at the origin by noting that the ellipse approaches a circle as the foci move closer together. If the foci F and F' coincide, the ellipse becomes a circle and $b = a = r$ in Formula 4. Thus, the equation of a circle is

$$\frac{x^2}{r^2} + \frac{y^2}{r^2} = 1$$

Multiplying both sides of this equation by r^2, we obtain

$$x^2 + y^2 = r^2 \tag{6}$$

as the standard form of the equation of a circle with center at the origin and radius equal to r.

Example 7. Find the equation of the circle centered at the origin and radius of 4.

Solution.   Substituting $r = 4$ into Formula 6

$$x^2 + y^2 = 16$$

is the desired equation.

EXERCISES

In Exercises 1 to 10, discuss each of the given equations and sketch their graphs.

1. $x^2 + y^2 = 25$
2. $4x^2 + 9y^2 = 36$
3. $16x^2 + 4y^2 = 16$
4. $3x^2 + 3y^2 = 27$
5. $3x^2 + y^2 = 9$
6. $25x^2 + 4y^2 = 100$
7. $2x^2 + 3y^2 - 24 = 0$
8. $5x^2 + 20y^2 = 20$
9. $9x^2 + 4y^2 = 4$
10. $4x^2 + y^2 = 25$

In Exercises 11 to 20, find the equation of the ellipses having the given properties. Sketch each curve.

11. Vertices at $(\pm 4, 0)$, minor axis 6.
12. Vertices at $(0, \pm 1)$, minor axis 1.
13. Vertices at $(0, \pm 5)$, semiminor axis $\frac{3}{2}$.
14. Vertices at $(\pm 6, 0)$, semiminor axis 2.
15. Major axis 10, foci at $(\pm 4, 0)$.
16. Major axis 10, foci at $(0, \pm 3)$.
17. Foci at $(\pm 1, 0)$, semimajor axis 4.
18. Vertices at $(0, \pm 7)$, foci at $(0, \pm \sqrt{28})$.
19. Vertices at $(\pm \frac{3}{2}, 0)$, one point of the curve at $(1, 1)$.
20. Vertices at $(\pm 3, 0)$, one point of the curve at $(\sqrt{3}, 2)$.

In Exercises 21 to 24, solve the given system of equations graphically.

21. $\dfrac{x^2}{4} + y^2 = 4$
    $2y + 3x = 0$

22. $\dfrac{x^2}{9} + \dfrac{y^2}{9} = 1$
    $y = e^{x+2}$

23. $y = \sin \frac{1}{2}x$
    $x^2 + 4y^2 = 4$

24. $y^2 - 12x = 0$
    $y^2 + 9x^2 = 9$

25. An elliptical cam with a horizontal major axis of 10 in. and a minor axis of 3 in. is to be machined by a numerically controlled vertical mill. Find the equation of the ellipse to be used in programming the control device.
26. An elliptical cam having the equation $9x^2 + y^2 = 81$ revolves against a push rod. What is the maximum travel of the push rod?

12.5   THE HYPERBOLA

The final conic which we shall consider is the hyperbola.

Definition:
A hyperbola is the set of all points in a plane the difference of whose distances from two fixed points in the plane is constant.

In Figure 12.11 we have drawn a hyperbola with foci at $F(c, 0)$ and $F'(-c, 0)$. The origin is then at the midpoint between the foci, which corresponds to the *center* of the hyperbola. The points $V(a, 0)$ and $V'(-a, 0)$ are called the *vertices*, and the segment VV' is called the

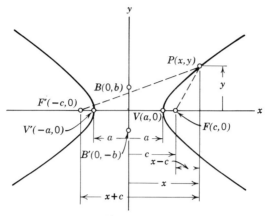

Figure 12.11

transverse axis of the hyperbola. Clearly the length of the transverse axis is $2a$. The segment BB', which is perpendicular to the transverse axis at the center of the hyperbola, is called the *conjugate axis* and has a length of $2b$. As we will see, the conjugate axis has an important relation to the curve even though it does not intersect the curve.

To find the algebraic equation of a hyperbola, we consider a point $P(x, y)$ on the hyperbola. Then by definition

$$F'P - FP = 2a$$

which, in turn, can be written

$$\sqrt{(x+c)^2 + y^2} - \sqrt{(x-c)^2 + y^2} = 2a \tag{7}$$

Using the same procedure here that we used for the ellipse, Equation (7) can be reduced to

$$\frac{x^2}{a^2} - \frac{y^2}{c^2 - a^2} = 1$$

Or, letting $b^2 = c^2 - a^2$, the equation of the hyperbola becomes

$$\frac{x^2}{a^2} - \frac{y^2}{b^2} = 1 \tag{8}$$

which like the ellipse is symmetric about both axes and the origin. Letting $y = 0$, the x-intercepts of equation are found to be $x = \pm a$. Additional information regarding the shape of the hyperbola can be obtained by solving Equation 8 for y. Performing the necessary algebraic operations, we get

$$y = \pm \frac{b}{a}\sqrt{x^2 - a^2}$$

from which it is evident that the curve does not exist for $x^2 < a^2$. Consequently, the hyperbola consists of two separate curves or branches — one to the right of $x = a$ and a similar one to the left of $x = -a$.

The shape of the hyperbola is dictated by two straight lines called the *asymptotes* of the hyperbola. The asymptotes of a hyperbola are the extended diagonals of the rectangle formed by drawing lines parallel to the coordinate axes through the end-points of both the transverse axis and the conjugate axis. Referring to Figure 12.12, we see that the slope of the diagonals of this rectangle are

$$m = \pm \frac{b}{a}$$

and, therefore, the asymptotes are given by the lines

$$y = \pm \frac{b}{a}x$$

We now want to show that these lines are asymptotes of the hyperbola; that is, that the hyperbola approaches arbitrarily close to the lines as x increases without bound. Solving Equation 8 for y^2, we can write it in

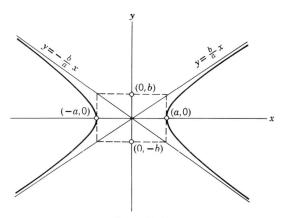

Figure 12.12

the form

$$y^2 = \frac{b^2 x^2}{a^2}\left(1 - \frac{a^2}{x^2}\right)$$

or, taking the square root of both sides

$$y = \pm \frac{b}{a} x \sqrt{1 - \frac{a^2}{x^2}} \qquad (9)$$

Now consider the limit of the right member as x increases without bound. Under this condition a^2/x^2 approaches zero as a limit and, therefore,

$$\lim_{x \to \infty} \pm \frac{b}{a} x \sqrt{1 - \frac{a^2}{x^2}} = \pm \frac{b}{a} x$$

which means that the hyperbola $x^2/a^2 - y^2/b^2 = 1$ approaches the lines $y = \pm (b/a) x$ as a limit as $x \to \infty$.

If we begin with the foci of the hyperbola on the y-axis and the center at the origin, the standard form of the equation of a hyperbola becomes

$$\frac{y^2}{a^2} - \frac{x^2}{b^2} = 1 \qquad (10)$$

The vertices of this hyperbola are on the y-axis; *the positive term always indicates the direction of the transverse axis.* Notice that the standard form of the hyperbola, like that for the ellipse, demands the coefficients of x^2 and y^2 be in the denominator, and the number on the right hand side be 1. In the case of the ellipse, the major axis is determined by the larger of the two denominators. In the case of the hyperbola, the transverse axis is determined by the sign of the term, *not* by the magnitude of the denominator.

To sketch the hyperbola, first draw the rectangle through the extremities of the transverse and conjugate axes and extend the diagonals of the rectangle. Then draw the hyperbola so that it passes through the vertex and comes closer and closer to the extended diagonals as x moves away from the origin.

Example 8.   Discuss and sketch the graph of $4x^2 - y^2 = 16$.

Solution.   Dividing by 16, we have

$$\frac{x^2}{4} - \frac{y^2}{16} = 1$$

which is the equation of a hyperbola with center at the origin and foci on the x-axis. It has vertices at $(\pm 2, 0)$ and its conjugate axis extends from $(0, 4)$ to $(0, -4)$. The foci are found from the equation $b^2 = c^2 - a^2$. Thus

$$c = \sqrt{a^2 + b^2} = \sqrt{4 + 16} = \sqrt{20} = 2\sqrt{5}$$

and the foci are located at $(\pm 2\sqrt{5}, 0)$. Plotting these points and drawing the rectangle and its extended diagonals, the hyperbola shown in Figure 12.13 is obtained.

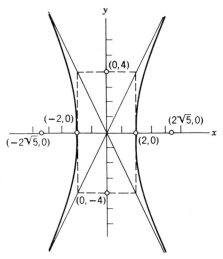

Figure 12.13

Example 9. Discuss and sketch the graph of $y^2/13 - x^2/9 = 1$.

Solution. This hyperbola has a vertical transverse axis with vertices at $(0, \sqrt{13})$ and $(0, -\sqrt{13})$. The extremes of the conjugate axis are then $(3, 0)$ and $(-3, 0)$. The foci are located at $(0, \sqrt{22})$ and $(0, -\sqrt{22})$. This information is used to sketch the hyperbola in Figure 12.14.

Example 10. Determine the equation of the hyperbola centered at the origin with foci at $(\pm 6, 0)$ and a transverse axis 8 units long.

Solution. Here $a = 4$ and $c = 6$. Since $c^2 = a^2 + b^2$, we have $b^2 = c^2 - a^2 = 36 - 16 = 20$. Substituting $a^2 = 16$ and $b^2 = 20$ into Equation 8, we get

$$\frac{x^2}{16} - \frac{y^2}{20} = 1$$

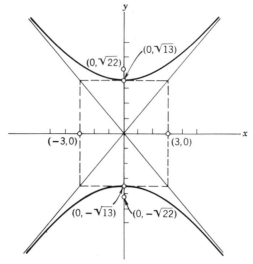

Figure 12.14

Example 11. The path of an alpha particle moving toward an atomic nucleus is described by a hyperbola. How close will an alpha particle come to a nucleus located at the origin of the coordinate system, if the equation of the path of the alpha particle is $x^2 - y^2 = 0.01$?

Solution. The vertex is the closest point on a hyperbola to its center so we want to find the distance from the center to the vertex. The equation can be written

$$\frac{x^2}{0.01} - \frac{y^2}{0.01} = 1$$

which indicates that the vertices are $(0.1, 0)$ and $(-0.1, 0)$. That is, the alpha particle comes within 0.1 units of the nucleus. It is interesting to note that the transverse and conjugate axes have the same length in this problem. Hyperbolas of this type are called *equilateral* hyperbolas. Obviously, the asymptotes have slopes of ± 1.

EXERCISES

In Exercises 1 to 10, discuss the properties of the graph of each of the given equations and then sketch the graph.

1. $x^2 - y^2 = 16$ 4. $9x^2 - y^2 = 9$
2. $y^2 - x^2 = 9$ 5. $4y^2 - 25x^2 = 100$
3. $4x^2 - 9y^2 = 36$ 6. $3x^2 - 3y^2 = 9$

7. $4x^2 - 16y^2 = 25$ **9.** $y^2 + 1 = x^2$
8. $4y^2 - x^2 = 9$ **10.** $x^2 - 25 = 5y^2$

In Exercises 11 to 20, find the equation of the hyperbolas having the given properties. Sketch each curve.

11. Vertices at $(\pm 4, 0)$, foci at $(\pm 5, 0)$
12. Vertices at $(0, \pm 5)$, foci at $(0, \pm 3)$
13. Conjugate axis 4, vertices at $(0, \pm 1)$
14. Conjugate axis 1, vertices at $(\pm 4, 0)$
15. Transverse axis 6, foci at $(\pm \frac{7}{2}, 0)$
16. Transverse axis 3, foci at $(\pm 2, 0)$
17. Vertices at $(0, \pm 4)$, asymptotes $y = \pm (1/2)x$
18. Vertices at $(\pm 3, 0)$, asymptotes $y = \pm 2x$
19. Vertices at $(0, \pm 3)$, one point of the curve $(2, 7)$
20. Vertices at $(\pm 3, 0)$, one point of the curve $(7, 2)$

12.6 TRANSLATION OF AXES

The equation of a circle centered at the origin has the form $x^2 + y^2 = r^2$. It is sometimes desirable to be able to write the equation of a circle centered at some point in the plane other than the origin. This can be done by writing the equation of the circle with respect to a new pair of axes that are parallel to the original axes. The process of changing from one pair of axes to another in this manner is called a *translation of axes*. In this section, we shall investigate the effect of a translation of axes on the form of the standard equations of the conic sections (Figure 12.15).

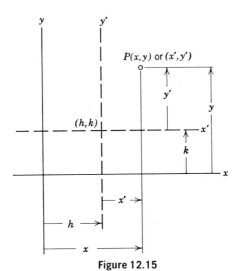

Figure 12.15

Consider a point $P(x,y)$ in the xy-coordinate plane. Under a translation of axes the coordinates of P are changed to agree with a new pair of axes. To see how this happens, let us construct an x'-axis parallel to the x-axis and a y'-axis parallel to the y-axis as shown in Figure 12.15. Let the intersection of the new axes be at the point (h,k) in the original coordinate system. Points referred to the x'-axis and y'-axis will be designated by a prime $(')$ notation. Thus the coordinates of the point P, with respect to the translated axes, are (x',y'). In order to transform (x,y) into (x',y'), the relation between the new and old axes must be known. From Figure 12.15, we see that the two coordinate systems are related by the equations:

$$x = x' + h$$
$$y = y' + k \tag{11}$$

or

$$x' = x - h$$
$$y' = y - k \tag{12}$$

These equations are called the equations for the translation of axes.

Now let us consider a circle with its center at (h,k) in the xy-coordinate plane as shown in Figure 12.16. If (h,k) is taken as the intersection of the $x'y'$-axes, the equation of the circle referred to these axes can be written in the form

$$(x')^2 + (y')^2 = r^2$$

In order to express this equation in xy-coordinates, we apply Equation 12.

Thus
$$(x-h)^2 + (y-k)^2 = r^2 \tag{13}$$

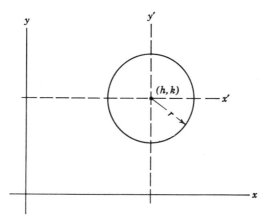

Figure 12.16

is the standard equation of a circle of radius r centered at a point (h,k).

The procedure used in obtaining the standard form of the equation of a circle centered at a point (h, k) can be extended to the other conic sections when they have been displaced from the origin.

The equation of the parabola, with vertex at the origin having a vertical axis, is given by $x^2 = 4ay$. Hence, the equation of the same parabola with its vertex at (h,k) can be written in terms of the $x'y'$-coordinate plane as

$$(x')^2 = 4ay'$$

or. using the translating Equation 12 as

$$(x-h)^2 = 4a(y-k) \tag{14}$$

Similarly, the parabola with a horizontal axis and vertex at (h, k) is given by

$$(y-k)^2 = 4a(x-h) \tag{15}$$

By a similar procedure, the standard equations of an ellipse centered at a point (h, k) are seen to be

$$\frac{(x-h)^2}{a^2} + \frac{(y-k)^2}{b^2} = 1 \tag{16}$$

if the major axis is horizontal, and

$$\frac{(x-h)^2}{b^2} + \frac{(y-k)^2}{a^2} = 1 \tag{17}$$

if it is vertical.

Finally, the standard equations of the hyperbola centered at a point (h,k) are seen to be

$$\frac{(x-h)^2}{a^2} - \frac{(y-k)^2}{b^2} = 1 \tag{18}$$

if the transverse axis is horizontal, and

$$\frac{(y-k)^2}{a^2} - \frac{(x-h)^2}{b^2} = 1 \tag{19}$$

if it is vertical.

Example 12. Write the equation of the ellipse centered at $(2, -3)$ with horizontal axis 10 units and vertical axis 4 units (Figure 12.17).

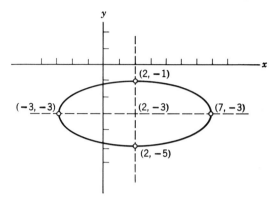

Figure 12.17

Solution.   Since the longer of the two axes is the horizontal axis, we conclude that Equation 16 should be used with $a=5$ and $b=2$. Also. $h=2$ and $k=-3$. Making these substitutions, we have as the equation of the ellipse

$$\frac{(x-2)^2}{25}+\frac{(y+3)^2}{4}=1$$

Example 13.   Write the equation of the parabola whose directrix is the line $y=-2$ and whose vertex is located at $(3, 1)$. (See Figure 12.18.)

Solution.   The parabola has a vertical axis and opens upward since the directrix is horizontal and lies below the vertex. The vertex lies midway between the focus and the directrix so that $a=3$. Letting $h=3, k=1$,

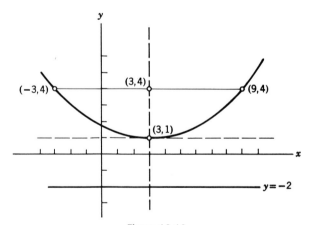

Figure 12.18

and $a = 3$ in Equation 14. the equation of the parabola is

$$(x-3)^2 = 12(y-1)$$

Example 14.   Discuss and sketch the graph of the hyperbola

$$\frac{(x+5)^2}{16} - \frac{(y-2)^2}{9} = 1$$

Solution.   The hyperbola is centered at $(-5, 2)$. By Equation 18 it has a horizontal transverse axis with vertices at $(-9, 2)$ and $(-1, 2)$. The end-points of the conjugate axis are located at $(-5, 5)$ and $(-5, -1)$. Also the foci are $(-10, 2)$ and $(0, 2)$ since $c = \sqrt{9 + 16} = 5$. The graph appears in Figure 12.19.

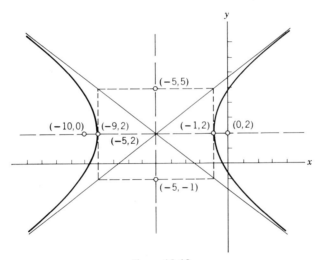

Figure 12.19

EXERCISES

In Exercises 1 to 6, write the equations of the *parabolas* having the given properties. Sketch each graph.

1. Vertex at $(3, 1)$, focus at $(5, 1)$
2. Vertex at $(-2, 3)$, focus at $(-2, 0)$
3. Directrix $y = 2$, vertex at $(1, -1)$
4. Directrix $x = -1$. vertex at $(0, 4)$

5. End points of right chord $(2, 4)$ and $(2, 0)$, opening to the right
6. End points of right chord $(-1, -1)$ and $(5, -1)$, opening upward

In Exercises 7 to 12, write the equations of the *ellipses* having the given properties. Sketch each graph.

7. Major axis 8, foci at $(5, 1)$ and $(-1, 1)$
8. Minor axis 6, vertices at $(2, -1)$ and $(10, -1)$
9. Minor axis 2, vertices at $(\frac{1}{2}, 0)$ and $(\frac{1}{2}, -8)$
10. Semimajor axis 3/2, foci at $(1, 1)$ and $(1, -1)$
11. Vertices at $(-6, 3)$ and $(-2, 3)$, foci at $(-5, 3)$ and $(-3, 3)$
12. Center at $(1, -3)$, major axis 10, minor axis 6, vertical axis

In Exercises 13 to 18, write the equations of the *hyperbolas* having the given properties. Sketch each graph.

13. Center at $(-1, 2)$, transverse axis 7, conjugate axis 8, vertical axis
14. Center at $(3, 0)$, transverse axis 6, conjugate axis 2, horizontal axis
15. Vertices at $(5, 1)$ and $(-1, 1)$, foci at $(6, 1)$ and $(-2, 1)$
16. Vertices at $(2, \pm 4)$, conjugate axis 2
17. Vertices at $(-4, -2)$ and $(0, -2)$, asymptotes $m = \pm \frac{1}{2}$
18. Vertices at $(3, 3)$ and $(5, 3)$, asymptotes $m = \pm 3$

Write the equation of the indicated family of curves.

19. Circles with center on the x-axis
20. Parabolas with vertical axis and vertex on the x-axis
21. Parabolas with vertex and focus on the x-axis
22. Ellipses with center on the y-axis and horizontal major axis
23. Circles passing through the origin with center on the x-axis
24. Circles tangent to the x-axis

12.7 THE GENERAL SECOND DEGREE EQUATION

Each of the conic sections presented in this chapter can be described by a quadratic equation of the form

$$Ax^2 + By^2 + Cx + Dy + E = 0 \qquad (20)$$

where A, B, C, D, and E are constants. This is accomplished by expanding the standard form of the equation of the conic. For instance, the standard form of a circle is

$$(x - h)^2 + (y - k)^2 = r^2$$

If we expand this equation, we obtain

$$x^2 - 2hx + y^2 - 2ky + h^2 + k^2 - r^2 = 0$$

By substituting $C = -2h$, $D = -2k$, and $E = h^2 + k^2 - r^2$, this equation can be written

$$x^2 + y^2 + Cx + Dy + E = 0 \qquad (21)$$

Formula 21 is called the *general form* of the equation of a circle. Comparing (21) with (20), we see that (20) represents a *circle* when $A = B = 1$.

Expanding (16) and (17), we find that (20) represents an *ellipse* when $A \neq B$ but A and B have the same numerical sign.

Similarly, (20) represents a *hyperbola* when A and B have different numerical signs.

Finally, (20) represents a *parabola* when either $A = 0$ or $B = 0$, but not both.

If the equation of a conic is given in general form, it can be reduced to one of the standard forms by completing the square on x and y. Examples of the reduction of equations of the form (20) to standard form are given below.

Example 15. Discuss and sketch the graph of $9x^2 + 4y^2 - 72x - 8y + 144 = 0$.

Solution. This equation can be put into standard form by rearranging the terms and completing the square on the x terms and the y terms. Thus, the given equation may be rewritten in the form

$$9(x^2 - 8x \qquad) + 4(y^2 - 2y \qquad) = -144$$

Completing the square,

$$9(x^2 - 8x + 16) + 4(y^2 - 2y + 1) = -144 + 144 + 4$$
$$9(x - 4)^2 + 4(y - 1)^2 = 4$$

$$\frac{9(x - 4)^2}{4} + \frac{4(y - 1)^2}{4} = 1$$

$$\frac{(x - 4)^2}{4/9} + \frac{(y - 1)^2}{1} = 1$$

We conclude from this equation that the center of the ellipse is at $(4, 1)$, the major axis is parallel to the y-axis, the semimajor axis is $a = 1$, and the semiminor axis is $b = \frac{2}{3}$. From $c^2 = a^2 - b^2$, the focal distance is $c = \frac{\sqrt{5}}{3}$. The ellipse shown in Figure 12.20 is drawn with the aid of this information.

Example 16. Discuss and sketch the graph of

$$x^2 - 4y^2 + 6x + 24y - 43 = 0$$

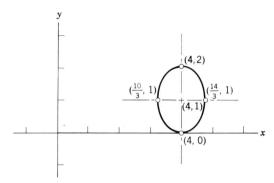

Figure 12.20

Solution.   This is the equation of a hyperbola since the coefficients of the x^2 and y^2 terms have unlike signs. In order to sketch the hyperbola, we must reduce the given equation to standard form by rearranging the terms and completing the square on the x-terms and the y-terms. Thus

$$x^2 - 4y^2 + 6x + 24y - 43 = 0$$

may be written

$$(x^2 + 6x \quad) - 4(y^2 - 6y \quad) = 43$$

Completing the square on each variable, we have

$$(x^2 + 6x + 9) - 4(y^2 - 6y + 9) = 43 + 9 - 36$$

$$(x+3)^2 - 4(y-3)^2 = 16$$

$$\frac{(x+3)^2}{16} - \frac{(y-3)^2}{4} = 1$$

The center of the hyperbola is the point $(-3, 3)$. The transverse axis is horizontal with vertices at $(1, 3)$ and $(-7, 3)$. The end points of the conjugate axis are located at $(-3, 5)$ and $(-3, 1)$. Finally, the foci are at $(-3 + 2\sqrt{5}, 3)$ and $(-3 - 2\sqrt{5}, 3)$, since $c = \sqrt{16 + 4} = 2\sqrt{5}$. The graph appears in Figure 12.21.

Example 17.   Discuss and sketch the graph of

$$2y^2 + 3x - 8y + 9 = 0$$

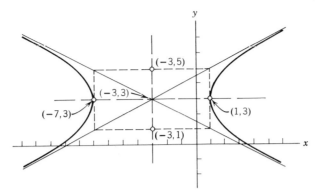

Figure 12.21

Solution.   By completing the square on the y variable this equation can be reduced to form (15). Thus

$$2y^2 + 3x - 8y + 9 = 0$$

$$2(y^2 - 4y \quad) = -3x - 9$$

$$2(y^2 - 4y + 4) = -3x - 9 + 8$$

$$2(y - 2)^2 = -3x - 1$$

$$2(y - 2)^2 = -3(x + \tfrac{1}{3})$$

$$(y - 2)^2 = -\frac{3}{2}\left(x + \frac{1}{3}\right)$$

This is the standard form of the equation of a parabola with horizontal axis and vertex at $(-1/3, 2)$. We see that $4a = -3/2$. so $a = -3/8$. There-fore. the focus is at $(-17/24, 2)$ and the end-points of the right chord are $(-17/24, 11/4)$ and $(-17/24, 5/4)$. The parabola. which opens to the left. is shown in Figure 12.22.

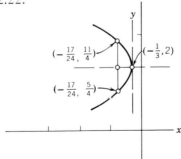

Figure 12.22

EXERCISES

Transform each of the following equations into standard form and sketch the graph.

1. $x^2 + y^2 + 4x + 6y + 4 = 0$
2. $x^2 - 2x - 8y + 25 = 0$
3. $9x^2 + 4y^2 + 18x + 8y - 23 = 0$
4. $2x^2 + 2y^2 - 4x - 16 = 0$
5. $x^2 - y^2 - 4x - 21 = 0$
6. $y^2 + 4y + 6x - 8 = 0$
7. $x^2 - 6x - 3y = 0$
8. $x^2 + 4y^2 + 8x = 0$

9. $2y^2 - 2y + x - 1 = 0$
10. $3x^2 + 4y^2 - 18x + 8y + 19 = 0$
11. $4x^2 - y^2 + 8x + 2y - 1 = 0$
12. $y^2 - 2x - 4y + 10 = 0$
13. $9x^2 + 4y^2 - 18x + 16y - 11 = 0$
14. $y^2 - 25x^2 + 50x - 50 = 0$
15. $x^2 - 2y^2 + 2y = 0$
16. $y^2 - y - 1/2x + 1/4 = 0$

13

Differential Equations

13.1 INTRODUCTION

The study of science and engineering is, to a large extent, concerned with rates of change. For this reason, equations involving derivatives are of considerable importance. An equation that involves at least one derivative of a function is called a *differential equation*. Usually the function is the unknown and it is our job to learn how to find the function or functions that satisfy a given differential equation. We shall limit our discussion to equations involving only first or second derivatives of a function.

We note in passing that it would probably be more correct to call a differential equation a *derivative* equation. Popular usage, however, allows us to consider the two equations

$$x\frac{dy}{dx} + y = 0$$

and

$$x\,dy + y\,dx = 0$$

to mean the same thing and, hence, we see the reason for the terminology "differential equation."

13.2 DIFFERENTIAL EQUATIONS OF FAMILIES OF CURVES

A set of curves obtained from a relation having one arbitrary constant is called a *one* parameter family of curves. Thus, the relation $y^2 = 4ax$ represents the family of parabolas with vertex at the origin and focus on the x-axis, as shown in Figure 13.1. One curve in the family is generated for each value of the arbitrary constant a.

The slope of any curve in this family can be found by differentiating $y^2 = 4ax$ with respect to x. If the differentiation is carried out so that we eliminate the arbitrary constant a in the equation, the result is called the *differential equation of the family of curves*.

To find the differential equation of a family of curves, we proceed as follows:

(1) Write the equation of the family if it is not already known.

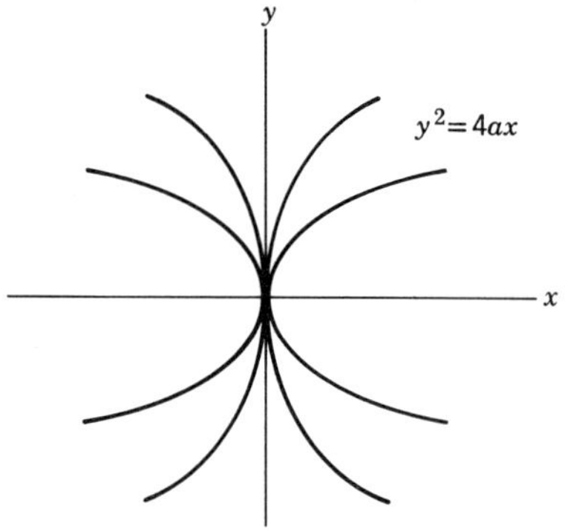

Figure 13.1

(2) Find an equation for dy/dx (involving usually y and x).

(3) Use the results in (1) and (2) to eliminate the arbitrary constant.

Example 1.   Find the differential equation of the family of curves in Figure 13.1.

Solution.   By differentiating $y^2 = 4ax$ implicitly with respect to x, we obtain

$$2yy' = 4a \qquad \text{or} \qquad a = \frac{yy'}{2}$$

We now substitute this value for a into the given equation to get

$$y^2 = 4 \left[\frac{yy'}{2} \right] x$$

which becomes

$$y^2 - 2xyy' = 0$$

This equation may also be written in terms of dx and dy as

$$y^2 dx - 2xy\, dy = 0$$

Essentially the same technique may be used to find the differential equation of *two* parameter families, that is, families involving two arbitrary constants. In this case, it is usually necessary to differentiate the given relation twice in order to eliminate the arbitrary constants. The differential equation of a two-parameter family will then involve y''.

Example 2.   Find the differential equation of the family of circles with center on the x-axis. (See Figure 13.2.)

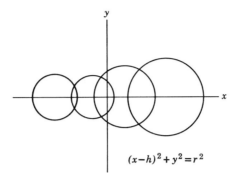

$$(x-h)^2+y^2=r^2$$

Figure 13.2

Solution.   The equation of this family is

$$(x-h)^2+y^2 = r^2$$

which has two constants (h and r) to be eliminated. The constant r is eliminated when we take the first derivative. Thus

$$2(x-h)+2yy' = 0$$

which may be written

$$yy'+x = h$$

Differentiating this equation, the constant h is eliminated and we have

$$yy''+(y')^2+1 = 0$$

as the differential equation of the indicated family of circles.

Example 3.   Find the differential equation of the family of parabolas with axis parallel to the x-axis having a fixed focal distance. (See Figure 13.3.)

Solution.   The equation of this family is

$$(y-k)^2 = 4a(x-h)$$

In this problem we seek to eliminate only the constants h and k, since the focal distance a is fixed. Differentiating the equation of the family with respect to x, we get

$$2(y-k)y' = 4a$$

or

$$(y-k)y' = 2a \qquad (1)$$

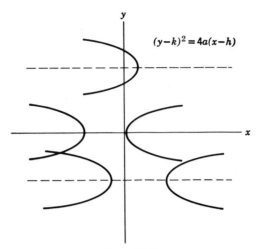

Figure 13.3

Differentiating again,

$$(y-k)y''+y'\cdot y' = 0$$

or

$$(y-k)y''+(y')^2 = 0 \tag{2}$$

We have already eliminated h from Equation (1) and (2) so all we need to do is eliminate k from these two. Solving (1) for k, we have

$$k = y - \frac{2a}{y'}$$

Inserting this into Equation (2), yields

$$\left(y-y+\frac{2a}{y'}\right)y''+(y')^2 = 0$$

or

$$(y')^3+2ay'' = 0$$

EXERCISES

Determine the differential equation of the indicated families of curves by eliminating the arbitrary constants. Sketch a few curves of each family.

1. Straight lines through the origin.
2. Straight lines with a y-intercept of 10.

3. Straight lines with a y-intercept equal to the slope.
4. Circles with center at origin.
5. Circles with center on the y-axis having a fixed radius.
6. Circles with center on the x-axis having a fixed radius.
7. Hyperbolas with center at the origin and with transverse axis equal to the conjugate axis and vertices on the x-axis.
8. Parabolas with vertex at the origin having a vertical axis.
9. Parabolas with vertex on the x-axis having a vertical axis and a fixed focal distance.
10. Parabolas with vertex on the y-axis having a horizontal axis and a fixed focal distance.
11. Parabolas with vertical axis and a fixed focal distance.
12. Parabolas with vertex and focus on the x-axis.
13. Parabolas with vertex and focus on the y-axis.
14. Circles with center on the y-axis.
15. Circles tangent to the x-axis having a fixed radius.

13.3 SOLUTIONS TO DIFFERENTIAL EQUATIONS

By a solution to a differential equation we mean a function which satisfies the differential equation over some interval. If $y = f(x)$ is a solution to a given differential equation, then the differential equation will be reduced to an identity if y, and the derivatives of y with respect to x, are replaced by $f(x)$ and its respective derivatives. We begin our discussion of solutions of differential equations by showing how the validity of a solution can be checked. This exercise is helpful in both understanding what is meant by a solution and demonstrating the kinds of solutions that can be expected.

Example 4. Show that $y = \sin x$ satisfies the differential equation $y'' + y = 0$.

Solution. Since $y = \sin x$, we have

$$y' = \cos x$$
$$y'' = -\sin x$$

Substituting these values into the given differential equation yields the identity

$$y'' + y = -\sin x + \sin x = 0$$

Hence, $y = \sin x$ is a solution.

Example 5. Show $y_1 = 1 + 5e^{-3x}$ is a solution to $y'' + 3y' = 0$.

Solution.
$$y_1 = 1 + 5e^{-3x}$$
$$y_1' = -15e^{-3x}$$
$$y_1'' = 45e^{-3x}$$

When we substitute y_1'' and y_1' for y'' and y' in the differential equation, we get

$$45e^{-3x} + 3(-15e^{-3x}) = 0$$

which is identically 0 and, hence, $1 + 5e^{-3x}$ is a solution.

Example 6.   Show that $xy + \ln y = c$ defines a function $y(x)$ that is implicitly a solution to $(xu + 1)u' + u^2 = 0$.

Solution.   We differentiate implicitly to find y'. Thus,

$$xy' + y + \frac{y'}{y} = 0$$

or

$$xyy' + y^2 + y' = 0$$

Rearranging the terms, we find that $y'(xy + 1) + y^2 = 0$ and, therefore, y and y' satisfy the given differential equation.

EXERCISES

In each of the following problems, show that the equation on the right is a solution of the corresponding differential equation.

Differential Equation	Solution
1. $\dfrac{dy}{dx} - 2x = 0$	$y = x^2 + c$
2. $\dfrac{dy}{dx} = 3x^2 + 2$	$y = x^3 + 2x + c$
3. $t\dfrac{ds}{dt} - 2s = 0$	$s = Kt^2$
4. $x\dfrac{dy}{dx} - 2y = -x$	$y = x + cx^2$
5. $\dfrac{dV}{dr} + V = \cos r - \sin r$	$V = \cos r + ce^{-r}$
6. $\dfrac{dy}{dx} - 2y = 3e^{2x}$	$y = (3x + C)e^{2x}$
7. $\dfrac{d^2y}{dx^2} - \dfrac{dy}{dx} - 2y = 0$	$y = Ae^{-x} + Be^{2x}$

8. $\dfrac{d^2s}{dt^2} + 16s = 0$ $s = C_1 \sin 4t + C_2 \cos 4t$

9. $\dfrac{d^3y}{dx^3} + 9\dfrac{dy}{dx} = 0$ $y = A + B \cos 3x + C \sin 3x$

10. $\dfrac{d^2s}{dt^2} - s = 1$ $s = C_1 e^t + C_2 e^{-t} - 1$

11. $\dfrac{d^2y}{dx^2} + \dfrac{dy}{dx} = -\cos x$ $y = c_1 + c_2 e^{-x} + \tfrac{1}{2}\cos x - \tfrac{1}{2}\sin x$

12. $\dfrac{d^2i}{dt^2} - \dfrac{di}{dt} - 2i = 4t$ $i = c_1 e^{-t} + c_2 e^{2t} + 1 - 2t$

13. $\dfrac{d^2v}{dt^2} = 9 \sin 3t$ $v = -\sin 3t + At + B$

14. $y'' - 4y' + 4y = 0$ $y = c_1 e^{2x} + c_2 x e^{2x}$

15. $\dfrac{d^3s}{dt^3} + 9\dfrac{ds}{dt} = 0$ $s = c_1 + c_2 \sin 3t + c_3 \cos 3t$

16. $\dfrac{d^3i}{dt^3} + 4\dfrac{di}{dt} = 15 \sin 3t$ $i = A + B \cos 2t + C \sin 2t + \cos 3t$

17. $(1 + ye^x)\dfrac{dy}{dx} + y^2 e^x = 0$ $\ln y + ye^x = C$

18. $(x + \cos y)y' + y = 0$ $xy + \sin y = K$

13.4 DIFFERENTIAL EQUATIONS OF THE FORM $y' = f(x)$

In the last section we showed how to verify that a particular function was a solution to a differential equation; however, we gave no clues as to how to find a solution. We are now ready to consider this problem. The technique of solving general differential equations is very difficult, but there are a few special techniques that we can study. Fortunately, these techniques apply to the more commonly occurring differential equations.

You have already been exposed to some rather elementary differential equations of the form $y' = f(x)$. In this case. we found that the solution is

$$y = \int f(x)\, dx + C$$

Example 7. Solve the differential equation

$$\frac{dy}{dx} = 3x^2$$

Solution. By antidifferentiation, we obtain

$$y = x^3 + C$$

as a solution to the given differential equation. Notice there are many solutions to this differential equation, each differing by a constant. We say we have a *family* of solutions since any member of the family satisfies the differential equation.

Example 8.   Solve $d^2y/dx^2 = \cos x$.

Solution.   Here we antidifferentiate once to obtain

$$\frac{dy}{dx} = \sin x + C$$

and then again to get the solution

$$y = -\cos x + Cx + C_1$$

We see that in this case we have a *two* parameter family of solutions.

The preceding examples are illustrative of the problems we will be dealing with in that they demonstrate the fact that *the number of arbitrary constants in a solution will always be equal to the highest order derivative appearing in the differential equation.*

13.5   DIFFERENTIAL EQUATIONS OF THE FORM $F(x)\,dx + G(y)\,dy = 0$

A differential equation which can be put into the form

$$F(x)\,dx + G(y)\,dy = 0 \tag{3}$$

is said to be a *separable* equation. Separable equations can be solved by finding the antiderivative of each term. Thus, the solution of (3) is given by

$$\int F(x)\,dx + \int G(y)\,dy = C$$

In most cases, the equation is not immediately in form (3) and it is necessary to perform some algebraic manipulation to actually separate the variables. Never try to apply the separable technique unless the coefficient of dx is a function of x only and that of dy is a function of y only.

Example 9.   Solve the equation $3(y^2 + 1)\,dx + 2xy\,dy = 0$

Solution.   The variables of this equation can be separated by dividing each term by $x(y^2 + 1)$. Performing this division, we get

$$\frac{3dx}{x} + \frac{2y\,dy}{y^2 + 1} = 0$$

Antidifferentiating each term, the solution is

$$3 \ln x + \ln (y^2 + 1) = C \tag{4}$$

The solution obtained in (4) is correct, but it can be written more compactly by using the properties of logarithms. Hence

$$\ln x^3 + \ln (y^2 + 1) = C$$
$$\ln x^3 (y^2 + 1) = C \tag{5}$$

is a more compact solution. An additional simplification can be made on Equation 5 by writing the arbitrary constant in the form $C = \ln C_1$, so that

$$\ln x^3 (y^2 + 1) = \ln C_1$$

and

$$x^3 (y^2 + 1) = C_1$$

This form is desirable because of its compactness, but it must be pointed out that any of the above forms are proper solutions to the given differential equation. In most cases, the form of the solution is a matter of individual preference.

Example 10.   Solve the equation $(1 - \cos \theta) \, dr = -r \sin \theta \, d\theta$.

Solution.   Separation of variables is accomplished in this equation by dividing by $r(1 - \cos \theta)$. Thus

$$\frac{dr}{r} + \frac{\sin \theta \, d\theta}{1 - \cos \theta} = 0$$

Antidifferentiating, we obtain as the desired solution

$$\ln r + \ln (1 - \cos \theta) = \ln C$$
$$\ln r (1 - \cos \theta) = \ln C$$
$$r (1 - \cos \theta) = C$$

In many applied problems we are not interested in the entire family of solutions, but in one *particular* solution. Usually we are given additional information, such as an *initial* or *boundary* condition, to enable us to pick one function from the entire family of solutions.

Example 11.   Find the particular solution for $dy/dx = xy^2 e^x$ given that $y = 2$ when $x = 0$.

Solution.   Separating variables, we have

$$\frac{dy}{y^2} = xe^x \, dx$$

Antidifferentiation of the first term is immediate, but the formula for integration by parts must be used to antidifferentiate the second term. (See Exercise 4, Section 11.4.) Thus

$$-\frac{1}{y} - xe^x + e^x = C$$

is the equation for the family of solutions. Letting $x=0$ and $y=2$ in this equation, we have

$$-\frac{1}{2} - 0 + 1 = C$$

and

$$C = \frac{1}{2}$$

The particular solution satisfying the given boundary conditions is then

$$-\frac{1}{y} - xe^x + e^x = \frac{1}{2}$$

or

$$y = \frac{2}{2e^x - 2xe^x - 1}$$

Separable differential equations arise in many of the physical laws used in science and technology. For instance, quantities that change at a rate which is proportional to the quantity itself are described by differential equations of this type.

Example 12.   When the switch is closed in the RC circuit in Figure 13.4, there is an initial charging current (transient current) in the capacitor. The transient current i in a capacitor changes at a rate equal to $(-1/RC)i$ where RC is the time constant of the circuit. Find the expression for the current in the capacitor at any time t if $i = V/R$ when $t=0$.

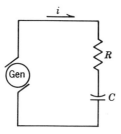

Figure 13.4

Solution.   The rate of change of the transient current is given by

$$\frac{di}{dt} = -\frac{1}{RC}\,i$$

Separating variables

$$\frac{di}{i} = -\frac{1}{RC}\,dt$$

Hence

$$\ln i + \ln K = -\frac{t}{RC}$$

$$\ln iK = -\frac{t}{RC}$$

$$i = K_1 e^{-t/RC}$$

Letting $t = 0$ and $i = V/R$ in this equation, we find $K_1 = V/R$ and, consequently,

$$i = \frac{V}{R}\,e^{-t/RC}$$

is the equation for the transient current.

Example 13.   A radioactive isotope decomposes at a rate proportional to its mass. Find the expression for the amount of the isotope remaining at any time if the initial mass is 100 mg. and the mass 2 years later is 75 mg.

Solution.   Let M be the mass of the isotope remaining after t years. Then the rate of decomposition is given by

$$\frac{dM}{dt} = -kM$$

Antidifferentiating and adding a constant in the form $\ln C$, we find

$$\ln M + \ln C = -kt$$

Simplifying,

$$\ln MC = -kt$$
$$MC = e^{-kt}$$
$$M = C_1 e^{-kt} \tag{6}$$

To find the value of C_1 we recall that $M = 100$ when $t = 0$. Substituting these values into Equation 6, we get $C_1 = 100$ and

$$M = 100\,e^{-kt} \tag{7}$$

The value of k may now be determined by substituting the boundary condition $t = 2$, $M = 75$ into Equation 7. Thus

$$75 = 100\, e^{-2k}$$

or

$$e^{-2k} = \frac{3}{4}$$

so

$$k = \frac{1}{2}\ln\frac{4}{3} = \frac{1}{2}(0.2852) = 0.1426$$

The mass of the isotope remaining after t years is then given by

$$M = 100\, e^{-0.14t}$$

EXERCISES

Obtain the family of solutions of each of the given differential equations.

1. $2y\, dx + 2x\, dy = 0$

2. $x^2\, dy + y^2\, dx = 0$

3. $xy\, dx + y^2\, dy = 0$

4. $xy\, dx + (1+x^2)\, dy = 0$

5. $\dfrac{ds}{dt} = ts^3$

6. $y\, dx + (3x+2)\, dy = 0$

7. $x(1-y)\, dx + (x^2-2)\, dy = 0$

8. $\cot y\, dx = x\, dy$

9. $dx + (1+x^2)y\, dy = 0$

10. $\dfrac{dp}{dt} = 2 - p$

11. $xy\, dx + e^x\, dy = 0$

12. $\dfrac{dq}{dt} + \dfrac{t\sin t}{q} = 0$

13. $dx + e^{x+y}\, dy = 0$

14. $v(t+3)\, dv + (v^2 - 2v - 3)\, dt = 0$

15. $(x^2-1)\, dy + x(y^2+5y+4)\, dx = 0$

16. $e^x\cos^2 y\, dx - (1+e^x)\, dy = 0$

In each of the following exercises, obtain the particular solution of the differential equation for the given boundary conditions.

17. $(y+2)\, dx + (x-4)\, dy = 0;$ $x = 2, y = 0$

18. $\dfrac{ds}{dt} = -\dfrac{t}{s};$ $t = -1, s = 3$

19. $xy\, dx + \sqrt{9+x^2}\, dy = 0;$ $x = 4, y = 1$

20. $\cos y\, dx + x \sin y\, dy = 0;$ $x = 3, y = \pi/3$
21. $(1 + x^2)\, dy + (1 + y^2)\, dx = 0;$ $x = 1, y = \sqrt{3}$
22. $dy + y\, dx = x\, dy;$ $x = 2, y = e$
23. The intensity of light I emitted by the phosphor of a television tube decreased at a rate proportional to the intensity I. Find the expression for the intensity of light being emitted at any time t if the initial intensity is $I = 25$ and the intensity at $t = 1$ is $I = 20$.
24. Newton's law of cooling states that the difference T in temperature between a body and its surrounding medium changes at a rate which is proportional to T. Find the expression for T at any time t if $T = 200$ deg when $t = 0$, and $T = 150$ deg when $t = 20$ sec.
25. A certain radioactive substance decays at a rate which is always equal to 25 percent of the mass of the substance. Find the formula for the mass of the substance remaining after t hours if $M = 200$ g when $t = 0$.
26. The transient current in a series RL circuit decreases at a rate equal to $-Ri/L$. Find the expression for the transient current if $i = V/R$ when $t = 0$.
27. The rate of change of atmospheric pressure P with altitude h is given by $dP/dh = -kP$, where k is a constant of proportionality. If the atmospheric pressure at ground level is 15 psi and at 10,000 ft is 10 psi, find the expression for the pressure as a function of altitude. What is the pressure at 3,000 ft?
28. It is found in chemistry that simple chemical conversions take place at a rate proportional to the amount of unconverted substance remaining. If initially there was 50 g of a substance and 1 hr later 20 g remained, how long will it take to convert 80 percent of the substance?
29. During an experiment, an object of mass m is moved through a resisting medium by a constant force F. The differential equation of this object is found to be

$$m \frac{dv}{dt} = F - kv$$

where k is a constant related to the resisting medium. Find the expression for the velocity of an object starting from rest if $m = 10, F = 150$, and $k = 1$.

13.6   DIFFERENTIAL EQUATIONS OF THE FORM $dy + P(x)y\, dx = Q(x)\, dx$

Differential equations which can be put into the form

$$dy + P(x)y\, dx = Q(x)\, dx \qquad (8)$$

are perhaps the most important ones that occur in applications. We choose not to give the general theory governing the solution to this equation but we will show the technique.

Equations such as (8) may be solved by multiplying the equation by an appropriate function of x called an *integrating factor*. The purpose of the integrating factor is to produce a function on each side of the equation for which we can find an antiderivative. The integrating factor that we shall use is the function

$$R(x) = e^{\int P(x)dx} \tag{9}$$

This factor makes the left-hand side of (8) an "exact" differential; that is, the left-hand side becomes

$$d(ye^{\int P(x)dx})$$

Therefore, all we must do is to find

$$\int Q(x)e^{\int P(x)dx}\, dx$$

The technique for solving differential equations that can be put into form (8) is now presented in summary form.

(1) By algebraic manipulation, put the differential equation into the form $dy + P(x)y\, dx = Q(x)\, dx$. Note that the coefficient of dy is 1.
(2) Identify $P(x)$ and compute the integrating factor $R(x)$ by Equation 9.
(3) Multiply both sides of the equation by $R(x)$.
(4) Antidifferentiate both sides and solve for y.

Example 14.   Solve $2(dy/dx) - 4y = 16e^x$.

Solution.   Applying the four-step procedure outlined above, we write the given equation as

$$dy - 2y\, dx = 8e^x\, dx \tag{10}$$

We see from this equation that $P(x) = -2$ and, therefore the integrating factor is

$$R(x) = e^{-\int 2dx} = e^{-2x}$$

Multiplying Equation 10 by e^{-2x}, we obtain

$$e^{-2x}dy - 2ye^{-2x}dx = 8e^{-x}dx$$

The left-hand side of this equation is the expression for the differential of the product ye^{-2x}, so we may write

$$d(ye^{-2x}) = 8e^{-x}dx$$

Now antidifferentiating both sides, we have the solution

$$ye^{-2x} = -8e^{-x} + C$$

or

$$y = -8e^x + Ce^{2x}$$

Example 15.   Solve $x(dy/dx) + 2y = (1/x) \cos x$.

Solution.   Writing this equation in standard form yields

$$dy + \frac{2}{x} y \, dx = \frac{1}{x^2} \cos x \, dx \tag{11}$$

Here, $P = 2/x$ and

$$R = e^{\int 2dx/x} = e^{2 \ln x} = x^2$$

You may wonder how $e^{2 \ln x} = x^2$. Since integrating factors of this type are fairly common, we will demonstrate the algebra involved. Let

$$R = e^{2 \ln x} = e^{\ln x^2}$$

Taking the natural logarithm of both sides, yields

$$\ln R = \ln e^{\ln x^2} = \ln x^2 \cdot \ln e$$

from which

$$\ln R = \ln x^2$$

and, therefore

$$R = x^2$$

Applying the integrating factor to Equation 11, we get

$$x^2 \, dy + 2xy \, dx = \cos x \, dx$$

The solution to this equation is then

$$x^2 y = \sin x + C$$

Example 16.   When an object of mass m moves through a resisting medium with acceleration dv/dt, the differential equation describing this motion is

$$m \frac{dv}{dt} + rv = F$$

where v is the velocity, r is a constant, and F is the force causing the motion. Find a formula for v if $F = t$.

Solution.   Substituting $F = t$ into the given equation we find

$$m\frac{dv}{dt} + rv = t$$

and, therefore,

$$dv + \frac{r}{m}v\,dt = \frac{t}{m}\,dt$$

Letting $P = r/m$, the integrating factor is

$$R = e^{\int r/m\,dt} = e^{rt/m}$$

Applying the integrating factor, we obtain the equation

$$e^{rt/m}\,dv + \frac{r}{m}e^{rt/m}\,vdt = \frac{t}{m}e^{rt/m}\,dt$$

or

$$d(ve^{rt/m}) = \frac{t}{m}e^{rt/m}\,dt$$

Using integration by parts on the right member of this equation, the solution is

$$ve^{rt/m} = \frac{t}{r}e^{rt/m} - \frac{m}{r^2}e^{rt/m} + C$$

$$v = \frac{t}{r} - \frac{m}{r^2} + Ce^{-rt/m}$$

The arbitrary constant C may readily be determined if the initial conditions of motion are known.

EXERCISES

Find the family of solutions of the following differential equations.

1. $dy + 2y\,dx = 4\,dx$

2. $2\,dy + 8xy\,dx = x\,dx$

3. $dy - 4y\,dx = 3\,dx$

4. $\dfrac{di}{dt} + i - 2 = 0$

5. $\dfrac{dy}{dx} + \dfrac{y}{x} = x + 1$

6. $x^2\, dy - 2xy\, dx = dx$

7. $(t + 2q)\, dt - dq = 0$

8. $(3x + xy)\, dx + dy = 0$

9. $(t^2 + 1)\, ds + 2ts\, dt = 3dt$

10. $(x^2 - 4)\, dy + xy\, dx = x\, dx$

11. $(x^2 + 1)\, dy - xy\, dx = 2x\, dx$

12. $dy + 2y\, dx = e^{-x}\, dx$

13. $\dfrac{dv}{dr} + v\tan r = \cos r$

14. $dy = (\csc x - y\cot x)\, dx$

15. $dy - y\, dx = 2e^{-x}\, dx$

16. $\dfrac{di}{dt} + i = e^{-t}$

17. $t\dfrac{ds}{dt} + s = t$

18. $(y\cot x - x)\, dx + dy = 0$

In each of the following problems, find the particular solution corresponding to the given boundary conditions.

19. $x\, dy - y\, dx = x^2 e^x\, dx$; $x = 1, y = 0$
20. $dy/dx + 2y = 4$; $x = 0, y = 1$
21. $dy/dx + y\cot x = -\cot x$; $x = \pi/3. y = 1$
22. $dr/d\theta + r\cot\theta = \sec\theta$; $\theta = \pi/3. r = 2$
23. Find the equation for the velocity of the object in a resisting medium if motion is produced by a force of $F = e^t$. (See Example 16.)
24. Evaluate the arbitrary constant in Exercise 23 if $v = 0$ when $t = 0$.
25. When a constant voltage V is applied to a resistance R with a capacitance C in series, the rate at which the capacitor is charged is given by

$$R\frac{dq}{dt} + \frac{q}{C} = V$$

Find the equation for the charge transferred q if the capacitor is initially discharged.
26. Using the results of Exercise 25 and the fact that $i = dq/dt$, find the equation for the current in the capacitor. Compare this with the result obtained in Example 12.
27. Referring to Exercise 25 again, show that for large values of t the charge on the plates of the capacitor is given by $q = CV$.

13.7   ORTHOGONAL TRAJECTORIES

An important application of differential equations is concerned with families of curves. Two families of curves are said to be *orthogonal* (or perpendicular) if every curve of one family intersects every curve of the other family at right angles. We can easily give the equation governing the relationship between two orthogonal families since the slopes of lines that intersect at right angles are negative reciprocals of each other. Thus, if $y = f(x)$ is one family and $y = g(x)$ is the other, then

$$f'(x) = -\frac{1}{g'(x)}$$

or using a different notation

$$\frac{dy}{dx} = -\frac{1}{dy_0/dx}$$

The family of curves orthogonal to a given set is said to be the *orthogonal trajectories* of the given family of curves. To find the orthogonal trajectories we proceed as follows.

(1) Find the differential equation of the given family by the methods of Section 13.2.
(2) Replace dy/dx by $-1/(dy_0/dx)$, and y by y_0. This is now the differential equation of the orthogonal trajectories.
(3) Solve the differential equation for the orthogonal trajectories.

One of the more important applications of finding orthogonal trajectories is in electromagnetic theory since the lines of flux of the electric field (that is, the direction of the field) are orthogonal to the equipotential curves. (See Figure 13.5.)

Example 17.   The equipotential lines of the electric field in Figure 13.5 can be shown by experimental methods to be the family of circles

$$x^2 + y^2 = r^2$$

Find the equation for the lines of flux in the field.

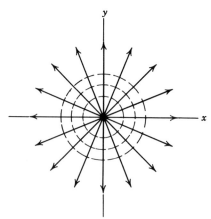

Figure 13.5

Solution.   The constant to be eliminated in the given equation is r^2. The differential equation of the equipotential curves is then

$$2x + 2y\frac{dy}{dx} = 0$$

Replacing dy/dx by $-1/(dy_0/dx)$ and y by y_0, the differential equation of the lines of flux is

$$x + y_0\left(-\frac{1}{dy_0/dx}\right) = 0$$

or

$$x\,dy_0 - y_0\,dx = 0$$

The solution of this equation is found by separating variables to be

$$\ln y_0 - \ln x = \ln m \qquad (m = \text{a constant})$$

or

$$y_0 = mx$$

That is, the lines of flux project radially from the origin.

Example 18.   Find the orthogonal trajectories of the family of parabolas with vertex at the origin and foci on the y-axis.

Solution.   The equation of the indicated family is

$$x^2 = 4a\,y$$

which can be written in the form

$$\frac{x^2}{y} = 4a$$

The differential equation of this family is then

$$x^2\frac{dy}{dx} - 2xy = 0$$

or

$$\frac{dy}{dx} = \frac{2y}{x}$$

Consequently, the differential equation of the orthogonal trajectories is

$$\frac{dy_0}{dx} = -\frac{x}{2y_0}$$

Solving by separation of variables, we have

$$\tfrac{1}{2}x^2 + y_0^2 = C$$

which is the family of ellipses shown in Figure 13.6.

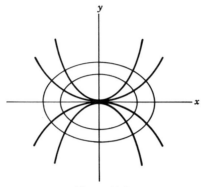

Figure 13.6

EXERCISES

Find the orthogonal trajectories of each of the following families of curves and sketch a few curves of each family.

1. Straight lines having a slope of 2.
2. Straight lines with y-intercept equal to the slope.
3. The parabolas with vertex on the x-axis having a vertical axis and a fixed focal distance.
4. The parabolas with vertex at $(0,0)$ and focus on the x-axis.
5. The hyperbolas $x^2 - y^2 = C$.
6. The cubics $y = Cx^3$.
7. The exponentials $y = Ce^x$.
8. The logarithms $y = \ln x + C$.
9. In the transfer of heat through a body, the points through which the temperature is constant are called *isotherms*. The orthogonal trajectories of the isotherms give the direction of heat flow and are consequently called lines of flow. Find the lines of flow in a long copper wedge if the isotherms are given by $y = kx$.
10. If the equipotential lines of the electric field of a single-phase induction motor are given by $c^2x^2 + y^2 = c^2$, find the equation for the lines of flux for the electric field.
11. In fluid mechanics, the streamlines of an object moving through a fluid can be considered to be orthogonal trajectories of the velocity equipotential lines of the object. If the velocity equipotential lines of the object are $2x^2 + y^2 = k^2$, what is the form of the streamlines?

13.8 DIFFERENTIAL EQUATIONS OF THE FORM $a_0 y'' + a_1 y' + a_2 y = 0$

We conclude this chapter with a discussion of differential equations that can be put in the form

$$a_0 y'' + a_1 y' + a_2 y = 0 \qquad (12)$$

where a_0, a_1, and a_2 are constants. You may wonder if such a restriction means that little or no application can be found for (12). The importance of this type of differential equation is attested to by the fact that all undriven electrical and mechanical systems are described by (12). For example, if R, L, and C are the basic constant components of a passive electrical network (Figure 13.7), then the current i in the circuit is given by

$$L\frac{d^2 i}{dt^2} + R\frac{di}{dt} + \frac{1}{C} i = 0$$

Similarly, the transverse movement y of a mechanical system (see Figure 13.8) with mass m, fluid damping C, and spring constant k is governed by

$$m\frac{d^2 y}{dt^2} + C\frac{dy}{dt} + ky = 0$$

Since Equation 12 involves the second derivative we would expect a two-parameter family of solutions. The technique of solving this equation rests with the tacit assumption that the solution is of the form

$$y = C_1 e^{m_1 x} + C_2 e^{m_2 x}$$

where m_1 and m_2 are constants to be determined. We choose to demonstrate this fact by solving a specific equation, namely,

$$\frac{d^2 y}{dx^2} - 4\frac{dy}{dx} + 3y = 0$$

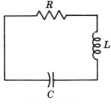

Figure 13.7

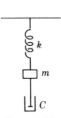

Figure 13.8

If $y = e^{mx}$ is a solution of this equation, then the left side will be identically zero when y and its derivatives are replaced with e^{mx} and its corresponding derivatives. Since

$$\frac{dy}{dx} = me^{mx}$$

and

$$\frac{d^2y}{dx^2} = m^2 e^{mx}$$

the given equation becomes

$$m^2 e^{mx} - 4me^{mx} + 3e^{mx} = 0$$

or

$$(m^2 - 4m + 3)e^{mx} = 0$$

The exponential function can never equal zero and, therefore, may be divided out of the equation to yield

$$m^2 - 4m + 3 = 0$$

which is called the *auxiliary equation*. The roots of this equation are $m = 1$ and $m = 3$ and, therefore, both $y_1 = e^x$ and $y_2 = e^{3x}$ are solutions and $y = C_1 e^x + C_2 e^{3x}$ is a solution also.

We see from this example that the solution to a differential equation of this type is ultimately dependent upon the solution to the auxiliary equation – which is purely algebraic. For differential equations of the form (12), the auxiliary equation will always be a quadratic and hence we will always be able to find its roots. The form of the solution to (12) is governed by the roots of the auxiliary equation of which there are three possibilities.

(1) The roots are real and distinct, that is, $m_1 \neq m_2$ as in the preceeding example. Then the solution is

$$y = C_1 e^{m_1 x} + C_2 e^{m_2 x}$$

(2) The roots are equal. Then the solution is

$$y = (C_1 + C_2 x)e^{mx}$$

(3) The roots are complex conjugates $m_1 = a + ib$, $m_2 = a - ib$. Then the solution is

$$y = e^{ax}(C_1 \cos bx + C_2 \sin bx)$$

We begin with several examples in which the roots of the auxiliary equation are real and distinct.

Example 19.   Solve $\dfrac{d^2 s}{dt^2} - 3\dfrac{ds}{dt} = 0$.

Solution.   The auxiliary equation is

$$m^2 - 3m = 0$$
$$m(m - 3) = 0$$

The roots of this equation are clearly $m = 0$ and $m = 3$. The general solution is then

$$s = c_1 e^0 + c_2 e^{3t} = c_1 + c_2 e^{3t}$$

Example 20.   Find the particular solution for $2(d^2 y/dx^2) + 5(dy/dx) - 3y = 0$, if $y = 1$ and $dy/dx = -2$ when $x = 0$.

Solution.   The auxiliary equation is

$$2m^2 + 5m - 3 = 0$$
$$(2m - 1)(m + 3) = 0$$

from which

$$m_1 = \tfrac{1}{2} \quad \text{and} \quad m_2 = -3$$

The general solution is then

$$y = c_1 e^{\frac{1}{2}x} + c_2 e^{-3x} \tag{13}$$

Since there are two arbitrary constants to be evaluated, we must establish a second equation containing c_1 and c_2. This can readily be obtained by differentiating the general solution. Thus

$$\frac{dy}{dx} = \frac{1}{2}c_1 e^{\frac{1}{2}x} - 3c_2 e^{-3x} \tag{14}$$

Applying the given conditions to Equations 13 and 14, we get the equations

$$1 = c_1 + c_2$$

and

$$-2 = \tfrac{1}{2}c_1 - 3c_2$$

Solving these two equations simultaneously yields

$$c_1 = \frac{2}{7} \quad \text{and} \quad c_2 = \frac{5}{7}$$

and, therefore, the particular solution is

$$y = \frac{2}{7} e^{1/2x} + \frac{5}{7} e^{-3x}$$

When the auxiliary equation has repeated real roots, the solution is written in the form

$$y = c_1 e^{mx} + c_2 x e^{mx} = (c_1 + c_2 x) e^{mx} \qquad (15)$$

Example 21. Find the family of solutions for $y'' - 6y' + 9y = 0$ and verify the solution.

Solution. Here the auxiliary equation is

$$m_2 - 6m + 9 = 0$$
$$(m - 3)^2 = 0$$

from which

$$m = 3, 3$$

Substituting into (15), we have

$$y = c_1 e^{3x} + c_2 x e^{3x} = (c_1 + c_2 x) e^{3x}$$

To verify this solution we note that

$$y' = (3c_1 + c_2 + 3c_2 x) e^{3x}$$
$$y'' = (9c_1 + 6c_2 + 9c_2 x) e^{3x}$$

Substituting these values into the given equation, we have

$$(9c_1 + 6c_2 + 9c_2 x) e^{3x} - 6(3c_1 + c_2 + 3c_2 x) e^{3x} + 9(c_1 + c_2 x) e^{3x} = 0$$

which is identically zero.

The third possible type of solution occurs when the auxiliary equation has complex roots of the form $m = a \pm ib$. In this case, the family of solutions is given by

$$y = e^{ax} (c_1 \sin bx + c_2 \cos bx) \qquad (16)$$

Example 22.   Find the family of solutions for

$$\frac{d^2y}{dx^2} - 2\frac{dy}{dx} + 5y = 0$$

Solution.   The auxiliary equation is

$$m^2 - 2m + 5 = 0$$

The roots of this equation are found by the quadratic formula to be

$$m = \frac{2 \pm \sqrt{-16}}{2} = 1 \pm 2i$$

Hence, the general solution of the given differential equation is

$$y = e^x (c_1 \cos 2x + c_2 \sin 2x)$$

where c_1 and c_2 are the arbitrary constants.

Example 23.   Find the particular solution for $(d^2\theta/dt^2) + 4\theta = 0$ if $\theta = 3$ when $t = 0$, and $d\theta/dt = -4$ when $t = \pi/2$.

Solution.   In this case the auxiliary equation is

$$m^2 + 4 = 0$$

The roots of this equation are

$$m = \pm 2i$$

so that the general solution is

$$\theta = c_1 \cos 2t + c_2 \sin 2t$$

Substituting the conditions $\theta = 3$ when $t = 0$ into this equation requires that

$$3 = c_1 \cos 0 + c_2 \sin 0$$

or

$$c_1 = 3$$

Similarly, the condition that $d\theta/dt = -4$ when $t = \pi/2$ requires that

$$-4 = -2c_1 \sin \pi + 2c_2 \cos \pi$$

or

$$c_2 = 2$$

Therefore

$$\theta = 3 \cos 2t + 2 \sin 2t$$

Example 24.   Consider a body of mass M attached to a spring as in Figure 13.9. If the mass is displaced through a distance x from its equilibrium position and released with no initial velocity, its equation of motion is given by

$$M\frac{d^2x}{dt^2} + kx = 0 \qquad\qquad (17)$$

where k is a constant related to the stiffness of the spring. Letting $p^2 = k/M$, Equation 17 may be written

$$\frac{d^2x}{dt^2} + p^2x = 0 \qquad\qquad (18)$$

The motion defined by Equation 18 is called *simple harmonic motion*; it is characterized by the fact that the acceleration is proportional to the displacement and of opposite direction. Find the expression for the displacement of the mass.

Solution.   Here, the auxiliary equation is $m^2 + p^2 = 0$. Solving for m we have

$$m = \pm pi$$

and, therefore,

$$x = A \cos pt + B \sin pt$$

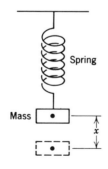

Spring

Mass

Figure 13.9

EXERCISES

Find the family of solutions of each of the following problems.

1. $\dfrac{d^2y}{dx^2} + \dfrac{dy}{dx} = 0$

13. $4y'' + 4y' + y = 0$

2. $\dfrac{d^2s}{dt^2} - 4s = 0$

14. $\dfrac{d^2r}{d\theta^2} = 3\dfrac{dr}{d\theta} - 2r$

3. $3y'' + 6y' = 0$

15. $\dfrac{d^2s}{dt^2} + \dfrac{2}{3}\dfrac{ds}{dt} + \dfrac{1}{9}s = 0$

4. $\dfrac{d^2i}{dt^2} + \dfrac{di}{dt} - 6i = 0$

16. $4y'' - y = 0$

5. $\dfrac{d^2y}{dx^2} + 5\dfrac{dy}{dx} + 4y = 0$

17. $\dfrac{d^2q}{dt^2} - 3q = 0$

6. $\dfrac{d^2y}{dx^2} + 4y = 0$

18. $\dfrac{d^2r}{d\theta^2} + r = 0$

7. $\dfrac{d^2s}{dt^2} + 9s = 0$

19. $9\dfrac{d^2y}{dx^2} + y = 0$

8. $2y'' - 7y' + 3y = 0$

20. $\dfrac{d^2y}{dx^2} + 2\dfrac{dy}{dx} - y = 0$

9. $\dfrac{d^2v}{dt^2} - 6\dfrac{dv}{dt} + 9v = 0$

21. $\dfrac{d^2y}{dx^2} + \dfrac{dy}{dx} - y = 0$

10. $y'' + 4y' + 5y = 0$

22. $2\dfrac{d^2y}{dx^2} - 3\dfrac{dy}{dx} - y = 0$

11. $\dfrac{d^2x}{dt^2} - 2\dfrac{dx}{dt} + 2 = 0$

23. $\dfrac{d^2v}{ds^2} - 4\dfrac{dv}{ds} + 7v = 0$

12. $\dfrac{d^2v}{dt^2} + 3v = 0$

24. $y'' + 8y' + 25y = 0$

Find the particular solution of each of the differential equations which satisfies the given boundary conditions.

25. $\dfrac{d^2y}{dx^2} - 4y = 0$; when $x = 0$, $y = 0$ and $\dfrac{dy}{dx} = 2$

26. $\dfrac{d^2s}{dt^2} - 2\dfrac{ds}{dt} - 3s = 0$; when $t = 0$, $s = 0$ and $\dfrac{ds}{dt} = -4$

27. $y'' - 8y' + 16y = 0$; when $x = 0$, $y = 0$ and $y' = 1$

28. $y'' + 16y = 0$; when $x = 0$, $y = 2$ and $y' = 0$

29. $\dfrac{d^2s}{dt^2} + 9s = 0$; when $t = \pi/6$, $s = 2$ and $\dfrac{ds}{dt} = 0$

30. $\dfrac{d^2y}{dx^2} + 3\dfrac{dy}{dx} = 0$; when $x = 0$, $y = 2$ and $\dfrac{dy}{dx} = 6$

31. $\dfrac{d^2y}{dx^2} - 6\dfrac{dy}{dx} + 10y = 0$; when $x = 0$, $y = 2$ and $\dfrac{dy}{dx} = 1$

32. The elastic curve of a column satisfies approximately the differential equation

$$EI\dfrac{d^2y}{dx^2} + Py = 0$$

where E is the modulus of elasticity of the beam, I is its moment of inertia, P is the applied force, and y is the deflection of the beam caused by the force P. Find an equation for the deflection y, if E, I, and P are constants.

33. A mass of 10 gm is attached to a spring whose stiffness constant is $k = 0.1$. Find the expression for the displacement of the object if $x = 2$ and $dx/dt = 0$ when $t = 0$.

34. When a vibrating body is acted upon by a retarding force, the equation of motion becomes

$$\dfrac{d^2x}{dt^2} + 2a\dfrac{dx}{dt} + b^2x = 0$$

where a and b are constants. When $a > b$, the motion is non-oscillatory and the system is said to be *overdamped;* when $a = b$, the motion is non-oscillatory and the system is said to be *critically damped;* when $a < b$, the motion is said to be *damped oscillatory motion.* Name the type of damping associated with each equation and find the general solution.

(a) $\dfrac{d^2x}{dt^2} + 4\dfrac{dx}{dt} + 4x = 0$ (d) $\dfrac{d^2x}{dt^2} + 2\dfrac{dx}{dt} + 3x = 0$

(b) $\dfrac{d^2x}{dt^2} + 10\dfrac{dx}{dt} + 9x = 0$ (e) $\dfrac{d^2x}{dt^2} + \dfrac{dx}{dt} + x = 0$

(c) $3\dfrac{d^2x}{dt^2} + 6\dfrac{dx}{dt} + 3x = 0$

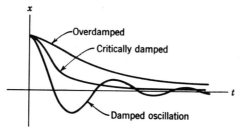

Figure 13.10

Figure 13.10 depicts each of the three conditions — an overdamped motion, a critically damped motion, and a damped oscillatory motion.

14

Calculus in Two Dimensions

14.1 INTRODUCTION

Our discussion to this point has been limited to consideration of functions with one independent variable. However, this is not an inherent limitation; sometimes the value of a variable that we encounter will depend upon more than one quantity. For example, the volume of an ideal gas is a function of both temperature and pressure. Specifically the volume V is given by

$$V = k\frac{T}{P}$$

where T is the temperature, P is the applied pressure, and k is a constant.

Functions of two independent variables are generally represented in much the same way as are functions of one real variable. The functional value is designated by $f(x, y)$ which means that, if the values of x and y are known, $f(x, y)$ is also determined.

Example 1. Let $f(x, y) = 2x^2 + y^2 + xy$. Find $f(3, 2)$, $f(u^2, v)$, and $f(y, x)$.

Solution. $\quad f(3, 2) = 2(3)^2 + (2)^2 + (3)(2)$
$$= 2(9) + 4 + 6 = 28$$
$$f(u^2, v) = 2(u^2)^2 + v^2 + (u^2)(v)$$
$$= 2u^4 + v^2 + u^2v$$
$$f(y, x) = 2y^2 + x^2 + xy$$

14.2 GRAPH OF $z = f(x, y)$

A function of one variable may be graphically represented by a curve in the plane. One axis represents the value of the independent variable and the other represents the functional value. To graphically represent a function of two variables, three mutually perpendicular axes are used — two to represent the values of the independent variables (the x and y variables) and the third to represent the value of the function (the "z" variable). Such a system is called a three dimensional rectangular coordinate system.

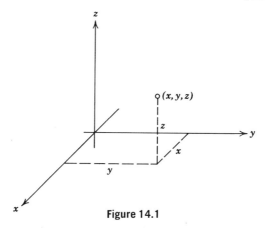

Figure 14.1

In Figure 14.1, the y and z coordinate axes are to be regarded as lying in the plane of the paper while the x-axis is regarded as perpendicular to this plane. The planes formed by three coordinate axes are designated as the xy-plane, the xz-plane, and the yz-plane. The coordinate axes are directed line segments with the positive x-axis toward the reader, the positive y-axis to the right, and the positive z-axis upward. A point is then located in space by giving an x, y, and z value as shown in Figure 14.1.

A three dimensional rectangular coordinate system divides the space into eight *octants* but usually, when sketching the graph of a function, we are content with a "first octant sketch", that is, where x, y, and z are all greater than zero. The sketch usually turns out to be some kind of surface although it could be just a set of disconnected points.

To make a first octant sketch, one would in a naive way probably attempt to plot all the points (x, y, z) where $z = f(x, y)$. It will not take long for you to realize that this point plotting technique does not readily give even a remote idea of the shape of the graph of the function. A method which is of some value is that of sketching the curves formed by the intersection of the surface and planes parallel to one of the coordinate planes. If you sketch enough of these curves, you can in many cases get a rough idea of the shape of the surface. The following examples will show the general idea.

Example 2. Make a first octant sketch of $z = x^2 + 2y^2$.

Solution. Here if we hold z fixed and allow x and y to vary, we obtain a family of curves — one curve for each value of z that we choose. From Chapter 12 we see that each of these curves is an ellipse. If we sketch a few of them, we get the surface of Figure 14.2. In addition, it is helpful to note that when $x = 0$, $z = 2y^2$ and when $y = 0$, $z = x^2$.

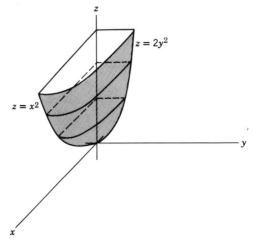

Figure 14.2

Example 3. Sketch the graph of the function $V = k(T/P)$.

Solution. To sketch the graph of this function, we can hold T fixed and allow P to vary. This will yield a family of curves: one curve for each value of T that we choose. If we now consider all possible values of T and all possible values of P, the resulting figure will be a surface in space as shown in Figure 14.3. The volume of the gas for any given pair of values of T and P will be equal to the height of the surface above the TP-plane.

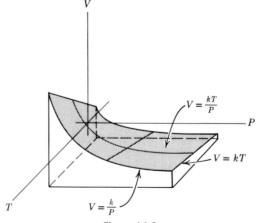

Figure 14.3

Example 4.   Make a first octant sketch of $f(x, y) = x^2$.

Solution.   Since y is not a part of the defining rule, the curve $z = x^2$ is obtained for *each* value of y. See Figure 14.4.

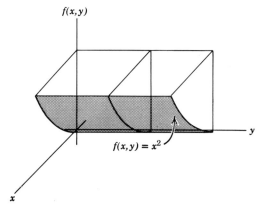

Figure 14.4

EXERCISES

1. Given $f(x, y) = x^2 + 4xy$; find $f(1, 3)$.
2. Given $g(u, v) = \sin uv$; find $g(\tfrac{1}{2}, \pi)$.
3. Given $f(x, y) = x \ln x^2 y$; find $f(a^2, b)$.
4. Given $H(r, s) = r^2 s + 3s^4 + 1$; find $H(-1, -1)$.

Sketch the graph of the following functions. Use only the first octant for your sketch.

5. $z = x + 2y$ **9.** $z = x^2 + y^2$
6. $z = 3x$ **10.** $z = 5 - x - y$
7. $z = y^2$ **11.** $z = 8 - 2x - y$
8. $z = 4 - x^2$ **12.** $z = x^3$

14.3   PARTIAL DERIVATIVES

The concepts of limit and continuity are readily extended to functions of two real variables although the meaning is different.

The extension of the definition of the derivative to $z = f(x, y)$ yields two different derivatives called *partial derivatives*. One derivative is

$$z_x = \lim_{\Delta x \to 0} \frac{f(x + \Delta x, y) - f(x, y)}{\Delta x}$$

which is called the *partial derivative of f(x, y) with respect to x*; and the other is

$$z_y = \lim_{\Delta y \to 0} \frac{f(x, y + \Delta y) - f(x, y)}{\Delta y}$$

which is called the *partial derivative of f(x, y) with respect to y*.

The notation for z_x is also written $\partial f(x, y)/\partial x$, $\partial z/\partial x$, or $f_x(x, y)$ and similarly for the partial derivative of z with respect to y.

By virtue of the definition of z_x, the process of finding its value is simply one of treating y as a constant and computing the derivative with respect to x in the usual manner. All the rules of differentiation previously developed for functions of one variable are valid. Hence, no new differentiation formulas need be learned.

Example 5.   Find the partial derivatives of $z = x^2y^3 + x \cos y$.

Solution.   Treating y as a constant and differentiating with respect to x, we obtain for the partial derivative of z with respect to x

$$\frac{\partial z}{\partial x} = 2xy^3 + \cos y$$

Similarly, we get

$$\frac{\partial z}{\partial y} = 3x^2y^2 - x \sin y$$

for the partial derivative of z with respect to y. In this case x is treated as a constant.

Example 6.   Find $\partial z/\partial x$ if $z = x^2 e^{xy}$.

Solution.   The function $x^2 e^{xy}$ is a product function with respect to x. Therefore, treating y as a constant, the desired partial derivative is

$$\frac{\partial z}{\partial x} = x^2 \cdot y e^{xy} + 2x \cdot e^{xy} = x e^{xy}(xy + 2)$$

Example 7.   The volume of an ideal gas is $V = (KT/P)$. Find the rate of change of volume with respect to pressure if the temperature remains constant during the process.

Solution.   Treating T as a constant and differentiating, we obtain

$$\frac{\partial V}{\partial P} = -\frac{KT}{P^2}$$

14.4   GEOMETRIC INTERPRETATION

Consider a function $z = f(x, y)$ whose graph is the surface shown in Figure 14.5a and b. The geometric interpretation of the partial derivatives of the function can be given with reference to this figure. The partial derivative $\partial z/\partial x$ treats y as a constant — a fact that is interpreted to mean that the point P is constrained to move on the given surface along a path that is parallel to the xz-plane. (This is the same path that would be described by cutting the surface with a plane which is held parallel to the xz-plane.) Hence, the partial derivative $\partial z/\partial x$ is interpreted as the slope of the curve of intersection cut from the surface by a plane parallel to the xz-plane. Figure 14.5a is a geometric representation of $\partial z/\partial x$ at the point $P(x, y, z)$.

The partial derivative of z with respect to y is represented in Figure 14.5b. Since x is treated as a constant in the partial derivative $\partial z/\partial y$, we interpret this partial derivative as the slope of the curve of intersection cut from the surface by a plane parallel to the yz-plane.

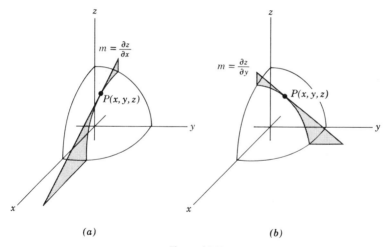

(a)   (b)

Figure 14.5

EXERCISES

Find both of the partial derivatives of the following functions.

1. $z = x^3 y^3 + x^2$
2. $z = y^{1/2} - xy^4$
3. $f(x, t) = x^2 \cos t$
4. $h = s^2 t^2 - \sin st$

5. $G(x, y) = e^{xy}$
6. $R = V \ln T + \cos T$
7. $F = 1 + 6uv$
8. $V = \frac{1}{2} \pi r^2 h$

9. $z = \sin x \cos y$

10. $f(x, y) = \dfrac{e^x}{\sin y}$

11. $A = r \arctan x$

12. $v = \sqrt{xy} + 3x$

13. $H(r, s) = \sec^2 rs$

14. $z = \sqrt{1 + xy}$

15. $Z = \sqrt{X^2 + R^2}$

16. $i = R \sin 2t - 2 \cos 2t$

17. $p = \sin^3 \omega t$

18. $\theta = \alpha^2 \tan \alpha\beta$

19. $v = e^t \sin \omega t$

20. $z = \dfrac{\ln xy}{\sqrt{x^2 + y^2}}$

21. The hydraulic diameter of a noncircular pipe is defined by the equation $D = 4A/P$ where A is the flow cross section and P is the wetted perimeter. Find the rate at which D varies: (a) as a function of A, and (b) as a function of P.

22. The torque acting on a single loop of wire in a magnetic field is given by $T = KI \cos \theta$, where K is a constant, I is the current in the coil, and θ is the angle that the coil makes with the magnetic field. Find the rate at which the torque changes as a function of the angle θ.

23. The work done by a moving body is given by $W = wv^2/2g$, where w is the weight of the body, v is its velocity, and g is the acceleration of gravity. Find the rate at which W changes as a function of the velocity.

24. The mutual conductance, g_m, of a triode electronic vacuum tube is defined as $g_m = \partial i_b/\partial e_c$, where i_b is the plate current, and e_c is the grid voltage. For a certain triode. the plate current is given by $i_b = (e_b + 10e_c)^{5/2} \times 10^{-6}$ amp. where e_b is the plate voltage. Find the expression for the mutual conductance of this triode.

25. What is the mutual conductance of the triode in Exercise 24 when $e_b = 350$ volts and $e_c = -25$ volts? Mutual conductance is measured in mhos, which is the name given to reciprocal ohms.

26. The stress x in. from the fixed end of a uniformly loaded cantilever beam is given by $s = (W/2ZL)(L - x)^2$, where W is the load, L is the length, and Z is the section modulus of the beam. Find $\partial s/\partial x$. If $W = 1000$ lb., $L = 50$ in., and $Z = 1.7$ in.^{3}, at what rate is the stress changing 30 in. from the fixed end.

27. If $f(x, y) = x \sin (2x + y)$, show that $f_x(0, \pi/2) = 1$.

28. If $f(s, t) = s^2 e^{2t}$, show that $f_t(2, 0) = 8$.

14.5 HIGHER PARTIAL DERIVATIVES

If we consider the function $z = f(x, y)$, the partial derivatives $\partial z/\partial x$ and $\partial z/\partial y$ are also functions of x and y. (Of course, functions of x and y include functions of x only and y only, as well as constant functions.) Therefore, it is possible to take partial derivatives of these partial derivatives which results in new functions called *partial derivatives of higher order*. For example, from $\partial z/\partial x$ we obtain

$$\frac{\partial}{\partial x}\left[\frac{\partial z}{\partial x}\right] = \frac{\partial^2 z}{\partial x^2} \tag{1}$$

which is called the *second partial derivative of z with respect to x*.

From a function of two variables there are four possible second partial derivatives — namely.

$$\frac{\partial}{\partial x}\left[\frac{\partial z}{\partial x}\right] = \frac{\partial^2 z}{\partial x^2} = f_{xx}(x, y) \tag{2}$$

$$\frac{\partial}{\partial y}\left[\frac{\partial z}{\partial x}\right] = \frac{\partial^2 z}{\partial y \partial x} = f_{xy}(x, y) \tag{3}$$

$$\frac{\partial}{\partial x}\left[\frac{\partial z}{\partial y}\right] = \frac{\partial^2 z}{\partial x \partial y} = f_{yx}(x, y) \tag{4}$$

$$\frac{\partial}{\partial y}\left[\frac{\partial z}{\partial y}\right] = \frac{\partial^2 z}{\partial y^2} = f_{yy}(x, y) \tag{5}$$

The partial derivatives (3) and (4) are called *mixed* partial derivatives. As we will see shortly, mixed partial derivatives have a convenient relationship to one another.

Example 8. Find all of the second partial derivatives of

$$z = x^3 y^2 + y^3 + xy$$

Solution. In order to find the second partial derivatives, we obtain the first partial derivatives

$$\frac{\partial z}{\partial x} = 3x^2 y^2 + y \qquad \text{and} \qquad \frac{\partial z}{\partial y} = 2x^3 y + 3y^2 + x$$

Hence

$$\frac{\partial^2 z}{\partial x^2} = 6xy^2$$

$$\frac{\partial^2 z}{\partial y \partial x} = 6x^2 y + 1$$

$$\frac{\partial^2 z}{\partial x \partial y} = 6x^2 y + 1$$

$$\frac{\partial^2 z}{\partial y^2} = 2x^3 + 6y$$

Example 9. Find all of the second partial derivatives of

$$f(x, y) = x^2 y^2 + x \sin y$$

Solution.   The first partial derivatives are

$$f_x(x, y) = 2xy^2 + \sin y \qquad \text{and} \qquad f_y(x, y) = 2x^2 y + x \cos y$$

Hence

$$f_{xx}(x, y) = 2y^2$$
$$f_{xy}(x, y) = 4xy + \cos y$$
$$f_{yx}(x, y) = 4xy + \cos y$$
$$f_{yy}(x, y) = 2x^2 - x \sin y$$

Notice that in both of these examples the mixed partial derivatives have identical values; that is, $f_{xy}(x, y) = f_{yx}(x, y)$. These results encourage the conjecture that the order in which mixed partial derivatives are formed is immaterial to the end result. It can be proved that this conjecture is true for functions whose successive derivatives are continuous functions of the variables involved. Fortunately, most functions encountered in physical problems satisfy this condition. We are therefore justified in using mixed partial derivatives interchangeably. This result can be extended to higher partial derivatives as long as the number of differentiations with respect to each independent variable is the same. For instance, the following third order mixed partial derivatives are equal:

$$f_{xxy} = f_{xyx} = f_{yxx} \qquad \text{and} \qquad f_{xyy} = f_{yxy} = f_{yyx}$$

EXERCISES

In Exercises 1 to 10, find all of the second partial derivatives.

1. $z = x^5 y^2 + \sqrt{x}$
2. $i = v\sqrt{r}$
3. $f(x, y) = x^2 \tan y$
4. $p = \sin x \cos y$
5. $Z = \sqrt{R^2 + X^2}$
6. $v = e^{-RC}$
7. $h = \sin st + s^2$
8. $w = pe^{pt}$
9. $F = z \tan zy$
10. $R = \arc \sin st$

11. Find all of the third partial derivatives of the function $z = x^5 y^3 + y^2 - x^4$.
12. Show that $u = e^{-4t} \sin 2x$ satisfies the simple heat equation

$$\frac{\partial u}{\partial t} = \frac{\partial^2 u}{\partial x^2}$$

13. The voltage across the plates of a discharging capacitor is given by $v = Ve^{-t/RC}$, where V and C are constants. Find $\partial^2 v / \partial t \partial R$.

14. In fluid mechanics the volume flow rate associated with streamlines is called the *stream function* and is designated by u. It can be shown that the stream function of an ideal fluid must satisfy the equation

$$\frac{\partial^2 u}{\partial x^2} + \frac{\partial^2 u}{\partial y^2} = 0$$

which is known as Laplace's equation. Show that $u = \ln \sqrt{x^2 + y^2}$ is a valid stream function of an ideal fluid.

14.6 THE TOTAL DIFFERENTIAL

You may recall that the differential of a function $y = f(x)$ was defined by $dy = f'(x)\, dx$. The idea of a differential may be extended to functions of two variables by choosing an appropriate definition.

Consider a function of two variables $z = f(x, y)$. The *total differential*, dz, of this function is defined by

$$dz = \frac{\partial z}{\partial x}\, dx + \frac{\partial z}{\partial y}\, dy \tag{6}$$

where dx and dy may represent actual increments in x and y (that is. $dx = \Delta x$ and $dy = \Delta y$). in which case $\partial z/\partial x$ and $\partial z/\partial y$ represent the slope of the surface in the direction of the respective increments.

Intuitively, the product $(\partial z/\partial x)\, dx$ represents the variation in z due to a change dx in the variable x, and $(\partial z/\partial y)\, dy$ represents the variation in z due to a change dy in the variable y. The quantity dz is then equal to the variation due to a change in x alone plus the variation due to y alone. This explains why we call dz the total differential of the function.

To further solidify the concept of the total differential of a function, let us consider what happens to the area of a rectangle when its width w and its length l increase simultaneously. The area of the rectangle in Figure 14.6 can be written as

$$A = f(l, w) = w \cdot l$$

The total differential of the area is then given by

$$dA = \frac{\partial A}{\partial w}\, dw + \frac{\partial A}{\partial l}\, dl \tag{7}$$

Figure 14.6

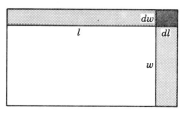

Using the fact that $A = wl$, we have

$$\frac{\partial A}{\partial w} = l \quad \text{and} \quad \frac{\partial A}{\partial l} = w$$

Substituting these values into Equation 7, we get

$$dA = l\,dw + w\,dl \tag{8}$$

This result has a geometric interpretation when related to the rectangle shown in Figure 14.6. Referring to this figure, we see that the first term in Equation 8 is equal to the increase in area due to an increase dw in width, under the assumption that the length remains constant. Similarly, the second term in Equation 8 is equal to the increase in area due to an increase dl in length assuming the width remains constant. In this light, Equation 8 is equal numerically to the sum of the lightly shaded areas shown in the figure.

The exact increase ΔA is given by

$$\begin{aligned}\Delta A &= (l + \Delta l)(w + \Delta w) - lw \\ &= l(\Delta w) + w(\Delta l) + (\Delta l)(\Delta w)\end{aligned} \tag{9}$$

where it is assumed that Δl and Δw are the same changes represented respectively by dl and dw in the above discussion. This is a valid assumption since both symbols represent a change in an independent variable. Comparing (8) and (9), it is clear that ΔA and dA differ by an amount equal to the area of the small rectangle in the upper right-hand corner of Figure 14.6. However, dA will be a good approximation to ΔA as long as dl and dw are both small. We conclude from this discussion that Equation 6 can be used to estimate small changes in the dependent variable in much the same way that we used differentials in connection with functions of only one independent variable.

Example 10.   A cylindrical piece of steel is initially 8-in. long and has a diameter of 8 in. During heat treating, the length and the diameter increase by 0.1 in. Find the approximate increase in the volume of the piece during heat treatment.

Solution.   The volume of a cylinder is given by

$$v = \pi r^2 h$$

where r is the radius and h is the height of the cylinder. Writing the total differential of this equation, we have

$$dv = \frac{\partial v}{\partial r}\,dr + \frac{\partial v}{\partial h}\,dh$$

But
$$\frac{\partial v}{\partial r} = 2\pi rh \quad \text{and} \quad \frac{\partial v}{\partial h} = \pi r^2$$

Therefore,
$$dv = 2\pi rh\,dr + \pi r^2 dh$$

Using the initial values of r and h and the values $dr = 0.05$ and $h = 0.1$ for the respective differentials, we get

$$dv = 2\pi (4)(8)(0.05) + \pi (4)^2(0.1)$$
$$= 10.05 + 5.03 = 15.08 \text{ in.}^3$$

The exact change in volume can be found by taking the difference in the volume when $r = 4.05$ and $h = 8.10$, and the volume when $r = 4$ and $r = 8$. The exact value of Δv is

$$\Delta v = 15.53 \text{ in.}^3$$

14.7 TOTAL DERIVATIVES

If the variables x and y in the function $z = f(x, y)$ are in turn functions of some other variable (say t), then we can assume from the form of the total differential that the derivative of z with respect to t is given by

$$\frac{dz}{dt} = \frac{\partial z}{\partial x}\frac{dx}{dt} + \frac{\partial z}{\partial y}\frac{dy}{dt} \tag{10}$$

This is referred to as the *total derivative* of the function because dz/dt includes the effects of both component rates dx/dt and dy/dt.

Example 11. If a coil of wire is placed in a uniform magnetic field of density B, and a current i sent through the coil, the resulting torque is given by

$$T = NBAi \cos \theta \tag{11}$$

where A is the area of the coil, N is the number of turns of wire, and θ is the angle the coil makes with the magnetic field. Find the expression for dT/dt if i and θ are changed simultaneously.

Solution. By Equation 10, dT/dt is

$$\frac{dT}{dt} = \frac{\partial T}{\partial i}\frac{di}{dt} + \frac{\partial T}{\partial \theta}\frac{d\theta}{dt}$$

$$= NBA \cos \theta \frac{di}{dt} - NBAi \sin \theta \frac{d\theta}{dt}$$

$$\frac{dT}{dt} = NBA \left[\cos \theta \frac{di}{dt} - i \sin \theta \frac{d\theta}{dt} \right]$$

EXERCISES

1. A triangle has an initial base $b = 10$ in. and an initial height $h = 4$ in. Find the approximate change in area that occurs when b is increased to 10.1 in. and h is increased to 4.01 in.

2. Find the exact change in the area of the triangle in Exercise 1.

3. The volume of a right circular cone is given by $V = \frac{1}{3}\pi r^2 h$. Find the expression for the total differential dV.

4. Assuming $r = 10$ and $h = 10$, find the approximate change in the volume of the right circular cone in Exercise 3 when $dr = 0.2$ and $dh = -0.1$.

5. The dynamic pressure of an object moving with a velocity v through a medium of density ρ is given by $P = \frac{1}{2}\rho v^2$. Find the formula that expresses the approximate error in P due to small errors in the measurement of ρ and v.

6. In designing a cone clutch, the normal pressure between the cone surfaces is found to be $P = S/\sin\alpha$, where S is the spring pressure and α is the angle that the clutch surface makes with the axis of the shaft. Find the formula for the approximate change in P caused by simultaneous small changes in S and α.

7. The velocity of an object sliding down an inclined plane can be expressed in the form $v = 32t\sin\alpha$, where t is the time in seconds and α is the angle of inclination of the plane. Find the equation for the approximate error in the velocity due to small errors in measuring t and α.

8. When an axial load is applied to a slender square column, there is a tendency for the column to buckle. The axial load that will cause failure is given by Euler's Formula, $F = \pi^2 d^4 E/12L^2$, where L is the length of the column, d is the length of its sides, and E is the modulus of elasticity. (a) What is the failure load of a column having $L = 300$ in., $d = 10$ in., and $E = 10^6$ lb/in? (b) By how much is this failure load in error if L may be in error by 0.2 in., and E may be in error by 600 lb/in?

9. When friction is ignored, the velocity of a freely falling object is given by $v = \sqrt{2gh}$, where g is the acceleration of gravity and h is the distance the object has fallen from rest. (a) Find the velocity of an object if $g = 32$ ft/sec^2 and $h = 100$ ft. (b) If the true value of g is 32.2 ft/sec^2, and if h can be measured to an accuracy of 0.1 ft, find the approximate error in the computed velocity.

10. The power in a resistor is given by $p = i^2 r$, where i is the current and r is the resistance. For a certain circuit $i = 2.25$ amps and $r = 10^4$ ohms. (a) Compute the power p for this circuit. (b) By how much may this answer be in error if i and r are subject to errors of 0.02 amps and 200 ohms, respectively?

11. Find the formula for the rate of change of volume of a cylinder which results when the radius and height of the cylinder change.

12. Suppose that the height of the cylinder in the preceding example is 100 in. and increases at a rate of 3 in./sec., and the radius is 10 in. and increases at 2 in./sec. What is the rate of change of the volume?

13. Work Exercise 12, but assume that the radius decreases at 2 in./sec.

14. As a ballistic missile rises through the atmosphere, the velocity of the missile and the atmospheric density are changing simultaneously. Find a formula for

the rate of change of dynamic pressure with respect to time due to changes in ρ and v. (See Exercise 5.)

15. During the compression stroke of a certain IC engine, the increase in the temperature of the gasoline-air mixture can be found by using $T = 0.01\ PV$. When the piston is $6°$ BTDC, the volume of the mixture is 20 in.^{3} and the pressure is 110 psi. (a) Find the increase in the temperature of the mixture at this point. (b) Find the rate at which the temperature is increasing if the volume is decreasing at a rate of 500 in.^{3}/sec and the pressure is increasing at a rate of 1000 psi/sec. (Assume dv/dt to be positive).

14.8   PARTIAL ANTIDIFFERENTIATION – ITERATED INTEGRALS

Consider the function $z = x^3y + y^2$. The partial derivative of z with respect to x is given by

$$\frac{\partial z}{\partial x} = 3x^2y$$

Now suppose we were given this partial derivative and told to reconstruct the original function. We would proceed by antidifferentiating $3\ x^2y$ with respect to x while holding y constant. Using the same notation for antidifferentiation that we have used previously, we obtain

$$z = \int 3x^2y\ dx = x^3y + G(y)$$

where $G(y)$ is some function of y only. Indeed, there was a y^2 term that dropped out when the partial derivative of z with respect to x was taken. Usually we can only reconstruct a function from its partial derivative to within an "arbitrary" function of one of the variables.

Example 12.   Find z if $z_y = 3x^2 + y^3$.

Solution. $z = \int (3x^2 + y^3)\ dy = 3x^2y + (y^4/4) + F(x)$. The arbitrary function in this case is a function of x only.

It is sometimes necessary to do the partial antidifferentiation more than once to reconstruct the function. Two such problems are solved in the following examples.

Example 13.   Find z, if $(\partial^2 z/\partial x^2) - x^2$.

Solution.   The first antidifferentiation yields

$$\frac{\partial z}{\partial x} = \int x^2\ dx = \frac{x^3}{3} + G(y)$$

and the second,

$$z = \int \left[\int x^2 dx \right] dx$$

$$= \int \left[\frac{x^3}{3} + G(y) \right] dx$$

$$= \frac{x^4}{12} + xG(y) + H(y)$$

Example 14.   Find z if $z_{xy} = x \sin xy$.

Solution.   Here the $\partial z / \partial x$ is given by

$$\frac{\partial z}{\partial x} = \int (x \sin xy) \, dy = -\cos xy + F(x)$$

Then z is

$$z = \int \left[\int x \sin xy \, dy \right] dx$$

$$= \int \left[-\cos xy + F(x) \right] dx$$

$$= -\frac{1}{y} \sin xy + \int F(x) \, dx + Q(y)$$

Antiderivatives of the type arising in Example 14 are usually written without brackets. Thus, we write

$$\int \int x \sin xy \, dy \, dx$$

with the understanding that this means to antidifferentiate with respect to y first and then with respect to x.

The same type of convention as described above is used for *repeated* or *iterated* definite integration. Iterated integrals are usually written in the form

$$\int_a^b \int_{f(x)}^{g(x)} F(x, y) \, dy \, dx \qquad \text{or} \qquad \int_c^d \int_{F(y)}^{G(y)} H(x, y) \, dx \, dy$$

Notice that the limits on the inner integral may be a function of x when dy is the differential and a function of y when dx is the differential.

Example 15.   Compute the iterated integral $\int_{-1}^3 \int_1^2 3x^2 y \, dx \, dy$.

Solution.   By definition, the notation means

$$\int_{-1}^3 \left[\int_1^2 3x^2 y \, dx \right] dy$$

Since
$$\int_1^2 3x^2y\,dx = [x^3y]_1^2 = 8y - y = 7y$$

and
$$\int_{-1}^3 7y\,dy = \left[\frac{7y^2}{2}\right]_{-1}^3 = \frac{63}{2} - \frac{7}{2} = 28$$

we have
$$\int_{-1}^3 \int_1^2 3x^2y\,dx\,dy = 28$$

Evaluation of iterated integrals will be important to us later in the chapter so we will do a few more examples to give you a little more proficiency.

Example 16.   Evaluate $\int_0^1 \int_0^2 (x+y)\,dy\,dx$.

Solution.
$$\int_0^1 \int_0^2 (x+y)\,dy\,dx = \int_0^1 \left[\int_0^2 (x+y)\,dy\right] dx$$
$$= \int_0^1 \left[xy + \frac{y^2}{2}\right]_0^2 dx$$
$$= \int_0^1 [2x+2]\,dx = [x^2 + 2x]_0^1 = 3$$

Example 17.   Evaluate $\int_{\pi/2}^\pi \int_0^1 x\sin y\,dx\,dy$.

Solution.
$$\int_{\pi/2}^\pi \int_0^1 x\sin y\,dx\,dy = \int_{\pi/2}^\pi \left[\int_0^1 x\sin y\,dx\right] dy$$
$$= \int_{\pi/2}^\pi \left[\frac{x^2}{2}\sin y\right]_0^1 dy = \int_{\pi/2}^\pi \frac{1}{2}\sin y\,dy$$
$$= \left[-\frac{1}{2}\cos y\right]_{\pi/2}^\pi = -\left[-\frac{1}{2} - 0\right] = \frac{1}{2}$$

Example 18.   Evaluate $\int_0^1 \int_0^{x^2} dy\,dx$.

Solution.
$$\int_0^1 \int_0^{x^2} dy\,dx = \int_0^1 \left[\int_0^{x^2} dy\right] dx = \int_0^1 [y]_0^{x^2} dx$$
$$= \int_0^1 x^2\,dx = \left[\frac{x^3}{3}\right]_0^1 = \frac{1}{3}$$

EXERCISES

Evaluate each of the following iterated integrals.

1. $\int_0^2 \int_0^3 xy \, dx \, dy$

9. $\int_1^3 \int_0^\pi x \sin xy \, dy \, dx$

2. $\int_0^2 \int_0^3 xy \, dy \, dx$

10. $\int_{-1}^1 \int_{-1}^1 e^{x+y} \, dx \, dy$

3. $\int_0^\pi \int_0^2 \cos y \, dx \, dy$

11. $\int_{-1}^2 \int_0^{y^2} x \, dx \, dy$

4. $\int_{-1}^2 \int_0^1 se^t \, dt \, ds$

12. $\int_0^2 \int_0^{\sqrt{y}} xe^{y^2} \, dx \, dy$

5. $\int_0^3 \int_1^2 3x^2y \, dx \, dy$

13. $\int_1^2 \int_v^{v^2} (v+2u) \, du \, dv$

6. $\int_1^4 \int_0^5 (x^2 + y^2) \, dx \, dy$

14. $\int_0^\pi \int_0^{\cos \theta} r \sin \theta \, dr \, d\theta$

7. $\int_{-1}^0 \int_0^4 (z^{1/2}y + 2) \, dz \, dy$

15. $\int_0^{\pi/2} \int_0^x \cos y \sin x \, dy \, dx$

8. $\int_0^2 \int_0^2 Te^{RT} \, dR \, dT$

16. $\int_0^9 \int_0^{\sqrt{25-x}} dy \, dx$

14.9   AREA AS AN ITERATED INTEGRAL

In Chapter 6 we showed that the area of a plane figure could be approximated by a summation process which led to the definite integral. We would now like to show how a plane area can be represented by an iterated integral.

Consider the plane area $ABCD$ shown in Figure 14.7. Divide the interval on the x-axis into n equal parts and draw a line from each point of division parallel to the y-axis. Let the width of each of these elements be Δx. Now do the same thing for the interval on the y-axis, representing the width of these elements by Δy. In this way the area $ABCD$ is divided into a network of elementary rectangles. Clearly, the area of each of these elementary rectangles is given by

$$\Delta A_{ij} = \Delta x_i \Delta y_j$$

If we now choose a typical Δx and sum all of the areas formed by the product of this Δx and the corresponding Δy's, the result will be the area

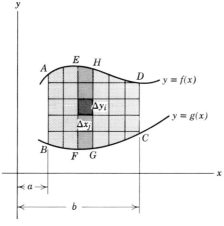

Figure 14.7

of a rectangle like $EFGH$ shown in the figure. This may be written symbolically as

$$\text{Area } EFGH \approx \sum_{i=1}^{n} \Delta y_i \Delta x_j = \left[\sum_{i=1}^{n} \Delta y_i \right] \Delta x_j \tag{12}$$

We may regard the exact area as the limit of (12) as $\Delta y \to 0$; that is,

$$\text{Area } EFGH = \left[\lim_{n \to \infty} \sum_{i=1}^{n} \Delta y_i \right] \Delta x_j \tag{13}$$

The Δy's in this equation are being summed from $y = g(x)$ to $y = f(x)$ and, therefore, by the definition of a definite integral we can write (13) as

$$\text{Area } EFGH = \left[\int_{g(x)}^{f(x)} dy \right] \Delta x_j \tag{14}$$

Observing that Area $EFGH$ is a typical element, we conclude that the desired area is the limit of the sum of these areas from $x = a$ to $x = b$ as $\Delta x \to 0$. Hence

$$\text{Area } ABCD = \lim_{m \to \infty} \sum_{j=1}^{m} \left[\int_{g(x)}^{f(x)} dy \right] \Delta x_j \tag{15}$$

or

$$\text{Area } ABCD = \int_{a}^{b} \left[\int_{g(x)}^{f(x)} dy \right] dx = \int_{a}^{b} \int_{g(x)}^{f(x)} dy \, dx \tag{16}$$

For areas like the one shown in Figure 14.8, it is more convenient to choose a typical Δy and sum horizontally first. Under these circumstances the area is given by

$$\text{Area } NMOP = \int_{c}^{d} \int_{G(y)}^{F(y)} dx \, dy$$

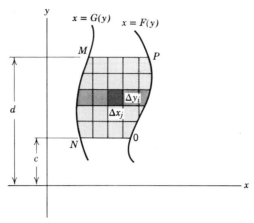

Figure 14.8

Example 19.   Use iterated integration to find the area bounded by the curves $y = x$ and $y = x^2$. (See Figure 14.9.)

Solution.   The area is given by

$$A = \int_a^b \int_{g(x)}^{f(x)} dy \, dx$$

The limits on the inner integral sign are $f(x) = x$ and $g(x) = x^2$. The limits on the outer integral can be found by solving the two equations simultaneously. Thus

$$x^2 = x$$
$$x^2 - x = 0$$
$$x = 0, 1$$

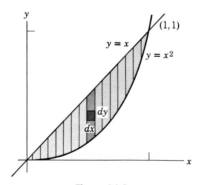

Figure 14.9

The numerical values of a and b are then $a = 0$, $b = 1$. The required area is then

$$A = \int_0^1 \int_{x^2}^x dy\, dx$$

$$= \int_0^1 [y]_{x^2}^x\, dx = \int_0^1 [x - x^2]\, dx$$

$$= \left[\frac{x^2}{2} - \frac{x^3}{3}\right]_0^1 = \left[\frac{1}{2} - \frac{1}{3}\right] = \frac{1}{6}$$

Example 20. Use iterated integration to find the area bounded by the two curves $y = 5x$ and $y = x^2$, and the straight lines $y = 1$ and $y = 4$. (See Figure 14.10.)

Solution. Here we use the iterated integral

$$A = \int_c^d \int_{G(y)}^{F(y)} dx\, dy$$

with $c = 1$, $d = 4$, $G(y) = y/5$, and $F(y) = y^{1/2}$. The required area is therefore

$$A = \int_1^4 \int_{y/5}^{y^{1/2}} dx\, dy = \int_1^4 [x]_{y/5}^{y^{1/2}}\, dy = \int_1^4 \left[y^{1/2} - \frac{y}{5}\right] dy$$

$$= \left[\frac{2y^{3/2}}{3} - \frac{y^2}{10}\right]_1^4 = \left[\frac{16}{3} - \frac{16}{10}\right] - \left[\frac{2}{3} - \frac{1}{10}\right] = \frac{19}{6}$$

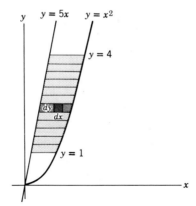

Figure 14.10

EXERCISES

Use iterated integration to find the area bounded by the given curves.

1. $y = 2 - x^2, y = x$ 6. $y = x - x^2, y = -x$
2. $y = x^2, y = 8x$ 7. $y^2 = 4x, x^2 = 4y$
3. $y = x^2, y = 4x, y = 1, y = 2$ 8. $y^2 = 8x, y = \frac{1}{2}x$
4. $y = x, y = -\frac{1}{2}x, x = 1, x = 3$ 9. $y = x, y = x^{3/2}$
5. $y = x^2 + 4x, y = x$ 10. $y = 3, y = x, x = -1, x = 2$

14.10   DOUBLE INTEGRATION

If f is a function of one real variable then $\int_a^b f(x)\, dx$ is a *number* defined as the limit of approximating sums. We now wish to show the analogous concept for functions of two real variables. Let $H(x, y)$ represent a function defined on some region of the xy-plane, for example the region shown in Figure 14.11.

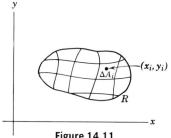

Figure 14.11

We subdivide the region R into elementary areas where the subdivision process may be carried out in *any* manner, not necessarily into rectangular regions parallel to the coordinate axes. Call the area of each of the sub-regions ΔA_i. Evaluate $H(x, y)$ at some point (x_i, y_i) in each of the sub-region and form the products $H(x_i, y_i)\Delta A_i$. Now if we sum these products, we shall have a sum analogous to that obtained for functions of one variable, that is,

$$\sum_{i=1}^n H(x_i, y_i)\, \Delta A_i \qquad (17)$$

Example 21.   Compute the sum in (17), if $H(x, y) = x^2 + 7y$, and R is the region bounded by the coordinate axes and the lines $x = 4$ and $y = 8$. Use four equal rectangular subregions and evaluate $H(x, y)$ at the midpoint of each elementary area. (See Figure 14.12.)

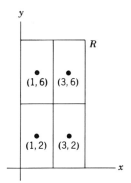

Figure 14.12

Solution. Each of the subregions shown in the figure have an area of $\Delta A_i = 2(4) = 8$. The functional values of H at the midpoints are

$$H(1, 2) = 1^2 + 7(2) = 15$$
$$H(3, 2) = 3^2 + 7(2) = 23$$
$$H(1, 6) = 1^2 + 7(6) = 43$$
$$H(3, 6) = 3^2 + 7(6) = 51$$

The desired sum is then

$$\sum_{i=1}^{4} H(x_i, y_i) \, \Delta A_i = (15 + 23 + 43 + 51)8 = 1056$$

As in the case of sums for functions of one variable, we are interested in the limit of (17) as n increases without bound, that is,

$$\lim_{n \to \infty} \sum_{i=1}^{n} H(x_i, y_i) \, \Delta A_i$$

If this limit exists we denote it by $\int_R \int H(x, y) \, dA$, and *we call it the double integral of the function H(x, y) over the region R.*

There are several elementary techniques for evaluating double integrals all based upon different methods of subdividing the region R. We will be content to show you what happens when the process used is that of subdividing the region into rectangles whose sides are parallel to the coordinate axes. In this case, the area of each subregion is given by

$$\Delta A_i = \Delta y_j \Delta x_i$$

and the sum $\Sigma \, H(x_i, y_i) \, \Delta A_i$ may be written as the double summation

$$\sum_{i=1}^{n} \sum_{j=1}^{m} H(x_i, y_j) \Delta y_j \Delta x_i$$

Suppose the region R is the one shown in Figure 14.7; then, if we let m approach infinity and n approach infinity independently, we arrive at the fact that

$$\int_R \int H(x, y) \, dA = \int_a^b \int_{g(x)}^{f(x)} H(x, y) \, dy \, dx \tag{18}$$

Similarly, if the region R is as shown in Figure 14.8, then we have

$$\int_R \int H(x, y) \, dA = \int_c^d \int_{G(y)}^{F(y)} H(x, y) \, dx \, dy \tag{19}$$

It is because double integrals can be evaluated in terms of iterated integrals that the two concepts are closely identified. Indeed, it is quite natural to refer to iterated integrals as double integrals.

Example 22.   Evaluate the double integral $\int_R \int (x^2 + 7y) \, dA$, where R is the same region described in Example 21.

Solution.   Using (18) to evaluate the given double integral, we have

$$\int_R \int (x^2 + 7y) \, dA = \int_0^4 \int_0^8 (x^2 + 7y) \, dy \, dx$$

$$= \int_0^4 \left[x^2 y + \frac{7y^2}{2} \right]_0^8 dx$$

$$= \int_0^4 (8x^2 + 224) \, dx$$

$$= \left[\frac{8x^3}{3} + 224x \right]_0^4$$

$$= \frac{512}{3} + 896 = \frac{3200}{3}$$

Example 23.   Express the area of the region R bounded by the curves $y = x$ and $y = x^2$ as a double integral. This is the same region that we considered in Example 19. (See Figure 14.13.)

Solution.   Since the area of R can be approximated by $\sum_{i=1}^{n} \Delta A_i$, we can express this area by the double integral $\int_R \int dA$. Using (18) to evaluate this double integral, we have

$$A = \int_R \int dA = \int_0^1 \int_{x^2}^x dy \, dx$$

which is the same iterated integral we used in Example 19.

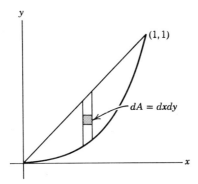

Figure 14.13

Example 24.   The pressure exerted on the triangular plate shown in Figure 14.14 increases with distance from the *y*-axis according to $P = 3x$ psi. Find the total force being exerted on the plate.

Solution.   The force being exerted on an element of area ΔA_i is

$$P(x_i)\Delta A_i,$$

so that the total force is given by the double integral $\int_R \int P(x)\, dA$. Thus

$$F = \int_R \int P(x)\, dA = \int_0^8 \int_0^y 3x\, dx\, dy$$

$$= \int_0^8 \left[\frac{3x^2}{2} \right]_0^y dy = \int_0^8 \frac{3y^2}{2}\, dy$$

$$= \left[\frac{y^3}{2} \right]_0^8 = 256\ \text{lb}$$

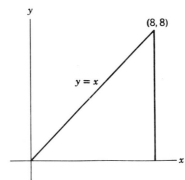

Figure 14.14

EXERCISES

Evaluate the given double integral over the indicated region. Draw each region.

1. $\int_R \int x \, dA$; R is bounded by $x = 0, y = 0, x = 2, y = 3$.

2. $\int_R \int y \, dA$; R is bounded by $x = 0, y = 0, x = 5, y = 4$.

3. $\int_R \int (x^2 + y^2) \, dA$; R is bounded by $x = 0, y = 0, x = 4, y = x$.

4. $\int_R \int xy \, dA$; R is bounded by $x = 0, y = 0, y = 4 - 2x$.

5. $\int_R \int (2x + 3y) \, dA$; R is bounded by $y = 2x$ and $y = x^2$.

6. $\int_R \int (3 + 4y) \, dA$; R is bounded by $y = 0$ and $y = 3x - x^2$.

7. Express the area bounded by $y = 4x - x^2$ and $y = x$ as a double integral and evaluate by (18).

8. The region R is bounded by $y = x^3, x = 2$, and $y = 0$. Find the area of R.

9. The moment of a region R about the x-axis is given by $\int_R \int y \, dA$. Find the moment about the x-axis of the region bounded by $y = x, y = x^2$.

10. Express the moment of a region R about the y-axis in terms of a double integral.

11. Using the expression in Exercise 10, find the moment of the region bounded by $y = x$ and $y = x^2$ about the y-axis.

12. The pressure on the triangular region formed by the lines $x = 0, y = 0, x = 4$, $y = \frac{1}{2}x$ increases with distance from the y-axis according to $P = x^2$ psi. Find the total force being applied to the region.

13. Find the total force being applied to the region in Exercise 12 if the pressure varies with distance from the x-axis according to $P = 3y$ psi.

14.11   VOLUME AS A DOUBLE INTEGRAL

We will conclude this chapter with a discussion of how an iterated integral, and thus a double integral, can be used to represent the volume under a surface.

Let $z = f(x, y)$ be a continuous function of x and y having as its graph the surface shown in Figure 14.15. Consider the volume V of the solid bounded by the given surface S, its projection on the xy-plane R and the vertical lines through the boundaries of the surfaces S and R. In the xy-plane, draw n lines parallel to the y-axis. Let the width of each interval intercepted by these lines on the x-axis be Δx. Now do the same thing with lines parallel to the x-axis representing the width of each of these elements

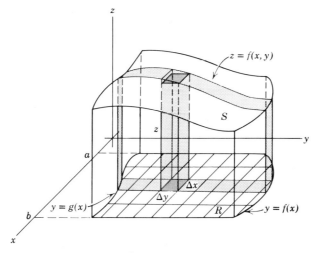

Figure 14.15

by Δy. In this way, the projection R in the xy-plane is divided into a net-work of elementary rectangles each having an area of $\Delta y \Delta x$. Through the lines in the xy-plane, pass planes parallel to the xz-plane and the yz-plane respectively. This divides the volume V into vertical rectangular columns. The volume of each of these rectangular columns can be written

$$V = z_{ij} \Delta y_i \Delta x_j$$

where z_{ij} is the height of each column. Now let us choose a typical Δx and sum the volumes of all of the rectangular columns corresponding to this Δx. The limit of this sum as $\Delta y \to 0$ is equal to the volume of a slice of the figure Δx units wide; that is,

$$V_y = \left[\lim_{\Delta y \to 0} \sum_{i=1}^{m} z_{ij} \Delta y_i \right] \Delta x_j \qquad (20)$$

Observing that the Δy's are summed between the lateral boundaries of the given surface, we may write (20) as

$$V_y = \left[\int_{g(x)}^{f(x)} z_j \, dy \right] \Delta x_j \qquad (21)$$

where $y = f(x)$ and $y = g(x)$ are the lateral boundaries of the given surface in the y direction. The notation $f(x)$ and $g(x)$ is used to indicate that, in general, these limits are functions of x. The desired volume is then the limit of the sum of the volume of these slices as $\Delta x \to 0$. Thus

$$V = \lim_{\Delta x \to 0} \sum_{j=1}^{n} \left[\int_{g(x)}^{f(x)} z_j \, dy \right] \Delta x_j \qquad (22)$$

$$= \int_{a}^{b} \int_{g(x)}^{f(x)} z \, dy \, dx \qquad (23)$$

The limits on the outer integral sign are the lateral boundaries of the given surface in the x-direction. The outer limits are always constants.

By a similar development, we may also write the volume as

$$V = \int_{c}^{d} \int_{G(y)}^{F(y)} z \, dx \, dy \qquad (24)$$

Example 25. Find the volume bounded by the plane surface $z = 4 - x - y$ and the first octant of the coordinate system. (See Figure 14.16.)

Solution.   This surface can be plotted by noting that:

when $x = 0$, $z = 4 - y$, which is a straight line in the yz-plane;
when $y = 0$, $z = 4 - x$, which is a straight line in the xz-plane;
when $z = 0$, $y = 4 - x$, which is a straight line in the xy-plane.

Drawing these lines and a plane surface through them, we get the figure shown at the top of the next pate. The required volume is then given by

$$V = \int_{0}^{4} \int_{0}^{4-x} (4 - x - y) \, dy \, dx$$

$$= \int_{0}^{4} \left[4y - xy - \frac{1}{2} y^2 \right]_{0}^{4-x} dx$$

$$= \int_{0}^{4} \left[4(4-x) - x(4-x) - \frac{1}{2}(4-x)^2 \right] dx$$

$$= \int_{0}^{4} \left[8 - 4x + \frac{1}{2} x^2 \right] dx = \left[8x - 2x^2 + \frac{x^3}{6} \right]_{0}^{4} = \frac{32}{3}$$

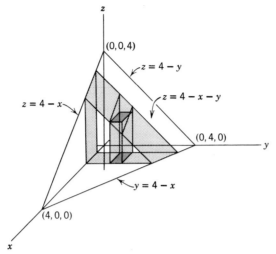

Figure 14.16

Example 26.   Find the volume under the surface $z = x^2 + y^2$ over the region R bounded by $y = 1$, $x = 0$, and $y = 3 - x$. (See Figure 14.17.)

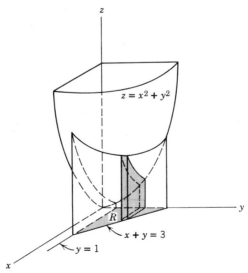

Figure 14.17

Solution.   The volume is given by

$$V = \int_R \int z\, dA = \int_1^3 \int_0^{3-y} (x^2 + y^2)\, dx\, dy$$

$$= \int_1^3 \left[\frac{x^3}{3} + xy^2\right]_0^{3-y} dx = \int_1^3 \left[\frac{(3-y)^3}{3} + (3-y)y^2\right] dy$$

$$= \left[-\frac{(3-y)^4}{12} + y^3 - \frac{y^4}{4}\right]_1^3$$

$$= \left(0 + 27 - \frac{81}{4}\right) - \left(-\frac{4}{3} + 1 - \frac{1}{4}\right) = \frac{22}{3}$$

EXERCISES

1. Find the volume bounded by the plane surface $z = 2 - x - y$ and the first octant of the coordinate system.
2. Find the volume in the first octant bounded by the plane surface $z = 1 - 2x - y$.
3. Find the volume under the plane surface $z = 3 + x + 4y$ over a region R, where R is bounded by $x = 0$, $y = 0$, $y = 2$, and $x = 3$.
4. Find the volume under the surface $z = x^2 + y^2$ over the region R bounded by $x = 0$, $y = 0$, and $y = 2 - x$.
5. Find the volume under the surface $z = x + y^2$ over the region R bounded by $y = x$ and $y = x^2$.
6. Find the volume under the surface $z = y$ over the region bounded by $y = 0$ and $y = 4x - x^2$.

15

Polar Coordinates

15.1 INTRODUCTION

The rectangular coordinate system was used exclusively in the first fourteen chapters of this book. Another coordinate system that is widely used in science and mathematics is the so-called *polar* coordinate system. In this system, the position of a point is determined by specifying a distance from a given point and the direction from a given line. Actually, this is not a new concept; we frequently use this system to describe the relative location of geographic points. Thus, when we say that Cincinnati is about 300 miles southeast of Chicago, we are, in fact, using polar coordinates.

15.2 POLAR COORDINATES

In establishing a frame of reference for the polar coordinate system, we begin by choosing a point O and extending a line from this point. The point O is called the *pole*, and the extended line is called the *polar axis*. The position of any point P in the plane is then determined if we know the distance OP, and the angle AOP as indicated in Figure 15.1. The distance OP is called the *radius vector* of P and is usually denoted by r. The angle AOP is called the *vectorial angle* and is denoted by θ. The coordinates of a point P are then written as the ordered pair (r,θ). Notice that the radius vector is the first element and the vectorial angle is the second.

Polar coordinates, like rectangular coordinates, are regarded as signed quantities. When stating the polar coordinates of a point, it is customary to use the following sign conventions.

Figure 15.1

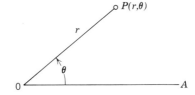

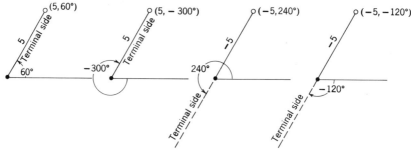

Figure 15.2

1. The radius vector is positive when measured on the terminal side of the vectorial angle and is negative when measured in the opposite direction.

2. The vectorial angle is positive when generated by a counter-clockwise rotation from the polar axis and negative when generated by a clockwise rotation.

The polar coordinates of a point determine the location of the point uniquely. However, the converse is not true, as we can see from Figure 15.2. Ignoring vectorial angles that are numerically greater than 360°, we have four pairs of coordinates that yield the same point. Thus, the pairs (5,60°), (5,−300°), (−5,240°), and (−5,−120°) represent the same point.

Polar coordinate paper is commercially available. As you can see in Figure 15.3, this paper consists of equally spaced concentric circles with radial lines extending at equal angles through the pole. While the use of polar coordinate paper is not mandatory, the student will find that it is very helpful in plotting polar curves. Several points are plotted in Figure 15.3 for illustrative purposes.

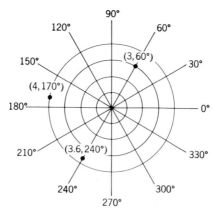

Figure 15.3

15.3 THE RELATIONSHIP BETWEEN POLAR COORDINATES AND RECTANGULAR COORDINATES

A coordinate system is simply a tool to make our work easier. The most desirable coordinate system is usually the one that gives the simplest description of the problem. In some cases, however, it is necessary to shift from one system to the other. It is, therefore, imperative that we have some knowledge of the relationship between the two systems.

The relationship between the polar coordinates of a point and the rectangular coordinates of the same point can be found by superimposing the rectangular coordinate system on the polar coordinate system so that the origin corresponds to the pole and the positive x-axis to the polar axis. Under these circumstances, the point P shown in Figure 15.4 has both (x,y) and (r,θ) as coordinates. The desired relationship is then an immediate consequence of triangle OMP. Hence, the equations

$$x = r \cos \theta \tag{1}$$

and

$$y = r \sin \theta \tag{2}$$

can be used to transform a rectangular equation into a polar equation.

Figure 15.4

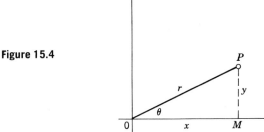

Example 1. Find the polar equation of the circle whose rectangular equation is $x^2 + y^2 = a^2$.

Solution. Substituting Equations (1) and (2) into the given equation, we have

$$r^2 \cos^2 \theta + r^2 \sin^2 \theta = a^2$$
$$r^2 (\cos^2 \theta + \sin^2 \theta) = a^2$$
$$r^2 = a^2$$
$$r = a$$

Perhaps this result was anticipated since a circle of radius a is the set of all points a units from a given point. It is clear from this result that a circle centered at the origin can be described most easily in polar form.

To make the transformation from polar coordinates into rectangular coordinates we use the following equations.

$$r = \sqrt{x^2 + y^2} \tag{3}$$

$$\sin \theta = \frac{y}{\sqrt{x^2 + y^2}} \tag{4}$$

$$\cos \theta = \frac{x}{\sqrt{x^2 + y^2}} \tag{5}$$

These equations are also derived from Figure 15.4.

Example 2. Transform the polar equation into a rectangular equation:

$$r = 1 - \cos \theta$$

Solution.   Substituting Equations 3 and 5 into the given equation, we have

$$\sqrt{x^2 + y^2} = 1 - \frac{x}{\sqrt{x^2 + y^2}}$$

$$x^2 + y^2 = \sqrt{x^2 + y^2} - x$$

$$x^2 + y^2 + x = \sqrt{x^2 + y^2}$$

Example 3. Show that $r = 1/(1 - \cos \theta)$ is the polar form of a parabola.

Solution.   Here our work is simplified if we multiply both sides of the given equation by $1 - \cos \theta$ before making the substitution. Thus

$$r - r \cos \theta = 1$$

Substituting Equations 3 and 5, we get

$$\sqrt{x^2 + y^2} - x = 1$$

Transposing x to the right and squaring both sides of the resulting equation, we get

$$x^2 + y^2 = x^2 + 2x + 1$$
$$y^2 = 2x + 1$$
$$y^2 = 2\left(x + \frac{1}{2}\right)$$

We recognize this as the standard form of a parabola having its vertex at $(-\frac{1}{2},0)$ and a horizontal axis.

EXERCISES

Plot each of the following points on polar coordinate paper.

1. $(5,30°)$

2. $(3.6,-45°)$

3. $(12,2\pi/3)$

4. $(0.5,220°)$

5. $(-7.1,14°)$

6. $(-2,7\pi/3)$

7. $(1.75,-200°)$

8. $(\sqrt{2},-311°)$

9. $(-5,-30°)$

10. $(5,150°)$

Convert each of the following rectangular equations into an equation in polar coordinates.

11. $2x+3y=6$

12. $y=x$

13. $x^2+y^2-4x=0$

14. $x^2-y^2=4$

15. $x^2+4y^2=4$

16. $xy=1$

17. $x^2=4y$

18. $y^2=16x$

Convert each of the following polar equations into an equation in rectangular coordinates.

19. $r=5$

20. $r=\cos\theta$

21. $r=10\sin\theta$

22. $r=2(\sin\theta-\cos\theta)$

23. $r=1+2\sin\theta$

24. $r\sin\theta=10$

25. $r=5/(1+\cos\theta)$

26. $r(1-2\cos\theta)=1$

15.4   PLOTTING CURVES OF POLAR EQUATIONS

A polar equation has a graph in the polar coordinate plane just as a rectangular equation has a graph in the rectangular coordinate plane. To draw the graph of a polar equation, we start by assigning values to θ and finding the corresponding values of r. The desired graph is then generated by plotting the ordered pairs (r,θ) and connecting them with a smooth curve.

Example 4.   Sketch the graph of the equation $r=1+\cos\theta$.

Solution.   Using increments of 45° for θ, we obtain the indicated table.

θ	r
0	2.00
45°	1.71
90°	1.00
135°	0.29
180°	0.00
225°	0.29
270°	1.00
315°	1.71
360°	2.00

The curve obtained by connecting these points with a smooth curve is called a *cardioid* (Figure 15.5).

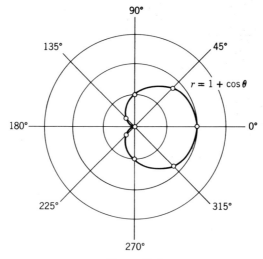

Figure 15.5

Example 5.   Sketch the graph of the equation $r = 4 \sin \theta$.

Solution.   We find the following table for values of r corresponding to the indicated values of θ. Drawing a smooth curve through the plotted points, we have the *circle* shown in Figure 15.6.

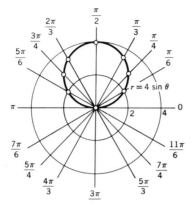

Figure 15.6

θ	r
0	0
$\pi/6$	2
$\pi/4$	$2\sqrt{2}$
$\pi/3$	$2\sqrt{3}$
$\pi/2$	4
$2\pi/3$	$2\sqrt{3}$
$3\pi/4$	$2\sqrt{2}$
$5\pi/6$	2
π	0

Notice that θ varies only from 0 to π radians. If we allow θ to vary from 0 to 2π radians, the graph will be traced out twice; once for $0 \leqslant \theta \leqslant \pi$ and again for $\pi < \theta \leqslant 2\pi$. The student should demonstrate this for himself by plotting points in the interval $\pi < \theta \leqslant 2\pi$.

Example 6. Sketch the graph of the equation $r = 1 - 2\cos\theta$.

Solution. Here we will assign values to θ in increments of 30°. The ordered pairs (r,θ) are then given in the table for the interval $0 \leqslant \theta \leqslant 180°$. This curve is called a *limacon* (Figure 15.7).

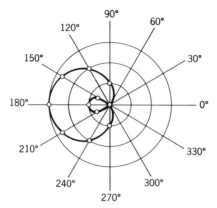

Figure 15.7

θ	r
0	−1.00
30°	−0.73
60°	0.00
90°	1.00
120°	2.00
150°	2.73
180°	3.00
210°	2.73
240°	2.00
270°	1.00
300°	0.00
330°	−0.73
360°	−1.00

Example 7.   Sketch the graph of the polar equation $r = 4 \cos 3\theta$.

Solution.   Substituting values for θ, we obtain the table opposite.

This curve is called a *three-leaved rose* (Figure 15.8). Notice that although 3θ is the argument, the curve is formed by plotting θ versus r. Also, we note that the curve retraces itself when θ assumes values greater than 180°.

In some cases it is to our advantage to express the polar equation in rectangular coordinates. This is especially true for curves which are straight lines or which become asymptotes to straight lines.

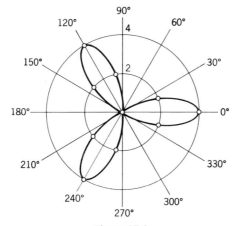

Figure 15.8

	$r = 4 \cos 3\theta$	
θ	3θ	r
0	0	4.00
20°	60°	2.00
30°	90°	0.00
40°	120°	−2.00
60°	180°	−4.00
80°	240°	−2.00
90°	270°	0.00
100°	300°	2.00
120°	360°	4.00
140°	420°	2.00
150°	450°	0.00
160°	480°	−2.00
180°	540°	−4.00

Example 8.   Determine the shape of the graph of $r = 4 \sin \theta$ by transforming it into rectangular coordinates.

Solution.   This curve has already been sketched by the method of point plotting and was said to be a circle; however, you might wonder how one could recognize this as a circle. When rectangular coordinates are used, the equation becomes

$$x^2 + y^2 = 4y$$

which by a technique similar to Example 15 in Chapter 12 becomes

$$x^2 + (y - 2)^2 = 4$$

Which is a circle of radius 2 centered at (0,2), as we said earlier.

Example 9.   Determine the shape of the graph of $4r \cos \theta = 1$ by transforming it into rectangular coordinates.

Solution.   Using the method of point plotting yields very inconclusive information. (Try it!) However, the Cartesian equation of the curve is $4x = 1$, or $x = \frac{1}{4}$, which is a line parallel to the y-axis.

EXERCISES

Sketch the graph of each of the following equations.

1. $r = 5.6$
2. $r = \sqrt{2}$
3. $\theta = \pi/3$
4. $\theta = 170°$
5. $r = 2 \sin \theta$
6. $r = 0.5 \cos \theta$
7. $r \sin \theta = 1$
8. $r \cos \theta = -10$
9. $r = 1 + \sin \theta$
10. $r = 1 - \cos \theta$
11. $r = \sec \theta$
12. $r = -\sin \theta$
13. $r = 4 \sin 3\theta$ (three-leaved rose)
14. $r = 2 - \sin \theta$ (limacon)
15. $r = 2 \cos 2\theta$ (four-leaved rose)
16. $r = \sin 2\theta$ (four-leaved rose)
17. The radiation pattern of a particular two-element antenna is a cardioid of the form $r = 100 \, (1 + \cos \theta)$. Sketch the radiation pattern of this antenna.
18. The radiation pattern of a certain antenna is given by $r = 1/(2 - \cos \theta)$. Plot this pattern.
19. By transforming the polar equation in Exercise 18 into rectangular coordinates, show that the indicated radiation pattern is elliptical.
20. The feedback diagram of a certain electronic tachometer can be approximated by the curve $r = \frac{1}{2}\theta$. Sketch the feedback diagram of this tachometer from $\theta = 0$ to $\theta = 7\pi/6$.

15.5   CURVILINEAR MOTION

When a particle travels in a curved path, it is sometimes convenient to express its position in polar coordinates. The position of the particle, P, in Figure 15.9 is then specified at any instant by giving the values of r and θ as functions of time.

The linear velocity of P, which is of interest to us, can be resolved into two components which are parallel and perpendicular to the radius vector. The component that is parallel to the radius vector is denoted v_r, and is called the *radial* component of the velocity v. The perpendicular compon-

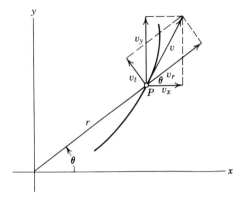

Figure 15.9

ent is denoted v_t, and called the *transverse* component of v. These components of velocity are more useful in some instances than horizontal and vertical components. They can be obtained from the horizontal and vertical components of velocity in the following manner. From Figure 15.9 the rectangular coordinates of P, in terms of polar coordinates, are

$$x = r \cos \theta \quad \text{and} \quad y = r \sin \theta$$

Differentiation of the above equations yields the rectangular components of velocity. Since r and θ are both functions of t, the expression on the right of each equation is differentiated as a product of two functions of t. Using $v_x = dx/dt$, and $v_y = dy/dt$, we have

$$v_x = r(-\sin \theta)\frac{d\theta}{dt} + \cos \theta \frac{dr}{dt} \tag{6}$$

and

$$v_y = r \cos \theta \frac{d\theta}{dt} + \sin \theta \frac{dr}{dt} \tag{7}$$

Next we see from Figure 15.10 that the x and y components of the velocity of P can be resolved into radial and transverse components by considering the radial and transverse directions as the reference for the coordinate system. The algebraic sum of the transverse components of v_x and v_y gives the transverse component of velocity as

$$v_t = -v_x \sin \theta + v_y \cos \theta$$

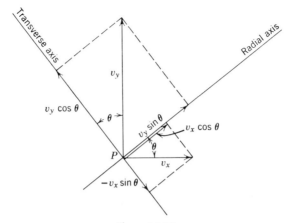

Figure 15.10

Substituting Equations 6 and 7 into this equation, we obtain

$$v_t = r \sin^2 \theta \, \frac{d\theta}{dt} - \sin \theta \cos \theta \, \frac{dr}{dt} + r \cos^2 \theta \, \frac{d\theta}{dt} + \sin \theta \cos \theta \, \frac{dr}{dt}$$

$$= r \frac{d\theta}{dt} \, (\sin^2 \theta + \cos^2 \theta)$$

$$v_t = r \frac{d\theta}{dt} \tag{8}$$

Also, we have

$$v_r = v_x \cos \theta + v_y \sin \theta$$

$$= \cos^2 \theta \, \frac{dr}{dt} - r \sin \theta \cos \theta \, \frac{d\theta}{dt} + \sin^2 \theta \, \frac{dr}{dt} + r \sin \theta \cos \theta \, \frac{d\theta}{dt}$$

$$v_r = \frac{dr}{dt} \tag{9}$$

The magnitude of the resultant velocity is then

$$v = \sqrt{v_t{}^2 + v_r{}^2} \tag{10}$$

The radial and transverse components of velocity change with time, which means that there is an acceleration in both directions. By a procedure which is very similar to the one used above for finding the radial and transverse components of velocity, we find the radial and transverse components of acceleration to be

$$a_r = -r\left(\frac{d\theta}{dt}\right)^2 + \frac{d^2r}{dt^2} \tag{11}$$

$$a_t = r\frac{d^2\theta}{dt^2} + 2\frac{dr}{dt}\frac{d\theta}{dt} \tag{12}$$

Also, the resultant acceleration is given by

$$a = \sqrt{a_r^2 + a_t^2} \tag{13}$$

Example 10. A particle moves along the curve $r = \sin\theta$ with a constant angular velocity of 2 rad/sec. Find v_r and v_t when $\theta = \pi/6$, if r is measured in ft.

Solution. By applying Equations 8 and 9, we have

$$v_r = \frac{dr}{dt} = \frac{d}{dt}[\sin\theta] = \cos\theta\,\frac{d\theta}{dt}$$

and

$$v_t = r\frac{d\theta}{dt} = \sin\theta\,\frac{d\theta}{dt}$$

Substituting $\theta = \pi/6$ and $d\theta/dt = 2$ into these two equations, we have

$$v_r = \cos\frac{\pi}{6}\,(2) = 0.866\,(2) = 1.732\,\text{ft/sec}$$

$$v_t = \sin\frac{\pi}{6}\,(2) = 0.500\,(2) = 1.000\,\text{ft/sec}$$

In addition, the velocity tangent to the curve at this point is given by

$$v = \sqrt{v_r^2 + v_t^2} = \sqrt{(1.732)^2 + (1.000)^2} = 2\,\text{ft/sec}$$

Example 11. Find the equations for the radial and transverse components of acceleration when the particle is moving in a circular path about the pole. (See Figure 15.11.)

Solution. When a particle moves in a circular path, the radius vector is constant and, therefore,

$$v_r = \frac{dr}{dt} = 0$$

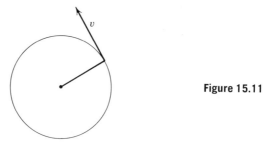

Figure 15.11

The velocity of an object moving in a circle is then equal to the transverse component of velocity alone. For this reason, when we talk about pure circular motion, we may drop the subscript t and simply write

$$v = r\frac{d\theta}{dt} = r\omega$$

where ω is frequently used to represent the angular velocity $d\theta/dt$. By substituting $dr/dt = 0$ and $d^2r/dt^2 = 0$ into Equations 11 and 12, the radial and transverse components of acceleration become

$$a_r = -r\left(\frac{d\theta}{dt}\right)^2 \tag{14}$$

$$a_t = r\frac{d^2\theta}{dt^2} \tag{15}$$

The minus sign in (14) indicates that a_r is directed inward along the radius vector.

The student may wonder why there is a radial component of acceleration when the radial component of velocity is zero. This is explained by the fact that an acceleration occurs when either the magnitude or direction of the velocity vector changes. The radial acceleration in (14) is a result of the continually changing direction of the velocity vector of P. This acceleration is referred to as the *normal* acceleration and is frequently written in the form

$$a_n = r\omega^2 \tag{16}$$

The negative sign is eliminated by assuming that accelerations directed toward the point of rotation are positive.

EXERCISES

1. A charged particle moves along a path that is given by the spiral $r = e^\theta$. Find the radial and transverse components of the velocity of the particle when $\theta = \pi$, if the particle is moving counter-clockwise at a constant rate of 1 radian/sec.
2. Find the magnitude of the velocity of the particle in Exercise 1.
3. Find the radial and transverse components of the acceleration of the particle in Exercise 1 and also find the resultant acceleration.
4. Find the radial and transverse components of the velocity of a particle when $\theta = \pi/2$, if its path is described by $r = 2\theta$, and if it is moving counter-clockwise at a constant rate of 2 radians/sec.
5. A particle moves along the curve $r = 3 \sin 2\theta$ in such a way that the transverse component of velocity is always 2 ft/sec. Find the radial component of velocity when $\theta = \pi/8$.
6. The equation of the center line of a certain fixed cam is $r = 1 + \cos \theta$. The follower slides along the cam in a counter-clockwise direction with a constant angular velocity of 0.5 radians/sec. Find the radial and transverse components of the follower velocity when $\theta = 30°$.
7. The follower of a fixed cam moves along the path $r = \sin \theta$ in a counter-clockwise direction. When the angle θ is 30°, the radial component of velocity is 15 in./sec. Find (a) the angular velocity of the follower at this point, (b) the transverse velocity, and (c) the total velocity.

15.6   AREA IN POLAR COORDINATES

The area bounded by a polar curve can be found by integration. Consider the area bounded by the polar curve $r = f(\theta)$, and the radial lines $\theta = \alpha$ and $\theta = \beta$ as shown in Figure 15.12. The area of this figure can be found by dividing the area into the circular sectors, and then taking the sum of the areas of these sectors as the number of sectors increases without bound.

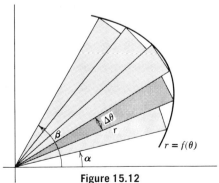

Figure 15.12

We begin our construction by dividing the central angle of the given area into n equal angles $\Delta\theta$ and then drawing circular arcs as shown in Figure 15.12. The area of a circular sector is given in geometry as $A = \frac{1}{2}r^2\theta$, where r is the radius of the circle and θ is the central angle of the sector. The area of any one of the elementary circular sectors is then given by

$$\Delta A_i = \frac{1}{2}r_i^2\Delta\theta$$

The approximate area bounded by $r = f(\theta)$, $\theta = \alpha$, and $\theta = \beta$ is

$$A_n = \sum_{i=1}^{n} \frac{1}{2}r_i^2\,\Delta\theta$$

The exact area is then given by the definition of the definite integral

$$A = \lim_{n\to\infty} \sum_{i=1}^{n} \frac{1}{2}r_i^2\,\Delta\theta = \frac{1}{2}\int_{\alpha}^{\beta} r^2\,d\theta \qquad (17)$$

Example 12.   Find the area bounded by the circle $r = 3\sin\theta$.

Solution.   This area is shown in Figure 15.13. The element of area is given by

$$\Delta A_i = \frac{1}{2}r_i^2\,\Delta\theta = \frac{9}{2}\sin^2\theta\,\Delta\theta$$

so that the desired area is given by

$$A = \frac{9}{2}\int_0^{\pi}\sin^2\theta\,d\theta = \frac{9}{2}\int_0^{\pi}\frac{1}{2}(1-\cos 2\theta)\,d\theta$$

$$= \frac{9}{4}\left[\theta - \frac{1}{2}\sin 2\theta\right]_0^{\pi} = \frac{9\pi}{4}$$

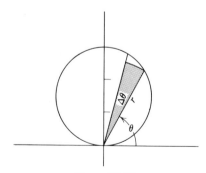

Figure 15.13

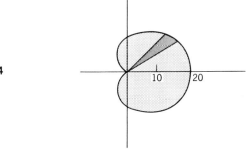

Figure 15.14

Example 13. The radiation pattern of a two element antenna is a cardioid of the form $r = 20 (1 + \cos \theta)$, where r is measured in miles. Find the area covered by this antenna. (See Figure 15.14.)

Solution. The area of this antenna pattern is given by the integral

$$A = \frac{1}{2} \int_0^{2\pi} [20(1 + \cos \theta)]^2 \, d\theta = 200 \int_0^{2\pi} (1 + 2 \cos \theta + \cos^2 \theta) \, d\theta$$

$$= 200 \left[\theta + 2 \sin \theta + \frac{\theta}{2} + \frac{1}{4} \sin 2\theta \right]_0^{2\pi}$$

$$A = 200 [2\pi + \pi] = 600\pi = 1884 \text{ sq mi}$$

EXERCISES

In Exercises 1 to 4, find the area enclosed by the graphs of the given equations. Sketch each graph.

1. $r = 10, \theta = 0, \theta = \pi/6$.

2. $r = \sin \theta, \theta = \pi/6, \theta = \pi/4$

3. $r = \theta, \theta = 45°, \theta = 180°$

4. $r = \tan \theta, \theta = 30°, \theta = 45°$

In Exercises 5 to 12, find the area enclosed by the graph of the given equation. Sketch each graph.

5. $r^2 = 4 \sin 2\theta$

6. $r^2 = \cos 2\theta$

7. $r = 7 \cos \theta$

8. $r = 8 \sin \theta$

9. $r = 3 \cos 2\theta$

10. $r = 2 \sin 2\theta$

11. $r = 3(1 - \sin \theta)$

12. $r = 2 + \cos \theta$

13. The directional pattern of certain types of microphones is given by the cardioid $r = a(1 + \cos \theta)$, where a is a constant. Find the area of this pattern in terms of the constant a.

16

Infinite Series

16.1 SEQUENCES

The half-life of a radioactive substance is the time it takes for half of the original atoms to undergo radioactive transformation. In the case of radium, the half-life is 1620 years; this means that if we start with 100 g of radium, we will have 50 g at the end of 1620 years, 25 g after 3240 years, 12.5 g after 4860 years, etc. The mass of radium remaining after successive half-lifes may then be written

$$100, 50, 25, 12.5 \tag{1}$$

Of course, this succession of half-lifes can be extended to include as many half-life values as we choose. Notice that the mass after each half-life is one half its previous value, so that (1) can also be written as

$$100, \frac{100}{2^1}, \frac{100}{2^2}, \frac{100}{2^3}, \cdots, \frac{100}{2^n}, \cdots \tag{2}$$

This example shows how a mathematical quantity called a *sequence* arises naturally in scientific work. In mathematics, a sequence of terms is a special type of function defined only for the positive integers. In the example above, the general rule for the sequence is given by

$$a_n = \frac{100}{2^n}$$

where each term of the sequence can be obtained by simply substituting successive integer values for n. Normally the general rule is either given or can be inferred from the information given.

Example 1. The sequence defined by the rule $a_n = 2n - 1$ gives the terms $a_1 = 1$, $a_2 = 3$, $a_3 = 5$, etc. Thus, the indicated sequence is

$$1, 3, 5, \ldots$$

Inspection of this sequence reveals that any two succeeding terms differ by a constant. Sequences that have the property $a_{n+1} - a_n = d$ are called

376

arithmetic sequences. The given sequence is therefore arithmetic with $d = 2$.

Example 2.   A sequence is called *geometric* if the ratio of any two succeeding terms is a constant, that is if $a_{n+1}/a_n = r$. The sequence $1, \frac{1}{2}, \frac{1}{4}, \frac{1}{8}, \ldots$ is geometric with $r = \frac{1}{2}$. Inspection of this sequence suggests that the general term is $a_n = 1/2^{n-1}$.

Example 3.   The sequence defined by the rule $a_n = (-1)^n$ has the terms

$$a_1 = -1, a_2 = 1, a_3 = -1, \text{etc.}$$

Example 4.   The sequence defined by the rule $a_n = \sum_{k=1}^{n} (\frac{1}{2})^{k-1}$ has the terms

$$a_1 = 1$$
$$a_2 = 1 + \frac{1}{2}$$
$$a_3 = 1 + \frac{1}{2} + \frac{1}{4}$$
$$\cdot$$
$$\cdot$$
$$\cdot$$

Example 5.   Find the general term of the sequence

$$-\tfrac{1}{2}, 0, \tfrac{1}{4}, \tfrac{2}{5}, \ldots$$

Solution.   We observe that the numerator of the first term is -1, of the second term is 0, of the third term is 1, of the fourth term is 2. This suggests that the numerator of the general term is specified by $n-2$. The denominators appear to be one more than the number of the term, so that $n+1$ represents the denominator of the general term. The expression

$$a_n = \frac{n-2}{n+1}$$

is therefore the general term of the sequence.

In working with sequences, products of successive integers beginning with 1 occur quite often. To handle this, we introduce the notation $n!$ to denote the product

$$n! = 1 \times 2 \times 3 \times \cdots \times (n-1) \times n$$

and we refer to this product as *n factorial*. Thus, the value of 4 factorial is

$$4! = 1 \times 2 \times 3 \times 4 = 24$$

Because of circumstances that arise in formulating the general term of a sequence it is convenient to define factorial zero to be equal to one; that is

$$0! = 1$$

Example 6.   Find the general term of the sequence

$$\frac{x}{1}, \frac{x^2}{1}, \frac{x^3}{1 \cdot 2}, \frac{x^4}{1 \cdot 2 \cdot 3}, \cdots$$

Solution.   The numerator of the general term is obviously x^n. The denominator can be described by $(n-1)!$ since the denominator of the first term is $0!$, of the second term is $1!$, of the third term is $2!$, etc. Hence, the general term is given by

$$a_n = \frac{x^n}{(n-1)!}$$

Example 7.   Find the general term of the sequence

$$\frac{1}{\sqrt{3}}, \frac{-1}{\sqrt{5}}, \frac{1}{\sqrt{7}}, \frac{-1}{\sqrt{9}}, \cdots$$

Solution.   To account for the alternate positive and negative terms of the sequence we use $(-1)^{n+1}$. This is positive when n is odd and negative when n is even. We also observe that the denominator can be expressed by $\sqrt{2n+1}$. The general term is therefore given by

$$a_n = \frac{(-1)^{n+1}}{\sqrt{2n+1}}$$

EXERCISES

In Exercises 1 to 4, write the first four terms of each sequence.

1. $a_n = \dfrac{1}{2n-1}$

3. $a_n = \dfrac{2^n}{(n-1)!}$

2. $a_n = \dfrac{n}{2^{n-1}}$

4. $a_n = \dfrac{x^{n-1}}{\sqrt{2n+1}}$

Discover by inspection the general term of the indicated sequence.

5. $1, 2, \dfrac{3}{2!}, \dfrac{4}{3!}, \cdots$

6. $1, \dfrac{1}{2}, \dfrac{1}{4}, \dfrac{1}{8}, \cdots$

7. $1, \dfrac{1}{3!}, \dfrac{1}{5!}, \dfrac{1}{7!}, \ldots$

11. $\dfrac{x^2}{2}, \dfrac{x^4}{4}, \dfrac{x^6}{8}, \dfrac{x^8}{16}, \ldots$

8. $1, \dfrac{1}{4}, \dfrac{1}{9}, \dfrac{1}{16}, \ldots$

12. $x, \dfrac{x^3}{3!}, \dfrac{x^5}{5!}, \dfrac{x^7}{7!}, \ldots$

9. $\dfrac{1}{3}, \dfrac{2}{5}, \dfrac{3}{7}, \dfrac{4}{9}, \ldots$

13. $1, x, \dfrac{x^2}{2!}, \dfrac{x^3}{3!}, \ldots$

10. $\dfrac{1}{\sqrt{2}}, \dfrac{1}{\sqrt{3}}, \dfrac{1}{\sqrt{4}}, \dfrac{1}{\sqrt{5}}, \ldots$

14. $\dfrac{e}{\sqrt{2}}, \dfrac{e^2}{\sqrt{4}}, \dfrac{e^3}{\sqrt{6}}, \dfrac{e^4}{\sqrt{8}}, \ldots$

16.2 CONVERGENCE AND DIVERGENCE

When we worked with general functions of the form $f(x)$, we were concerned with limiting values; with sequences, our main concern is with asymptotic behavior for large values of n.

The sequence $1, \frac{1}{2}, \frac{1}{3}, \ldots$ is specified by the general term $a_n = 1/n$. Since the general term is close to zero for large n, we say that the limit of the sequence is zero. Similarly, since $a_n = 3 - 1/\sqrt{n}$ is close to 3 when n is large, we say that the limit of the sequence defined by this general term is 3. Accordingly, we make the following definition of the limit of a sequence.

Definition

Let a_n be the nth term of a sequence. If a_n approaches some specific number L as n increases without bound, then L is said to be the limit of the sequence and we write

$$\lim_{n \to \infty} a_n = L$$

In this case the sequence is said to be *convergent* and *converges* to L; otherwise the sequence is said to be *divergent*.

Example 8. The sequence $a_n = 1/(n+1)$ converges to zero, since

$$\lim_{n \to \infty} a_n = \lim_{n \to \infty} \frac{1}{n+1} = 0$$

Example 9. The sequence $a_n = n/(n+1)$ converges to 1, since

$$\lim_{n \to \infty} a_n = \lim_{n \to \infty} \frac{n}{n+1} = \lim_{n \to \infty} = \frac{1}{1 + (1/n)} = 1$$

Example 10. The sequence $a_n = n^2$ diverges, since

$$\lim_{n \to \infty} a_n = \lim_{n \to \infty} n^2 = \infty$$

Example 11.   The sequence $a_n = (-1)^n$ diverges since $\lim_{n \to \infty} (-1)^n$ does not exist but alternates between 1 and -1.

Examples 10 and 11 are important because they show that a sequence can diverge for two basic reasons: (1) the value of the terms may increase without bound as n increases without bound, or (2) the general term may fail to approach a definite limit as n increases without bound.

Sometimes, even though the general rule is given, it is very difficult to determine the convergence or divergence of a sequence. This is especially true when the general rule is given in terms of a sum.

Example 12.   Show that the sequence $1, 1 + \frac{1}{2}, 1 + \frac{1}{2} + \frac{1}{4}, \ldots$ converges to 2.

Solution.   This is the sequence described in Example 4, so the general term is

$$a_n = \sum_{k=1}^{n} \left(\frac{1}{2}\right)^{k-1}$$

However, a_n is not in a form where we can easily take the limit as n increases without bound. The key to this problem is to recognize a_n as the sum of n terms of a geometric progression with $a = 1$ and $r = \frac{1}{2}$. Then, recalling from algebra that the sum of the first n terms of a geometric progression is $a(1 - r^n)/(1 - r)$, we can write

$$a_n = \sum_{k=1}^{n} \left(\frac{1}{2}\right)^{k-1} = \frac{1 - (1/2)^n}{1 - (1/2)} = 2\left[1 - \left(\frac{1}{2}\right)^n\right]$$

Since

$$\lim_{n \to \infty} a_n = \lim_{n \to \infty} 2\left[1 - \left(\frac{1}{2}\right)^n\right] = 2$$

the sequence converges to 2.

EXERCISES

In each of the following exercises the general term of the sequence is given. Write the first four terms of the sequence and then investigate the convergence or divergence of each sequence.

1. $a_n = \dfrac{2}{n}$

2. $a_n = 3 - \dfrac{2}{n}$

3. $a_n = \dfrac{n}{n+3}$

4. $a_n = \dfrac{n+1}{2n-1}$

5. $a_n = 5 + (-1)^n$

6. $a_n = \dfrac{n^2 - 2}{n^2 + 2}$

7. $a_n = \dfrac{(-1)^n}{n}$

8. $a_n = \dfrac{3n}{n^2 + 3}$

9. $a_n = \dfrac{n^3}{n^2 + 5}$

10. $a_n = (-1)^n + \dfrac{1}{n}$

11. $a_n = \displaystyle\sum_{k=1}^{n} \left(\frac{1}{3}\right)^k$

12. $a_n = \displaystyle\sum_{k=1}^{n} \left(\frac{1}{4}\right)^{k-1}$

16.3   SERIES

We are frequently interested in the sum of the terms of a sequence. When the sequence is finite, the problem of finding the sum consists simply of finding the algebraic sum of all of the terms in the sequence. It is always possible to find such a sum, although it is sometimes impractical. However, the problem is not so simple when we are dealing with an infinite sequence, since no matter how many terms we sum there are always more terms remaining to be summed. It is soon apparent that the "sum" of an infinite number of terms of a sequence does not exist in the sense that we ordinarily use for sum. We therefore define what we mean by an *infinite sum*.

Definition
An expression of the form

$$a_1 + a_2 + a_3 + \cdots + a_n + \cdots = \sum_{n=1}^{\infty} a_n \qquad (3)$$

is called an *infinite series*. This definition only gives a name to an endless summation, it does not give it a meaning.

We can arrive at a reasonable meaning for an infinite series by associating with the series in (3) a sequence $S_1, S_2, \ldots, S_n$ which is defined by

$$S_1 = a_1$$
$$S_2 = a_1 + a_2$$
$$S_3 = a_1 + a_2 + a_3$$
$$\cdot$$
$$\cdot$$
$$\cdot$$
$$S_n = a_1 + a_2 + a_3 + \cdots + a_n$$

The sequence $S_1, S_2, S_3, \ldots$ is called the *sequence of partial sums* for the series $a_1 + a_2 + a_3 + \ldots$. Each term of the sequence of partial sums is said to be a finite series of terms of the original series.

Example 13.   Find the sequence of partial sums of the series $\sum\limits_{n=1}^{\infty} n = 1 + 2 + 3 + \cdots$.

Solution.   The sequence of partial sums for this series is

$$S_1 = 1$$
$$S_2 = 1 + 2$$
$$S_3 = 1 + 2 + 3$$
$$\cdot$$
$$\cdot$$
$$\cdot$$
$$S_n = 1 + 2 + 3 + \cdots + n$$

As with any sequence, we are interested in whether a sequence of partial sums converges or diverges. In the case of a sequence of partial sums we must determine the existence or nonexistence of

$$\lim_{n\to\infty} S_n$$

or, another way to write the same thing

$$\lim_{n\to\infty} \sum_{k=1}^{n} a_k \tag{4}$$

The limit in (4) has come to be written $\sum\limits_{k=1}^{\infty} a_k$, and it is called the sum of the infinite series. You must realize that $\sum\limits_{k=1}^{\infty} a_k$ is not the sum of infinitely many terms, but merely a notation for the limit of the partial sums of the series $a_1 + a_2 + a_3 + \cdots$. If the limit in (4) exists, the infinite series is said to converge; otherwise, it is said to diverge.

Example 14.   Find the sum of the infinite series $\sum\limits_{k=1}^{\infty} \left(\frac{1}{2}\right)^{k-1}$ if it exists.

Solution.   We resist the temptation to think of adding infinitely many terms together by first writing the partial sum

$$S_n = \sum_{k=1}^{n} \left(\frac{1}{2}\right)^{k-1}$$

Then, as in Example 12, we can write

$$S_n = \sum_{k=1}^{n} \left(\frac{1}{2}\right)^{k-1} = \frac{1-(1/2)^n}{1-(1/2)} = 2\left[1 - \left(\frac{1}{2}\right)^n\right]$$

Therefore

$$\lim_{n\to\infty} S_n = \lim_{n\to\infty} 2\left[1 - \left(\frac{1}{2}\right)^n\right] = 2$$

Example 15.   Show that the sum of the infinite series $\sum_{k=1}^{\infty} (-1)^{k+1}$ does not exist.

Solution.   Here

$$S_n = 1 - 1 + 1 - 1 + \cdots + (-1)^{n+1}$$

The value of S_n depends upon whether n is odd or even. When n is odd, $S_n = 1$; and when n is even, $S_n = 0$. Consequently, $\lim_{n\to\infty} S_n$ does not exist, and we say the series diverges.

Intuitively you might suspect that a series where the terms decrease as n increases will always converge. However, this would be wrong. As the next example shows, the partial sums of a sequence can become arbitrarily large even though the limit of the nth term is zero.

Example 16.   By examining the sequence of partial sums, show that the series $\sum_{k=1}^{\infty} \frac{1}{\sqrt{k}}$ diverges.

Solution.   The general term of this series, which is $a_n = 1/\sqrt{k}$, is easily seen to go to zero as n increases without bound. However, the partial sum

$$S_n = \frac{1}{\sqrt{1}} + \frac{1}{\sqrt{2}} + \frac{1}{\sqrt{3}} + \cdots + \frac{1}{\sqrt{n}}$$

increases without bound as n increases. To see why, note that the smallest term in the sum is the last. It is therefore true that

$$S_n > \frac{1}{\sqrt{n}} + \frac{1}{\sqrt{n}} + \frac{1}{\sqrt{n}} + \cdots + \frac{1}{\sqrt{n}} = n\left(\frac{1}{\sqrt{n}}\right) = \sqrt{n}$$

Since $\sqrt{n}$ clearly increases as n increases, so must S_n. But this means the series diverges.

The message in Example 16 is that you cannot study infinite series using the concepts of finite addition you learned in grade school. If you are to understand convergence and divergence of infinite series, it is essential that you distinguish between finite addition and infinite addition.

The problem of finding the "sum" of an infinite series $\sum\limits_{n=1}^{\infty} a_n$ is usually dependent upon our ability to find a general formula for the partial sum S_n, which does not involve summation. (See Example 14.) Unfortunately the formulation of S_n into a rule not involving a sum is impractical for most series and, therefore, other techniques are needed. There are tests which will tell us whether a series converges or diverges merely by examining the terms of the series. None of these tests, however, will tell us the value to which the series converges.

In the next three sections we shall consider some of the more popular tests for convergence and divergence. Generally, the analysis of an infinite series is quite difficult, so we restrict ourselves in the next section to series of positive constants; in Section 16.5 we allow the terms to be of different signs.

16.4   THE INTEGRAL TEST

In this section we show a test which basically compares the terms of an infinite series with a function. This test, which is called the integral test, uses the convergence or divergence of the improper integral of a function to tell precisely what the infinite series does.

Consider a series of positive terms $\sum\limits_{n=1}^{\infty} a_n$ decreasing in size, such that $\lim\limits_{n\to\infty} a_n = 0$. Assume the terms of the series are represented graphically in Figure 16.1. Let $f(x)$ be a continuous function that agrees with the terms of the series on the positive integers. Then the graph of $f(x)$ would look something like the smooth curve in Figure 16.1.

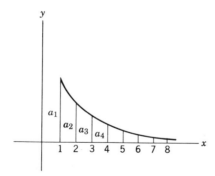

Figure 16.1

The basis of the integral test is suggested by the diagram. By comparing the area under the curve $y = f(x)$ from 1 to n with the sum $a_1 + a_2 + \cdots + a_n$, it can be shown that the series can converge only if the area under the curve approaches a limit as $n \to \infty$; otherwise, the series will diverge.

The Integral Test
Let $f(n)$ denote the general term of the series of positive terms

$$a_1 + a_2 + a_3 + \cdots + a_n + \cdots$$

If the function $f(x)$ is always positive and decreasing for $x \geq 1$, then the series converges if the integral

$$\int_1^\infty f(x)\, dx$$

exists, and diverges if it does not exist.

The function $f(x)$ referred to in the integral test is formed by simply replacing n with x in the expression for the general term of the series.

Note that the integral test allows us to discover whether or not a series is convergent, but it does not yield the numerical value to which the series converges. This, however, is no hindrance, since it is usually only the fact that a series converges that is important.

We will use the following examples to explain the application of the integral test.

Example 17.   Prove that this series converges

$$1 + 1/3^2 + 1/5^2 + 1/7^2 + \cdots$$

Solution.   The general term for this series is $f(n) = 1/(2n-1)^2$, so that $f(x) = 1/(2x-1)^2$. Throughout the interval $x \geq 1$, the function $f(x) = 1/(2x-1)^2$ is positive and decreases as x increases. The required conditions are satisfied so the integral test may be used. This yields the integral

$$\int_1^\infty \frac{dx}{(2x-1)^2} = \lim_{b \to \infty} \int_1^b \frac{dx}{(2x-1)^2} = \lim_{b \to \infty} \left[-\frac{1}{2(2x-1)} \right]_1^b$$

$$= \lim_{b \to \infty} \left[-\frac{1}{2(2b-1)} + \frac{1}{2} \right] = \frac{1}{2}$$

Since this integral exists, the given series converges. Keep in mind that the integral test only establishes convergence, it does not yield the value to which the series converges. Thus, the numerical value of the integral plays no essential role in the test.

Example 18.   Using the integral test, show that the series

$$\sum_{n=1}^{\infty} \frac{1}{n} = 1 + 1/2 + 1/3 + \cdots$$

diverges. This series is known as the harmonic series.

Solution.   Here, the general term is given by $f(n) = 1/n$, so $f(x) = 1/x$. The function $f(x) = 1/x$ is positive and decreasing as x increases and, therefore, the integral test is applicable to this series. Thus

$$\int_{1}^{\infty} \frac{dx}{x} = \lim_{p \to \infty} \int_{1}^{p} \frac{dx}{x} = \lim_{p \to \infty} [\ln x]_{1}^{p} = \lim_{p \to \infty} [\ln p] = \infty$$

Since this integral does not exist, the harmonic series diverges.

EXERCISES

Each of the series in Exercises 1 to 4 is a geometric series. Use Equation 4 and the definition of convergence to tell whether the series is convergent or divergent and, if it is convergent, find its sum.

1. $\displaystyle\sum_{n=1}^{\infty} \left(\frac{1}{3}\right)^{n-1}$

3. $\displaystyle\sum_{n=1}^{\infty} \frac{3}{2^{n-1}}$

2. $\displaystyle\sum_{n=1}^{\infty} \left(\frac{3}{2}\right)^{n-1}$

4. $\displaystyle\sum_{n=1}^{\infty} \left(\frac{4}{5}\right)^{n-1}$

Using the integral test, investigate the convergence of the following series.

5. $1 + \dfrac{1}{2^3} + \dfrac{1}{3^3} + \dfrac{1}{4^3} + \cdots + \dfrac{1}{n^3} + \cdots$

6. $\dfrac{1}{2} + \dfrac{1}{4} + \dfrac{1}{6} + \dfrac{1}{8} + \cdots + \dfrac{1}{2n} + \cdots$

7. $1 + \dfrac{1}{\sqrt{2}} + \dfrac{1}{\sqrt{3}} + \dfrac{1}{\sqrt{4}} + \cdots + \dfrac{1}{\sqrt{n}} + \cdots$

8. $1 + \dfrac{1}{2^{3/2}} + \dfrac{1}{3^{3/2}} + \dfrac{1}{4^{3/2}} + \cdots + \dfrac{1}{n^{3/2}} + \cdots$

9. $\dfrac{1}{e} + \dfrac{1}{e^2} + \dfrac{1}{e^3} + \dfrac{1}{e^4} + \cdots + \dfrac{1}{e^n} + \cdots$

10. $\dfrac{1}{3^2} + \dfrac{1}{5^2} + \dfrac{1}{7^2} + \dfrac{1}{9^2} + \cdots + \dfrac{1}{(2n+1)^2} + \cdots$

11. $\dfrac{1}{2} + \dfrac{2}{5} + \dfrac{3}{10} + \dfrac{4}{17} + \cdots + \dfrac{n}{n^2+1} + \cdots$

12. $\dfrac{1}{2} + \dfrac{1}{5} + \dfrac{1}{10} + \dfrac{1}{17} + \cdots + \dfrac{1}{n^2+1} + \cdots$

13. $\dfrac{e}{1+e} + \dfrac{e^2}{1+e^2} + \dfrac{e^3}{1+e^3} + \dfrac{e^4}{1+e^4} + \cdots + \dfrac{e^n}{1+e^n} + \cdots$

14. $\dfrac{1}{4} + \dfrac{2}{25} + \dfrac{3}{100} + \cdots + \dfrac{n}{(1+n^2)^2} + \cdots$

15. The series $\displaystyle\sum_{n=1}^{\infty} \dfrac{1}{n^p} = 1 + \dfrac{1}{2^p} + \dfrac{1}{3^p} + \cdots + \dfrac{1}{n^p} + \cdots$

is called the p-series. Use the integral test to prove that: (a) the series converges for $p > 1$, and (b) the series diverges for $p = 1$ and $p < 1$.

16.5 THE RATIO TEST

Another test for convergence is the so-called *ratio test*. The ratio test is based upon a comparison of the given series with some geometric series. In the ratio test we determine the ratio of two succeeding terms *in the limit*, which for a geometric series is always the same. The following example shows how to determine the convergence or divergence of a general geometric series.

Example 19. Determine the values of R for which the geometric series $\displaystyle\sum_{k=1}^{\infty} R^k$ is convergent.

Solution. We write the sequence of partial sums

$$S_n = \sum_{k=1}^{n} R^k$$

By the formula for the sum of a geometric series,

$$S_n = \frac{1 - R^{n+1}}{1 - R} = \frac{1}{1-R} - \frac{R^{n+1}}{1-R}, \qquad R \neq 1$$

Since R is a fixed number, we have

$$\lim_{n \to \infty} S_n = \frac{1}{1-R} \qquad \text{if } |R| < 1$$

$$= \infty \qquad \text{if } |R| > 1$$

If $R = 1$, then the original series becomes $\sum\limits_{k=1}^{\infty} (1)^k$ which obviously diverges.

The ratio test tells us that the same techniques we used to determine the convergence or divergence of a geometric series holds for nongeometric series, with one exception. When the limiting value of the ratio is 1, the test fails.

The Ratio Test

Consider the infinite series

$$a_1 + a_2 + a_3 + \cdots + a_n + a_{n+1} + \cdots$$

of constant terms, with either mixed or like signs. Let a_n and a_{n+1} represent any two consecutive terms of the series, and take the limit of the absolute value of the ratio of a_{n+1} to a_n as n increases. Denoting this by

$$R = \lim_{n \to \infty} \left| \frac{a_{n+1}}{a_n} \right|$$

it can be proved that

(1) When $R < 1$, the series is convergent.
(2) When $R > 1$, the series is divergent.
(3) When $R = 1$, the test fails.

If the ratio test yields $R = 1$, we cannot make a decision as to whether the series is convergent or divergent. When this happens, the series must be investigated for convergence by another test, such as the integral test.

Example 20.   Use the ratio test to investigate the convergence or divergence of the series

$$-\frac{1}{2} + \frac{2}{2^2} - \frac{3}{2^3} + \cdots + \frac{(-1)^n n}{2^n} + \cdots$$

Solution.   Here $a_n = [(-1)^n n]/(2^n)$ and $a_{n+1} = [(-1)^{n+1}(n+1)]/2^{n+1}$. The test ratio is then given by

$$\left| \frac{a_{n+1}}{a_n} \right| = \left| \frac{((-1)^{n+1}(n+1)/2^{n+1})}{(-1)^n n/2^n} \right| = \frac{n+1}{2^{n+1}} \cdot \frac{2^n}{n} = \frac{n+1}{2n} = \frac{1}{2} + \frac{1}{2n}$$

Taking the limit as $n \to \infty$, we have

$$R = \lim_{n \to \infty} \left(\frac{1}{2} + \frac{1}{2n} \right) = \frac{1}{2}$$

Since $R < 1$, we conclude that the given series converges.

Example 21.   Use the ratio test to investigate the convergence or divergence of the series

$$1 + \frac{1}{2^2} + \frac{1}{3^2} + \cdots + \frac{1}{n^2} + \cdots$$

Solution.   Here $a_n = 1/n^2$ and $a_{n+1} = 1/(n+1)^2$, so that

$$R = \lim_{n \to \infty} \left(\frac{1/(n+1)^2}{1/n^2} \right) = \lim_{n \to \infty} \left(\frac{n^2}{n^2 + 2n + 1} \right) = \lim_{n \to \infty} \left(\frac{1}{1 + 2/n + 1/n^2} \right) = 1$$

When $R = 1$, the test fails to determine whether the series is convergent or divergent. However, the series can be shown to be convergent by applying the integral test with $f(x) = 1/x^2$. Thus

$$\int_1^\infty \frac{dx}{x^2} = \lim_{b \to \infty} \int_1^b x^{-2} dx = \lim_{b \to \infty} \left[-\frac{1}{x} \right]_1^b = \lim_{b \to \infty} \left[-\frac{1}{b} + 1 \right] = 1$$

Therefore, since the integral exists, the given series converges.

Example 22.   Use the ratio test to show that this series is divergent:

$$\frac{1!}{\sqrt{2}} + \frac{2!}{\sqrt{3}} + \frac{3!}{\sqrt{4}} + \cdots + \frac{n!}{\sqrt{n+1}} + \cdots$$

Solution.   The nth term of this series is $a_n = n!/\sqrt{n+1}$, and the $(n+1)^{\text{st}}$ term is $a_{n+1} = (n+1)!/\sqrt{n+2}$, so that

$$\frac{a_{n+1}}{a_n} = \frac{(n+1)!}{\sqrt{n+2}} \cdot \frac{\sqrt{n+1}}{n!} = \frac{(n+1)!}{n!} \cdot \sqrt{\frac{n+1}{n+2}}$$

The test ratio can be simplified by noting that

$$\frac{(n+1)!}{n!} = \frac{(n+1) \times n \times (n-1) \times \cdots \times 2 \times 1}{n \times (n-1) \times \cdots \times 2 \times 1} = n+1$$

Making this simplification and at the same time dividing the numerator and denominator of the radicand by n, we have

$$R = \lim_{n \to \infty} \left((n+1) \sqrt{\frac{1 + (1/n)}{1 + (2/n)}} \right) = \infty$$

Since $R > 1$, the series is divergent.

Example 23.   Use the ratio test to show that this alternating series is divergent:

$$\frac{2}{3^2} - \frac{2^2}{5^2} + \frac{2^3}{7^2} - \frac{2^4}{9^2} + \cdots + \frac{(-1)^{n+1}2^n}{(2n+1)^2} + \cdots$$

Solution.   Here

$$|a_n| = \frac{2^n}{(2n+1)^2} \quad \text{and} \quad |a_{n+1}| = \frac{2^{n+1}}{[2(n+1)+1]^2} = \frac{2^{n+1}}{(2n+3)^2}$$

Therefore

$$R = \lim_{n\to\infty}\left|\frac{a_{n+1}}{a_n}\right| = \lim_{n\to\infty} \frac{2^{n+1}}{(2n+3)^2} \cdot \frac{(2n+1)^2}{2^n} = \lim_{n\to\infty} 2 \cdot \left(\frac{2n+1}{2n+3}\right)^2$$

$$= \lim_{n\to\infty} 2 \cdot \left(\frac{2+1/n}{2+3/n}\right)^2 = 2$$

Since $R > 1$, we conclude that the series is divergent.

16.6   THE ALTERNATING SERIES TEST

A series whose terms are alternately positive and negative is called an alternating series. There is a rather easy test to determine the convergence of this kind of series.

The Alternating Series Test
Let

$$a_1 - a_2 + a_3 - a_4 + \cdots + a_n - a_{n+1} + \cdots$$

represent an alternating series in which the absolute value of each term is eventually less than the preceding term; that is, eventually,

$$|a_{n+1}| < |a_n|$$

Then the series will be convergent if

$$\lim_{n\to\infty} a_n = 0$$

Example 24.   Show that the alternating series $1 - 1/2^2 + 1/3^2 - 1/4^2 + \cdots$ is convergent.

Solution.   Here $|a_n| = 1/n^2$ and $|a_{n+1}| = 1/(n+1)^2$. Therefore

$$|a_{n+1}| < |a_n|$$

for all values of n. Furthermore

$$\lim_{n\to\infty} \frac{1}{n^2} = 0$$

and the series is convergent.

Example 25. Show that the series $\sum_{n=1}^{\infty} (-1)^n/n$ converges.

Solution. The absolute value of each term of this series is less than the previous one, and $\lim_{n\to\infty} 1/n = 0$, hence the series converges.

EXERCISES

Use the ratio test to investigate the convergence or divergence of the indicated series. If the ratio test fails, apply either the integral test or the alternating series test.

1. $1 + \dfrac{1}{2!} + \dfrac{1}{3!} + \dfrac{1}{4!} + \cdots$

9. $\dfrac{1}{e^{10}} + \dfrac{2}{e^{10}} + \dfrac{3}{e^{10}} + \dfrac{4}{e^{10}} + \cdots$

2. $\dfrac{1}{2} - \dfrac{2}{2^2} + \dfrac{3}{2^3} - \dfrac{4}{2^4} + \cdots$

10. $\dfrac{1}{\pi} - \dfrac{2}{\pi^2} + \dfrac{3}{\pi^3} - \dfrac{4}{\pi^4} + \cdots$

3. $\dfrac{3}{2} + \dfrac{4}{2^2} + \dfrac{5}{2^3} + \dfrac{6}{2^4} + \cdots$

11. $\sum_{n=1}^{\infty} \dfrac{(-1)^{n+1}}{\sqrt{n}}$

4. $\sum_{n=1}^{\infty} \dfrac{(-1)^{n+1}3n}{(n+1)^2}$

12. $\sum_{n=1}^{\infty} \dfrac{\sqrt{n-1}}{(n-1)!}$

5. $1 - \dfrac{1}{2} + \dfrac{1}{3} - \dfrac{1}{4} + \cdots$

13. $\sum_{n=1}^{\infty} \dfrac{\sqrt{2n}}{n!}$

6. $1 - \dfrac{1}{3!} + \dfrac{1}{5!} - \dfrac{1}{7!} + \cdots$

14. $\sum_{n=1}^{\infty} \dfrac{1}{\sqrt{n}}$

7. $\sum_{n=1}^{\infty} \dfrac{n}{(n-1)!}$

15. $\sum_{n=1}^{\infty} \dfrac{5}{n^{2/3}}$

8. $\sum_{n=1}^{\infty} \dfrac{(-1)^n n!}{e^n}$

16. $\sum_{n=1}^{\infty} \dfrac{n-1}{n+1}$

16.7 POWER SERIES

So far we have considered only series of constant terms. Since some of the more important types of infinite series involve terms that are functions of a variable, we wish to extend our discussion to include series of this type.

The most important series involving variable terms is known as *a power series*. Letting x denote the variable quantity, a power series is given by

$$c_0 + c_1 x + c_2 x^2 + c_3 x^3 + \cdots + c_n x^n + \cdots$$

where $c_0, c_1, c_2, \ldots, c_n, \ldots$ are constant coefficients of the variable x and are independent of the value of x.

In dealing with power series we must again consider the problem of convergence and divergence. However, the problem is now one of finding for what values of the variable x the series will converge. We find that a power series may converge for all values of x, or it may converge for some values of x but not for others. The set of values of x for which a power series converges is usually referred to as the *interval of convergence* of the series. In this article we will determine the interval of convergence of a power series by applying the ratio test to the given power series.

Example 26.   Find the interval of convergence of the series

$$x + 2x^2 + 3x^3 + \cdots + nx^n + \cdots$$

Solution.   Here $a_n = nx^n$, and $a_{n+1} = (n+1)x^{n+1}$, so that

$$R = \lim_{n \to \infty} \left| \frac{(n+1)x^{n+1}}{nx^n} \right| = \lim_{n \to \infty} \frac{n+1}{n} \cdot |x| = \lim_{n \to \infty} \left(1 + \frac{1}{n}\right) \cdot |x| = |x|$$

The ratio test reveals that the given series is convergent for all values of x for which

$$|x| < 1$$

that is, for the interval

$$-1 < x < 1$$

We also know from the ratio test that the given series is divergent for $|x| > 1$, but we do not know what happens when $|x| = 1$. The end points of the interval of convergence, in this case $x = \pm 1$, must be examined separately.

If $x = 1$ the given series becomes

$$1 + 2 + 3 + \cdots + n + \cdots$$

which is clearly a divergent series.

If $x = -1$, the given series becomes the alternating series

$$-1 + 2 - 3 + 4 + \cdots + (-1)^n n + \cdots$$

The terms of this series increase without bound as n increases without bound and, therefore, the series is divergent.

Since the given series diverges at both $x = 1$ and $x = -1$, we conclude that the interval of convergence for this series is

$$-1 < x < 1$$

Example 27.   Find the interval of convergence of the series

$$x + \frac{x^2}{\sqrt{2}} + \frac{x^3}{\sqrt{3}} + \cdots + \frac{x^n}{\sqrt{n}} + \cdots$$

Solution.   Here $a_n = x^n/\sqrt{n}$, and $a_{n+1} = x^{n+1}/\sqrt{n+1}$, so that

$$R = \lim_{n \to \infty} \left| \frac{x^{n+1}}{\sqrt{n+1}} \cdot \frac{\sqrt{n}}{x^n} \right| = \lim_{n \to \infty} \left| \left(\frac{n}{n+1} \right)^{1/2} \right| \cdot |x| = |x|$$

Therefore, the series converges for $|x| < 1$, that is, $-1 < x < 1$. The ratio test fails at $x = 1$ and $x = -1$, so we examine these points separately.
If $x = 1$, the given series becomes

$$1 + \frac{1}{\sqrt{2}} + \frac{1}{\sqrt{3}} + \cdots + \frac{1}{\sqrt{n}} + \cdots$$

which can be tested for convergence by using the integral test with $f(x) = 1/\sqrt{x}$. Thus

$$\int_1^\infty \frac{dx}{\sqrt{x}} = \lim_{t \to \infty} \int_1^t x^{-1/2} \, dx = \lim_{t \to \infty} [2x^{1/2}]_1^t = \lim_{t \to \infty} [2t^{1/2} - 2] = \infty$$

Since the limit does not exist, the given series diverges for $x = 1$.
If $x = -1$, the given series becomes

$$-1 + \frac{1}{\sqrt{2}} - \frac{1}{\sqrt{3}} + \cdots + \frac{(-1)^n}{\sqrt{n}} + \cdots$$

which is an alternating series with $|a_n| = 1/\sqrt{n}$ and $|a_{n+1}| = 1/\sqrt{n+1}$. Since $|1/\sqrt{n+1}| < |1/\sqrt{n}|$ for all positive n, and

$$\lim_{n \to \infty} \frac{1}{\sqrt{n}} = 0$$

the given series will be convergent for $x = -1$.
Summarizing the result of the various tests, we see that the interval of convergence for the given series is

$$-1 \leqslant x < 1.$$

Example 28.   Find the interval of convergence for the series

$$1 - x + \frac{x^2}{2!} - \frac{x^3}{3!} + \cdots + \frac{x^n}{n!} + \cdots$$

Solution.   Here $|a_n| = |x^n/n!|$ and $|a_{n+1}| = |x^{n+1}/(n+1)!|$, so that

$$R = \lim_{n \to \infty} \left| \frac{x^{n+1}}{(n+1)!} \cdot \frac{n!}{x^n} \right| = \lim_{n \to \infty} \left(\frac{1}{n} \right) \cdot |x| = 0$$

Since $R < 1$ for all finite values of x, we conclude that the interval of convergence is

$$-\infty < x < \infty$$

EXERCISES

Find the interval of convergence of each of the given power series.

1. $x - x^2 + x^3 - x^4 + \cdots$

10. $\frac{x}{2} + \frac{x^2}{2^2} + \frac{x^3}{2^3} + \frac{x^4}{2^4} + \cdots$

2. $2x + 3x^2 + 4x^3 + 5x^4 + \cdots$

11. $1 - x + \frac{x^2}{2^2} - \frac{x^3}{3^2} + \cdots$

3. $\sum\limits_{n=1}^{\infty} \frac{x^{n-1}}{n!}$

12. $\sum\limits_{n=1}^{\infty} 2^n x^{n-1}$

4. $\sum\limits_{n=1}^{\infty} \frac{x^n}{n}$

13. $\sum\limits_{n=1}^{\infty} 3^{n-1} x^n$

5. $\frac{1}{2} + \frac{x}{3} + \frac{x^2}{4} + \frac{x^3}{5} + \cdots$

14. $\frac{x^2}{2} - \frac{x^4}{4} + \frac{x^6}{6} - \frac{x^8}{8} + \cdots$

6. $\sum\limits_{n=1}^{\infty} \frac{(-1)^{n+1} x^n}{\sqrt{n}}$

15. $\sum\limits_{n=1}^{\infty} \frac{2^n x^n}{n!}$

7. $\sum\limits_{n=1}^{\infty} \frac{(-1)^{n+1} x^{2n-1}}{(2n-1)!}$

16. $\sum\limits_{n=1}^{\infty} (-1)^{n+1} \frac{x^{2n-1}}{2n-1}$

8. $\sum\limits_{n=1}^{\infty} \frac{n x^{2n+3}}{(n+1)^2}$

17. $1 + \frac{x}{e} + \frac{x^2}{e^2} + \frac{x^3}{e^3} + \cdots$

9. $1 - \frac{x^2}{2!} + \frac{x^4}{4!} - \frac{x^6}{6!} + \cdots$

18. $\sum\limits_{n=1}^{\infty} \frac{x^n}{n 2^n}$

17
Expansion of Functions

17.1 INTRODUCTION

In trigonometry we learn how to find the values of the trigonometric functions for some special angles such as $0°$, $30°$, $45°$, $60°$, $90°$, etc. It is then pointed out that the values of the trigonometric functions for other angles are available in table form. Have you ever wondered how the values in the trigonometric tables were computed? For instance, how would you find $\sin 39°$ without a table? Surprising as it may seem, tables of trigonometric functions of angles are constructed using power series.

As we observed in the previous chapter, for values of x within the interval of convergence the sum of the terms of a power series is a function of x. This suggests that a power series can be denoted by

$$f(x) = a_0 + a_1 x + a_2 x^2 + a_3 x^3 + \cdots \tag{1}$$

and raises the question of whether or not a general function can be represented by a power series expansion similar to (1). The connection between functions and power series is illustrated by the following example. Consider the function

$$f(x) = \frac{1}{1-x} \tag{2}$$

If we divide 1 by $1 - x$, according to the ordinary rules of algebra we obtain

$$\frac{1}{1-x} = 1 + x + x^2 + x^3 + \cdots \tag{3}$$

where the terms on the right are the terms of an infinite power series. By the ratio test we can show that this power series converges for $-1 < x < 1$. It is also true, although we shall not prove it, that this power series is actually equal to $1/(1-x)$ on this interval. You should note that the power series diverges outside this interval and hence it could not possibly represent $1/(1-x)$ for $|x| > 1$.

395

17.2 MACLAURIN SERIES

Now that we have shown that a function can be represented by an infinite power series, you may wonder whether or not similar type series can be found for other functions. The problem is primarily one of method. The method employed in the case of $f(x) = 1/(1-x)$ is restricted to functions that are rational fractions and, therefore, is unsuitable for more general types of functions. The question that we wish to answer is: Does a given function $f(x)$ have a power series expansion? We proceed by assuming that $f(x)$ does have a power series expansion of the form

$$f(x) = a_0 + a_1 x + a_2 x^2 + a_3 x^3 + a_4 x^4 + \cdots + a_n x^n + \cdots \tag{4}$$

If the power series on the right is to represent $f(x)$, we must be able to evaluate the constants $a_0, a_1, a_2, \ldots$ by some means. Analysis of Equation 4 reveals that a_0 can be evaluated by setting $x = 0$. Since all terms to the right of the a_0 term are zero when x equals zero, we have as the first coefficient

$$a_0 = f(0)$$

In order to evaluate the remaining coefficients, we make the assumption that Equation 4 can be differentiated term by term, as many times as we wish, and that each successive derivative is defined at $x = 0$. Under this assumption the first derivative of Equation 4 is

$$f'(x) = a_1 + 2a_2 x + 3a_3 x^2 + 4a_4 x^3 + \cdots + na_n x^{n-1} + \cdots \tag{5}$$

Letting $x = 0$ in (5), we have

$$a_1 = f'(0)$$

Similarly

$$f''(x) = 2 \cdot 1 a_2 + 3 \cdot 2 a_3 x + 4 \cdot 3 a_4 x^2 + \cdots + n(n-1) a_n x^{n-2} + \cdots \tag{6}$$

from which

$$2 \cdot 1 a_2 = f''(0)$$

or

$$a_2 = \frac{f''(0)}{2 \cdot 1} = \frac{f''(0)}{2!}$$

Differentiating again,

$$f'''(x) = 3 \cdot 2 \cdot 1 a_3 + 4 \cdot 3 \cdot 2 a_4 x + 5 \cdot 4 \cdot 3 a_5 x^2 + \cdots$$

Letting $x = 0$,

$$3 \cdot 2 \cdot 1 a_3 = f'''(0)$$

or

$$a_3 = \frac{f'''(0)}{3 \cdot 2 \cdot 1} = \frac{f'''(0)}{3!}$$

If we continue this process we see that each of the coefficients a_0, a_1, $a_2, \ldots$ can be evaluated, and that the general term a_n is given by

$$a_n = \frac{f^{(n)}(0)}{n!}$$

Note that $f^{(n)}(0)$ represents the nth derivative of the function $f(x)$ evaluated at $x = 0$. In using this notation it is understood that

$$f^{(0)}(x) = f(x)$$

Substituting these values back into Equation 4 we conclude that if a function $f(x)$ has a power series expansion, it will be of the form

$$f(x) = f(0) + f'(0)x + \frac{f''(0)x^2}{2!} + \frac{f'''(0)x^3}{3!} + \cdots + \frac{f^{(n)}(0)x^n}{n!} + \cdots \qquad (7)$$

which can be written more compactly as

$$f(x) = \sum_{n=0}^{\infty} \frac{f^{(n)}(0)x^n}{n!} \qquad (8)$$

Equation 8 is called the *Maclaurin series expansion* of the function $f(x)$. Keep in mind that for a function to have a Maclaurin series expansion, the function and all its derivatives must be defined at $x = 0$.

Since the Maclaurin series expansion of a function is a power series, it is necessary to indicate the interval of convergence. For all of the functions considered in this book, the Maclaurin series expansion of the function is valid for all values of x within the interval of convergence of the series.

Example 1.   Using Equation 8, verify the Maclaurin series of

$$f(x) = \frac{1}{1-x}$$

Solution.   Evaluating the function and its successive derivatives at $x = 0$, we have

$$
\begin{aligned}
f(x) &= (1-x)^{-1}, & f(0) &= 1 \\
f'(x) &= 1(1-x)^{-2}, & f'(0) &= 1 \\
f''(x) &= 2(1-x)^{-3}, & f''(0) &= 2 \\
f'''(x) &= 6(1-x)^{-4}, & f'''(0) &= 6
\end{aligned}
$$

Substituting these values into Equation 8, the Maclaurin series expansion is

$$\frac{1}{1-x} = \frac{1 \cdot x^0}{0!} + \frac{1 \cdot x^1}{1!} + \frac{2 \cdot x^2}{2!} + \frac{6 \cdot x^3}{3!} + \cdots$$

$$= 1 + x + x^2 + x^3 + \cdots$$

This agrees with the series expansion obtained in Equation 3. As indicated previously, this expansion is valid in the interval $-1 < x < 1$.

Example 2.   Find the Maclaurin series expansion of the function $f(x) = \sin x$.

Solution.   Here

$$
\begin{aligned}
f(x) &= \sin x, & f(0) &= 0 \\
f'(x) &= \cos x, & f'(0) &= 1 \\
f''(x) &= -\sin x, & f''(0) &= 0 \\
f'''(x) &= -\cos x, & f'''(0) &= -1 \\
f^{(4)}(x) &= \sin x, & f^{(4)}(0) &= 0 \\
f^{(5)}(x) &= \cos x, & f^{(5)}(0) &= 1
\end{aligned}
$$

Substituting these values into Equation 8, we have

$$f(x) = \sin x = x - \frac{x^3}{3!} + \frac{x^5}{5!} - \cdots + \frac{(-1)^{n+1} x^{2n-1}}{(2n-1)!} + \cdots$$

This series converges for $-\infty < x < \infty$. This was shown in Exercise 7, Section 16.7.

Example 3.   Find the first four nonzero terms of the Maclaurin series expansion of $f(x) = \sqrt{x+1}$.

Solution. Writing $f(x)$ in exponential form we have

$$f(x) = (x+1)^{1/2}, \qquad f(0) = 1$$
$$f'(x) = \tfrac{1}{2}(x+1)^{-1/2}, \qquad f'(0) = \tfrac{1}{2}$$
$$f''(x) = -\tfrac{1}{4}(x+1)^{-3/2}, \qquad f''(0) = -\tfrac{1}{4}$$
$$f'''(x) = \tfrac{3}{8}(x+1)^{-5/2}, \qquad f'''(0) = \tfrac{3}{8}$$

Thus, by Equation 8,

$$\sqrt{x+1} = 1 + \frac{x}{2} - \frac{x^2}{8} + \frac{3x^3}{48} - \cdots .$$

EXERCISES

In Exercises 1 to 4, find the first three non-zero terms of the Maclaurin series expansion of the given functions and determine the interval of convergence.

1. $f(x) = \cos x$ **3.** $f(x) = \ln(1+x)$
2. $f(x) = e^x$ **4.** $f(x) = \ln(1-x)$

In the remaining problems, find the first three nonzero terms of the Maclaurin series expansion of the given function, but do not evaluate the interval of convergence.

5. $f(x) = \sin 3x$ **11.** $f(x) = e^{-x}$
6. $f(x) = \sqrt{1+4x}$ **12.** $f(x) = \tan x$
7. $f(x) = \dfrac{1}{(1+x)^2}$ **13.** $f(x) = \sec x$
8. $f(x) = \dfrac{1}{\sqrt{1+x}}$ **14.** $f(x) = e^{\sin x}$
9. $f(x) = \arcsin x$ **15.** $f(x) = \ln \cos x$
10. $f(x) = \arctan x$ **16.** $f(x) = e^{x^2}$

17.3 OPERATIONS WITH SERIES

Once a series expansion of a function $f(x)$ has been established, it is possible to use this expansion to express $f(z)$ by simply substituting $x = z$ into the expansion. Thus the Maclaurin series expansion of $\sin 3x$ is given by

$$\sin 3x = (3x) - \frac{(3x)^3}{3!} + \frac{(3x)^5}{5!} - \cdots$$

$$= 3x - \frac{9}{2}x^3 + \frac{81}{40}x^5 - \cdots$$

As we pointed out earlier, it is desirable to know where a given power series is convergent. One reason for this is the fact that many of the operations that we know to be valid for finite summations can be extended immediately to infinite series of terms, provided the series is convergent. The operations of differentiation and integration are such operations.

Differentiation of power series

A convergent power series may be differentiated term by term, and the resulting series will be convergent. Not only will this series be convergent, but it will have the same interval of convergence as the series from which it was obtained.

Example 4.   Use the expansion for $\sin x$ to derive the expansion for $\cos x$.

Solution.   Beginning with

$$\sin x = x - \frac{x^3}{3!} + \frac{x^5}{5!} - \cdots$$

we differentiate both sides to obtain

$$\cos x = 1 - \frac{3x^2}{3!} + \frac{5x^4}{5!} - \cdots$$

$$= 1 - \frac{x^2}{2!} + \frac{x^5}{4!} - \cdots$$

Integration of power series

A convergent power series may be integrated term by term and the resulting series will have the same interval of convergence.

Example 5.   Use the expansion for $1/(1+x^2)$ to derive the expansion for arc tan x.

Solution.   The expansion for $1/(1+x^2)$ is given by

$$\frac{1}{1+x^2} = 1 - x^2 + x^4 - x^6 + \cdots \tag{9}$$

The integral of the left side of (9) yields

$$\int_0^x \frac{dv}{1+v^2} = [\text{arc tan } v]_0^x = \text{arc tan } x$$

By evaluating the indefinite integral from 0 to x we can avoid the necessity of adding a constant. Hence, if we integrate both sides of (9), we have

$$\text{arc tan } x = x - \frac{x^3}{3} + \frac{x^4}{4} - \cdots$$

EXERCISES

In Exercises 1 to 8, write the infinite series expansion of the given function by using one of the Maclaurin series obtained in the preceding article.

1. $\sin 2x$
2. $e^{\pi x}$
3. $\cos \frac{1}{2} x$
4. $\sin x^2$

5. $\arcsin 3x$
6. $\sin (x - a)$
7. $\cos \left(x - \frac{\pi}{6} \right)$
8. $\tan (x + \theta)$

In Exercises 9 to 12, use the given Maclaurin series to derive the Maclaurin series of the expression in brackets.

9. $\ln (1 + x) = x - \dfrac{x^2}{2} + \dfrac{x^3}{3} - \dfrac{x^4}{4} + \dots,$ $\left[\dfrac{1}{1+x} \right]$

10. $\tan x = x + \dfrac{x^3}{3} + \dfrac{2x^5}{15} + \dots,$ $[\sec^2 x]$

11. $\cos x = 1 - \dfrac{x^2}{2!} + \dfrac{x^4}{4!} - \dfrac{x^6}{6!} + \dots,$ $[\sin x]$

12. $\tan x = x + \dfrac{x^3}{3} + \dfrac{2x^5}{15} + \dots,$ $[\ln \cos x]$

17.4   APPROXIMATION BY SERIES

If a series expansion represents a function $f(x)$, then we can expect to obtain a fairly good approximation of the value of $f(x)$ by using the first few terms of the series. This technique is widely used in the computer field and is the basis of this article.

Consider the problem of evaluating $\sin x$, when $x = 0.5$ radians. (Notice that angles must always be expressed in radians when a trigonometric function is being evaluated by a power series.) Using the first three terms of the series expansion of $\sin x$, we have

$$\sin x = 0.5 - \frac{(0.5)^3}{3!} + \frac{(0.5)^5}{5!}$$

$$= 0.5 - 0.020833 + 0.000260$$

$$= 0.47943$$

It is possible to show that this approximation is accurate to five decimal places, but we shall not do so since the analysis of approximating "errors" is a course in itself.

Example 6. Use four terms of the Maclaurin series expansion for e^x to approximate $e^{0.1}$.

Solution

$$e^{0.1} = 1 + 0.1 + \frac{(0.1)^2}{2!} + \frac{(0.1)^3}{3!}$$

$$= 1.00000 + 0.10000 + 0.00500 + 0.00017$$

$$= 1.1051$$

This answer is accurate to the indicated number of decimal places.

Example 7. Find the value of $\int_0^1 \cos x^2 dx$.

Solution. The integral in this problem can not be evaluated by elementary methods. However, the value of the definite integral can be approximated by employing power series. Using the power series expansion for $\cos x^2$, we have

$$\int_0^1 \cos x^2 \, dx = \int_0^1 \left[1 - \frac{x^4}{2!} + \frac{x^8}{4!} - \cdots \right] dx$$

$$= \left[x - \frac{x^5}{10} + \frac{x^9}{216} - \cdots \right]_0^1$$

Considering the first three terms of the series, we get

$$\int_0^1 \cos x^2 \, dx \approx \left[x - \frac{x^5}{10} + \frac{x^9}{216} \right]_0^1 = [1.000 - 0.100 + 0.005] = 0.905$$

EXERCISES

In Exercises 1 to 8 calculate the value of the given expression by using the first three terms of its Maclaurin series expansion.

1. $\cos 0.1$
2. $\tan 0.5$
3. $\sin 5°$
4. $\cos 10°$

5. $e^{0.2}$
6. $\ln 1.2$
7. $\ln 0.9$
8. $e^{1.1}$

In Exercises 9 to 12 evaluate the indicated definite integrals using three terms of the Maclaurin series expansion of the integrand.

9. $\int_0^1 \sin x^2 \, dx$

10. $\int_0^1 e^{x^2} \, dx$

11. $\int_0^1 \ln \cos x \, dx$

12. $\int_0^1 \arctan x \, dx$

13. Maclaurin series are particularly useful in working with digital computers. By using Maclaurin series, a programmer can express a trigonometric function as a series of algebraic terms that can be evaluated by the machine. Write the first few terms of the Maclaurin series for $f(x) = \sin(\sin x)$.
14. Evaluate the function in Exercise 13 at $x = \frac{1}{2}$. Use three terms of the series.
15. The displacement of a simple harmonic oscillator varies according to $S = \sin 2\pi t$. Write the first three terms of a series that could be used to evaluate this function on a digital computer.
16. What is the displacement of the oscillator in Exercise 15 when $t = 0.05$? Use the first 2 terms of the series expansion for this purpose.

17.5   TAYLOR SERIES

An analysis of the convergence of the Maclaurin series expansion of a function to the value of the function reveals that the rate of convergence of a given series is dependent upon the magnitude of the value of x to be computed. The larger the value of x, the greater the number of terms needed to obtain a certain accuracy. For instance, when $x = 0.5$, three terms of the Maclaurin series expansion of $\sin x$ are needed to obtain an answer which is accurate to five decimal places, although five terms of the expansion are needed to insure the same degree of accuracy when $x = 1.0$.

For large values of x, the Maclaurin series expansion of a function may converge so slowly that it becomes useless for purposes of computation. To circumvent this problem, we introduce what is known as the *Taylor series* expansion of a function. The Taylor series involves the expansion of the function about some point $x = a$. In order to develop the Taylor series of a function $f(x)$, we assume that $f(x)$ can be represented by the power series

$$f(x) = b_0 + b_1(x-a) + b_2(x-a)^2 + b_3(x-a)^3 + \cdots \qquad (10)$$

where a is some constant and $b_0, b_1, b_2, \ldots$ are coefficients which are to be determined. Assuming that (10) may be differentiated term by term any number of times, we get the following equations

$$f'(x) = b_1 + 2b_2(x-a) + 3b_3(x-a)^2 + 4b_4(x-a)^3 + \cdots$$

$$f''(x) = 2b_2 + 3 \cdot 2b_3(x-a) + 4 \cdot 3b_4(x-a)^2 + \cdots$$

$$f'''(x) = 3 \cdot 2b_3 + 4 \cdot 3 \cdot 2b_4(x-a) + \cdots$$

$$\cdot \qquad \cdot$$
$$\cdot \qquad \cdot$$
$$\cdot \qquad \cdot$$

The coefficients $b_0, b_1, b_2, \ldots$ can now be evaluated by letting $x = a$ in the above equations. When $x = a$, any term containing $x - a$ as a factor will be zero. The desired coefficients are then

$$b_0 = f(a)$$
$$b_1 = f'(a)$$
$$b_2 = \frac{f''(a)}{2!}$$
$$b_3 = \frac{f'''(a)}{3!}$$
$$\cdot$$
$$\cdot$$
$$\cdot$$

Therefore, if $f(x)$ can be represented by a power series like (10), that series will have the form

$$f(x) = f(a) + f'(a)(x-a) + \frac{f''(a)}{2!}(x-a)^2 + \cdots + \frac{f^{(n)}(a)}{n!}(x-a)^n + \cdots$$

$$= \sum_{n=0}^{\infty} \frac{f^{(n)}(a)}{n!}(x-a)^n \qquad (11)$$

which is known as the Taylor series expansion of $f(x)$.

Before considering the problems of finding the Taylor series expansion of a function, let us note that the Taylor series reduces to the Maclaurin series when $a = 0$

Example 8.   Write the Taylor series expansion of $\sin x$ about the point $a = \pi/3$.

Solution

$$f(x) = \sin x, \qquad f(\pi/3) = \frac{\sqrt{3}}{2}$$

$$f'(x) = \cos x, \qquad f'(\pi/3) = \frac{1}{2}$$

$$f''(x) = -\sin x, \qquad f''(\pi/3) = -\frac{\sqrt{3}}{2}$$

$$f'''(x) = -\cos x, \qquad f'''(\pi/3) = -\frac{1}{2}$$

Substituting the values into (11), we have

$$\sin x = \frac{\sqrt{3}}{2} + \frac{1}{2}\left(x - \frac{\pi}{3}\right) + \frac{(-\sqrt{3}/2)(x-(\pi/3))^2}{2!} + \frac{(-1/2)(x-(\pi/3))^3}{3!} + \cdots$$

$$= \frac{\sqrt{3}}{2} + \frac{1}{2}\left(x - \frac{\pi}{3}\right) - \frac{\sqrt{3}}{4}\left(x - \frac{\pi}{3}\right)^2 - \frac{1}{12}\left(x - \frac{\pi}{3}\right)^3 + \cdots$$

Example 9. Evaluate sin 33° by use of the Taylor series expansion of sin x about the point $a = \pi/6$.

Solution. We choose to expand the Taylor series about $a = \pi/6$ (that is, 30°), because 30° is the closest angle to 33° for which we know the values of sine and cosine. Thus

$$f(x) = \sin x, \qquad f(\pi/6) = \frac{1}{2}$$

$$f'(x) = \cos x, \qquad f'(\pi/6) = \frac{\sqrt{3}}{2}$$

$$f''(x) = -\sin x, \qquad f''(\pi/6) = -\frac{1}{2}$$

and

$$\sin x = \frac{1}{2} + \frac{\sqrt{3}}{2}\left(x - \frac{\pi}{6}\right) - \frac{1}{4}\left(x - \frac{\pi}{6}\right)^2 - \cdots$$

To find the value of sin 33°, we convert 33° into radians and substitute this value into the above Taylor series.
Thus

$$x = 33° = 33\frac{\pi}{180} = \frac{11\pi}{60} \text{ radians}$$

Using the first three terms of the Taylor series, we obtain

$$\sin 33° = \frac{1}{2} + \frac{\sqrt{3}}{2}\left(\frac{11\pi}{60} - \frac{\pi}{6}\right) - \frac{1}{4}\left(\frac{11\pi}{60} - \frac{\pi}{6}\right)^2$$

$$= \frac{1}{2} + \frac{\sqrt{3}}{2}\left(\frac{\pi}{60}\right) - \frac{1}{4}\left(\frac{\pi}{60}\right)^2$$

$$= 0.50000 + (0.86603)(0.05236) - (0.25000)(0.05236)^2$$

$$= 0.5446$$

This answer is accurate to the indicated number of decimal places.

Example 10. Use the Taylor series derived in the preceding example to evaluate sin 27°.

Solution. In this problem

$$x = 27° = 27\frac{\pi}{180} = \frac{3\pi}{20} \text{ radians}$$

so that

$$\left(x - \frac{\pi}{6}\right) = \left(\frac{3\pi}{20} - \frac{\pi}{6}\right) = -\frac{\pi}{60} = -0.05236$$

Therefore, the value of $\sin 27°$ to four decimal places is given by

$$\sin 27° = 0.50000 - (0.86603)(0.05236) - (0.25000)(0.05336)^2$$
$$= 0.4540$$

EXERCISES

In Exercises 1 to 6, find the Taylor series expansion of the given function about the indicated point.

1. $\cos x$, $a = \pi/3$
2. $\cos x$, $a = 30°$
3. $\sin x$, $a = 45°$

4. $\tan x$, $a = \pi/4$
5. e^x, $a = 1$
6. $\ln x$, $a = 1$

In the remaining exercises evaluate the given expressions using Taylor series expansions found above. Use three terms of the series.

7. $\cos 55°$
8. $\sin 48°$
9. $\tan 42°$

10. $\cos 62°$
11. $e^{1.01}$
12. $e^{0.98}$

18

Fourier Series

18.1 INTRODUCTION

This chapter can be considered to be a continuation of the preceding one, since we are going to consider another method of representing functions by an infinite series. Instead of expanding the given function in a power series, we expand it in a trigonometric series of the form

$$f(x) = \frac{a_0}{2} + a_1 \cos x + a_2 \cos 2x + \cdots + a_k \cos kx + \cdots \qquad (1)$$

$$+ b_1 \sin x + b_2 \sin 2x + \cdots + b_k \sin kx + \cdots$$

The trigonometric series in (1) is called the *Fourier series* of $f(x)$ and is more commonly written in the form

$$f(x) = \frac{a_0}{2} + \sum_{k=1}^{\infty} (a_k \cos kx + b_k \sin kx) \qquad (2)$$

The coefficients a_0, a_k, and b_k in (2) are constants to be determined. As you would suspect, the values of a_k and b_k are dependent upon the function $f(x)$. The constant term in (2) is written $a_0/2$, instead of simply a_0, in order to maintain consistency in the formulas that we shall develop for the indicated coefficients.

Not every function can be represented by a trigonometric series of the type shown above; however, the class of functions that can is surprisingly large and certainly includes any that you may be likely to encounter in applications. The fact that Fourier series can be used to expand functions even when their graphs have sharp corners, or perhaps even a finite "break" or discontinuity, makes it a much more usable tool than the power series expansion.

18.2 PERIODIC FUNCTIONS

A function $f(x)$ is called periodic if there exists a constant T for which

$$f(x+T) = f(x) \qquad (3)$$

The constant T is called a *period* of the function $f(x)$. Clearly, the graph of a periodic function repeats itself every T units. The most familiar example of a periodic function is $\sin x$, which has a period $T = 2\pi$. This is denoted by

$$\sin (x + 2\pi) = \sin x$$

It is interesting to note that $T = 2\pi$ is also a period of each of the harmonics

$$\sin kx, \qquad (k = 2, 3, 4, \ldots)$$

This is explained by the fact that an integral multiple of a period is again a period. Thus, $\sin 2x$ has a period of π, but it also has a period of 2π since it repeats every 2π radians as well. We conclude from this that every sum of the form

$$\frac{a_0}{2} + \sum_{k=1}^{n} (a_k \cos kx + b_k \sin kx) \tag{4}$$

is a function of period 2π, because it is the sum of functions of period 2π. The constant term $a_0/2$ does not change the periodicity of the summation.

The analysis of many mechanical and electrical systems requires the use of periodic waves that are structurally more complex than simple sinusoidal waves. Figure 18.1 depicts a few of the commonly used wave forms. Each of these periodic waves can be represented by a Fourier series as is the case for all of the functions considered in this chapter. Our problem is to find the relative amounts of constant and sinusoidal terms necessary to produce a given periodic wave.

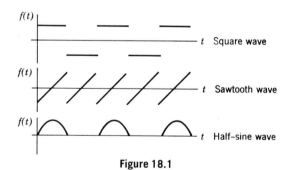

Figure 18.1

18.3   FOURIER SERIES OF FUNCTIONS OF PERIOD 2π

Let $f(x)$ be a function of period 2π. Then the Fourier series expansion of $f(x)$ can be written

$$f(x) = \frac{a_0}{2} + \sum_{k=1}^{\infty} (a_k \cos kx + b_k \sin kx) \tag{5}$$

The Fourier coefficients a_0, a_k, and b_k can be determined from a knowledge of $f(x)$ and the following basic assumptions.

(1) $f(x)$ is integrable over any period.
(2) $f(x)$ is equal to its Fourier series except, perhaps, at a finite number of points.
(3) The Fourier series of $f(x)$ is integrable term by term over any period, and the integral of the series is equal to the integral of $f(x)$.

Making use of the indicated assumptions, we can obtain the formulas to be used in evaluating the indicated Fourier coefficients.

Formula for a_0. To find the formula for a_0, we integrate both sides of (5) from 0 to 2π. Thus

$$\int_0^{2\pi} f(x)\,dx = \frac{a_0}{2}\int_0^{2\pi} dx + \sum_{k=1}^{\infty}\left(a_k\int_0^{2\pi}\cos kx\,dx + b_k\int_0^{2\pi}\sin kx\,dx\right) \qquad (6)$$

Beginning on the right of (6), we find that

$$\frac{a_0}{2}\int_0^{2\pi} dx = \frac{a_0}{2}[x]_0^{2\pi} = a_0\pi$$

Continuing on the right, we see that the value of the summation in (6) is zero; since for any positive integer k,

$$\int_0^{2\pi}\cos kx\,dx = \left[\frac{1}{k}\sin kx\right]_0^{2\pi} = 0$$

and

$$\int_0^{2\pi}\sin kx\,dx = \left[-\frac{1}{k}\cos kx\right]_0^{2\pi} = \left[-\frac{1}{k}(\cos 2k\pi - 1)\right] = 0$$

The second integral is zero because $\cos 2k\pi = 1$ for all integer values of k. Equation 6 can then be written as

$$\int_0^{2\pi} f(x)\,dx = a_0\pi$$

Solving this for a_0, we have the formula

$$a_0 = \frac{1}{\pi}\int_0^{2\pi} f(x)\,dx \qquad (7)$$

Formula for a_n. To find the formula for a_n (the coefficient of the nth harmonic) we multiply both sides of (5) by $\cos nx$ and integrate the result from 0 to 2π, that is

$$\int_0^{2\pi} f(x) \cos nx \, dx = \frac{a_0}{2} \int_0^{2\pi} \cos nx \, dx$$

$$+ \sum_{k=1}^{\infty} \left(a_k \int_0^{2\pi} \cos kx \cos nx \, dx \right.$$

$$\left. + b_k \int_0^{2\pi} \sin kx \cos nx \, dx \right) \tag{8}$$

Since n is the order of a harmonic, it is a positive integer; hence

$$\frac{a_0}{2} \int_0^{2\pi} \cos nx \, dx = \frac{a_0}{2} \left[\frac{1}{n} \sin nx \right]_0^{2\pi} = 0$$

The value of the integral

$$a_k \int_0^{2\pi} \cos kx \cos nx \, dx$$

depends upon whether $k = n$ or $k \neq n$. When $k = n$, the integral becomes

$$a_n \int_0^{2\pi} \cos^2 nx \, dx = \frac{a_n}{2} \int_0^{2\pi} (1 + \cos 2nx) \, dx$$

$$= \frac{a_n}{2} \left[x + \frac{\sin 2nx}{2n} \right]_0^{2\pi} = a_n \pi$$

When $k \neq n$, the integral can be evaluated by the trigonometric identity

$$\cos \alpha \cos \beta = \frac{1}{2} [\cos (\alpha + \beta) + \cos (\alpha - \beta)]$$

Thus

$$a_k \int_0^{2\pi} \cos kx \cos nx \, dx = \frac{a_k}{2} \int_0^{2\pi} [\cos (k+n)x + \cos (k-n)x] \, dx$$

$$= \frac{a_k}{2} \left[\frac{\sin (k+n)x}{k+n} + \frac{\sin (k-n)x}{k-n} \right]_0^{2\pi} = 0$$

Finally, the value of

$$b_k \int_0^{2\pi} \sin kx \cos nx \, dx$$

is found by using the trigonometric identity

$$\sin \alpha \cos \alpha = \frac{1}{2} [\sin (\alpha + \beta) + \sin (\alpha - \beta)]$$

When $k \neq n$, we have

$$b_k \int_0^{2\pi} \sin kx \cos nx\,dx = \frac{b_k}{2} \int_0^{2\pi} [\sin (k+n)x + \sin (k-n)x]\,dx$$

$$= \frac{b_k}{2} \left[-\frac{\cos (k+n)x}{k+n} - \frac{\cos (k-n)x}{k-n} \right]_0^{2\pi}$$

$$= \frac{b_k}{2} \left[\left(-\frac{1}{k+n} - \frac{1}{k-n} \right) - \left(-\frac{1}{k+n} - \frac{1}{k-n} \right) \right] = 0$$

and when $k = n$,

$$b_n \int_0^{2\pi} \sin nx \cos nx\,dx = \frac{b_n}{2} \int_0^{2\pi} \sin 2nx\,dx = 0$$

Hence, this integral is zero for any k and n. Applying the above results to (8), we obtain

$$\int_0^{2\pi} f(x) \cos nx\,dx = a_n\pi$$

The formula for a_n is then found to be

$$a_n = \frac{1}{\pi} \int_0^{2\pi} f(x) \cos nx\,dx \qquad (9)$$

Formula for b_n. To find the formula for b_n, we multiply both sides of (5) by $\sin nx$ and integrate the result from 0 to 2π. By performing an analysis similar to the one used to find a_n, we obtain the result

$$b_n = \frac{1}{\pi} \int_0^{2\pi} f(x) \sin nx\,dx \qquad (10)$$

Example 1.   Find the Fourier series for the function

$$f(x) = x, \qquad (0 \leqslant x < 2\pi)$$

$$f(x+2\pi) = f(x) \qquad (11)$$

   Solution.   Before finding the Fourier series of this function, let us draw its graph. First, we note that the graph of $f(x)$ is a straight line passing through the origin with a slope $m = 1$ on the interval $0 \leqslant x < 2\pi$.

The second condition states that the function is periodic with $T = 2\pi$. Hence, the given function is a sawtooth wave having a period of 2π. This is shown in Figure 18.2. With the graph of the function, let us find the Fourier coefficients. First we find a_0. Substituting (11) into (7), we have

$$a_0 = \frac{1}{\pi} \int_0^{2\pi} x\,dx = \frac{1}{\pi}\left[\frac{x^2}{2}\right]_0^{2\pi} = \frac{1}{\pi}\left[\frac{4\pi^2}{2}\right] = 2\pi \tag{12}$$

Next we find a_n by substituting (11) into (9) to get

$$a_n = \frac{1}{\pi} \int_0^{2\pi} x \cos nx\,dx$$

Integrating by parts, we find that

$$a_n = \frac{1}{\pi} \int_0^{2\pi} x \cos nx\,dx = \left[\frac{x \sin nx}{\pi n}\right]_0^{2\pi} - \frac{1}{\pi n} \int_0^{2\pi} \sin nx\,dx$$

$$= \frac{1}{\pi n^2}\left[\cos nx\right]_0^{2\pi} = 0 \tag{13}$$

Since a_n turns out to be zero for all n, we conclude that there are no cosine terms in the Fourier series. Finally, we find b_n by substituting (11) into (10) and integrating by parts

$$b_n = \frac{1}{\pi} \int_0^{2\pi} x \sin nx\,dx = \left[-\frac{x \cos nx}{\pi n}\right]_0^{2\pi} + \frac{1}{\pi n} \int_0^{2\pi} \cos nx\,dx$$

$$= \left[-\frac{x \cos nx}{\pi n} + \frac{1}{\pi n^2} \sin nx\right]_0^{2\pi}$$

$$b_n = -\frac{2}{n}$$

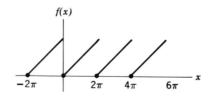

$f(x)$

-2π 2π 4π 6π x

Figure 18.2

The value of b_n, as we anticipated, is a function of the harmonic n. Hence, when

$$
\begin{aligned}
n &= 1, & b_1 &= -2 \\
n &= 2, & b_2 &= -1 \\
n &= 3, & b_3 &= -2/3 \\
n &= 4, & b_4 &= -1/2
\end{aligned}
$$

$$\cdot \qquad \cdot$$
$$\cdot \qquad \cdot$$
$$\cdot \qquad \cdot$$

Substituting $a_0 = 2\pi$, $a_n = 0$, and $b_n = -(2/n)$, $(n = 1, 2, 3 \ldots)$, into (5), we get

$$f(x) = \pi - \sum_{n=1}^{\infty} \frac{2}{n} \sin nx$$

$$= \pi - 2 \sin x - \sin 2x - \tfrac{2}{3} \sin 3x - \cdots \qquad (14)$$

It can be proven that a Fourier series converges to $f(x)$ at every point where the function is continuous. At points where finite discontinuities occur in $f(x)$, the series converges to a value equal to the average of the two extremes of the discontinuity. In the above problem, a finite discontinuity occurs at 2π and all multiples of 2π. The average of the extremes at each discontinuity is seen to be

$$\frac{1}{2}[2\pi + 0] = \pi$$

Therefore, the series (14) converges to π at the endpoint of each interval; a fact that can be verified by substituting $x = 2\pi$ into the series.

EXERCISES

In Exercises 1 to 6, sketch several periods of the given function and find its Fourier series expansion. Each function is defined on the interval $0 \leqslant x \leqslant 2\pi$ and $f(x + 2\pi) = f(x)$. (*Hint:* The required integrations can be effected most efficiently by using Table 5 of the Appendix.)

1. $f(x) = 2x$

4. $f(x) = \frac{1}{2}x^2$

2. $f(x) = \frac{1}{4}x$

5. $f(x) = \sin \frac{1}{2}x$

3. $f(x) = x^2$

6. $f(x) = x + \pi$

7. It is desired to use the exponential wave $v = e^t, 0 \le t \le 2\pi$, in the frequency control circuit of a radar. Find the Fourier series that represents this wave.

8. The forcing function being applied to a particular spring-mass system is given by $F = t^2 - t$ in the interval $0 \le t \le 2\pi$. Derive the Fourier series expansion for this function.

18.4   OTHER TYPES OF PERIODIC FUNCTIONS

It frequently happens that a wave cannot be described by a single equation throughout its entire period. As we will illustrate in the following examples, problems of this type are handled by writing separate integrals for each of the distinct parts making up one period of the wave.

Example 2.   (See Figure 18.3.) Find the Fourier series for the square wave

$$f(x) = 1, \qquad 0 \le x < \pi$$
$$= -1, \qquad \pi \le x < 2\pi$$

with

$$f(x + 2\pi) = f(x)$$

Solution.   The significance of this example is that the wave does not have the same equation throughout the interval from 0 to 2π. As indicated above, $f(x) = 1$ for the first half of the interval, and $f(x) = -1$ for the second. The coefficients are then found by writing separate integrals for each of the two parts. Hence to find a_0, we write

$$a_0 = \frac{1}{\pi} \int_0^{\pi} (1)\, dx + \frac{1}{\pi} \int_{\pi}^{2\pi} (-1)\, dx = \frac{1}{\pi} [x]_0^{\pi} - \frac{1}{\pi} [x]_{\pi}^{2\pi} = 0$$

The value for a_n is found similarly to be

$$a_n = \frac{1}{\pi} \int_0^{\pi} (1) \cos nx\, dx + \frac{1}{\pi} \int_{\pi}^{2\pi} (-1) \cos nx\, dx$$

$$= \frac{1}{\pi} \left[\frac{\sin nx}{n} \right]_0^{\pi} - \frac{1}{\pi} \left[\frac{\sin nx}{n} \right]_{\pi}^{2\pi} = 0$$

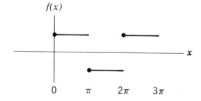

f(x)

x

0   π 2π 3π

Figure 18.3

Finally, b_n is given by

$$b_n = \frac{1}{\pi} \int_0^\pi (1) \sin nx \, dx + \frac{1}{\pi} \int_\pi^{2\pi} (-1) \sin nx \, dx$$

$$= \frac{1}{\pi} \left[-\frac{\cos nx}{n} \right]_0^\pi + \frac{1}{\pi} \left[\frac{\cos nx}{n} \right]_\pi^{2\pi}$$

$$= \frac{1}{\pi} \left[-\frac{\cos n\pi}{n} + \frac{1}{n} \right] + \frac{1}{\pi} \left[\frac{\cos 2n\pi}{n} - \frac{\cos n\pi}{n} \right]$$

The formula for b_n is more involved in this problem than it was in Example 1, because its value changes according to whether n is odd or even. This is caused by the fact that $\cos n\pi$ is equal to -1 when n is odd, and $+1$ when n is even. Hence, when n *is odd*, b_n is given by

$$b_n = \frac{1}{\pi} \left[\frac{1}{n} + \frac{1}{n} \right] + \frac{1}{\pi} \left[\frac{1}{n} + \frac{1}{n} \right] = \frac{4}{n\pi}$$

and when n *is even* by

$$b_n = \frac{1}{\pi} \left[-\frac{1}{n} + \frac{1}{n} \right] + \frac{1}{\pi} \left[\frac{1}{n} - \frac{1}{n} \right] = 0$$

We see from this that the Fourier series will contain only sine terms which are odd harmonics. Therefore, the desired Fourier series is

$$f(x) = \frac{4}{\pi} \left(\sin x + \frac{1}{3} \sin 3x + \frac{1}{5} \sin 5x + \cdots \right)$$

Example 3.   In electronics, a half-wave rectifier is a device that allows electric current to flow in only one direction. Consequently, if a simple alternating current is applied to the rectifier, current will flow during only half of the cycle. Find the Fourier series of the half-wave rectification of $i = \sin t$, which is depicted in Figure 18.4. This may be expressed mathematically as

$$\begin{aligned} i &= \sin t, & 0 \leqslant t < \pi \\ &= 0, & \pi \leqslant t \leqslant 2\pi \end{aligned}$$

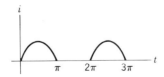

Figure 18.4

Solution.

$$a_0 = \frac{1}{\pi} \int_0^\pi \sin t \, dt = \frac{1}{\pi} [-\cos t]_0^\pi = \frac{1}{\pi} [1 + 1] = \frac{2}{\pi}$$

$$a_n = \frac{1}{\pi} \int_0^\pi \sin t \cos nt \, dt$$

Using Formula 45 from Table 5 (in the Appendix), we obtain

$$a_n = \frac{1}{\pi} \int_0^\pi \sin t \cos nt \, dt = -\frac{1}{2\pi} \left[\frac{\cos (1-n)t}{1-n} + \frac{\cos (1+n)t}{1+n} \right]_0^\pi$$

$$= -\frac{1}{2\pi} \left[\frac{\cos (1-n)\pi}{1-n} + \frac{\cos (1+n)\pi}{1+n} - \frac{1}{1-n} - \frac{1}{1+n} \right]$$

This is valid for all n, except $n = 1$. For $n > 1$, the value of a_n depends upon whether n is odd or even. When n is odd,

$$a_n = -\frac{1}{2\pi} \left[\frac{1}{1-n} + \frac{1}{1+n} - \frac{1}{1-n} - \frac{1}{1+n} \right] = 0$$

When n is even,

$$a_n = -\frac{1}{2\pi} \left[-\frac{1}{1-n} - \frac{1}{1+n} - \frac{1}{1-n} - \frac{1}{1+n} \right] = \frac{1}{\pi} \left[\frac{1}{1-n} + \frac{1}{1+n} \right]$$

For the case where $n = 1$, the integral becomes

$$a_1 = \frac{1}{\pi} \int_0^\pi \sin t \cos t \, dt = \frac{1}{\pi} \left[\frac{1}{2} \sin^2 t \right]_0^\pi = 0$$

Finally, the value of b_n is found by evaluating

$$b_n = \frac{1}{\pi} \int_0^\pi \sin t \sin nt \, dt$$

Using Formula 44 from Table 5, which is valid for $n > 1$, we get

$$b_n = \frac{1}{2\pi} \left[\frac{\sin (1-n)t}{1-n} - \frac{\sin (1+n)t}{1+n} \right]_0^\pi$$

$$= \frac{1}{2\pi} \left[\frac{\sin (1-n)\pi}{1-n} - \frac{\sin (1+n)\pi}{1+n} \right] = 0$$

Considering the separate case $n = 1$, the integral becomes

$$b_1 = \frac{1}{\pi} \int_0^\pi \sin t \sin t \, dt = \frac{1}{\pi} \int_0^\pi \sin^2 t \, dt = \frac{1}{2\pi} \left[t - \frac{\cos 2t}{2} \right]_0^\pi = \frac{1}{2}$$

Substituting the values of a_0, a_n, and b_n into (5), we get

$$i = \frac{1}{\pi} + \frac{1}{2}\sin t - \frac{2}{\pi}\left[\frac{1}{3}\cos 2t + \frac{1}{15}\cos 4t + \cdots\right]$$

as the Fourier series of the half-wave rectification of $\sin t$.

EXERCISES

In Exercises 1 to 8, sketch several periods of the given periodic function and find its Fourier series expansion.

1. $f(x) = 1,$ $0 \le x < \pi$
      $= 0,$ $\pi \le x < 2\pi$

2. $g(x) = 3,$ $0 \le x < \pi$
      $= -1,$ $\pi \le x < 2\pi$

3. $h(x) = x,$ $0 \le x < \pi$
      $= \pi,$ $\pi \le x < 2\pi$

4. $i = 2t,$ $0 \le t < \pi$
      $= 1,$ $\pi \le t < 2\pi$

5. $F(s) = 2,$ $0 \le s < \pi/2$
      $= 1,$ $\pi/2 \le s < \pi$
      $= 0,$ $\pi \le s < 2\pi$

6. $v = t,$ $0 \le t < \pi$
      $= 0,$ $\pi \le t < 2\pi$

7. $s(t) = t^2,$ $0 \le t < \pi/2$
      $= 0,$ $\pi/2 \le t < 2\pi$

8. $p(x) = \pi,$ $0 \le x < \pi$
      $= x,$ $\pi \le x < 2\pi$

9. In fatigue testing a specimen, the deforming force was applied alternately at two different levels. If the applied force is given by

$$f(t) = 1, \quad\quad 0 \le t \le \pi$$
$$= 2, \quad\quad \pi \le t < 2\pi$$

find the Fourier series that can be used to synthesize the necessary force.

10. The output of a full-wave rectifier is shown in Figure 18.5 when the input is a simple sinusoidal current. Find the Fourier series of the output if one period is described by

$$i = \sin t, \quad\quad 0 \le t < \pi$$
$$= -\sin t, \quad\quad \pi \le t \le 2\pi$$

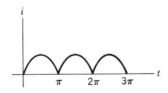

Figure 18.5

Appendix

Table 1   Natural Trigonometric Functions

Degree	Radian	Sine	Cosine	Tangent	Degree	Radian	Sine	Cosine	Tangent
0°	0.000	0.000	1.000	0.000	45°	0.785	0.707	0.707	1.000
1°	0.017	0.017	1.000	0.017	46°	0.803	0.719	0.695	1.036
2°	0.035	0.035	0.999	0.035	47°	0.820	0.731	0.682	1.072
3°	0.052	0.052	0.999	0.052	48°	0.838	0.743	0.669	1.111
4°	0.070	0.070	0.998	0.070	49°	0.855	0.755	0.656	1.150
5°	0.087	0.087	0.996	0.087	50°	0.873	0.766	0.643	1.192
6°	0.105	0.105	0.995	0.105	51°	0.890	0.777	0.629	1.235
7°	0.122	0.122	0.993	0.123	52°	0.908	0.788	0.616	1.280
8°	0.140	0.139	0.990	0.141	53°	0.925	0.799	0.602	1.327
9°	0.157	0.156	0.988	0.158	54°	0.942	0.809	0.588	1.376
10°	0.175	0.174	0.985	0.176	55°	0.960	0.819	0.574	1.428
11°	0.192	0.191	0.982	0.194	56°	0.977	0.829	0.559	1.483
12°	0.209	0.208	0.978	0.213	57°	0.995	0.839	0.545	1.540
13°	0.227	0.225	0.974	0.231	58°	1.012	0.848	0.530	1.600
14°	0.244	0.242	0.970	0.249	59°	1.030	0.857	0.515	1.664
15°	0.262	0.259	0.966	0.268	60°	1.047	0.866	0.500	1.732
16°	0.279	0.276	0.961	0.287	61°	1.065	0.875	0.485	1.804
17°	0.297	0.292	0.956	0.306	62°	1.082	0.883	0.469	1.881
18°	0.314	0.309	0.951	0.325	63°	1.100	0.891	0.454	1.963
19°	0.332	0.326	0.946	0.344	64°	1.117	0.899	0.438	2.050
20°	0.349	0.342	0.940	0.364	65°	1.134	0.906	0.423	2.145
21°	0.367	0.358	0.934	0.384	66°	1.152	0.914	0.407	2.246
22°	0.384	0.375	0.927	0.404	67°	1.169	0.921	0.391	2.356
23°	0.401	0.391	0.921	0.424	68°	1.187	0.927	0.375	2.475
24°	0.419	0.407	0.914	0.445	69°	1.204	0.934	0.358	2.605
25°	0.436	0.423	0.906	0.466	70°	1.222	0.940	0.342	2.748
26°	0.454	0.438	0.899	0.488	71°	1.239	0.946	0.326	2.904
27°	0.471	0.454	0.891	0.510	72°	1.257	0.951	0.309	3.078
28°	0.489	0.469	0.883	0.532	73°	1.274	0.956	0.292	3.271
29°	0.506	0.485	0.875	0.554	74°	1.292	0.961	0.276	3.487
30°	0.524	0.500	0.866	0.577	75°	1.309	0.966	0.259	3.732
31°	0.541	0.515	0.857	0.601	76°	1.326	0.970	0.242	4.011
32°	0.559	0.530	0.848	0.625	77°	1.344	0.974	0.225	4.332
33°	0.576	0.545	0.839	0.649	78°	1.361	0.978	0.208	4.705
34°	0.593	0.559	0.829	0.675	79°	1.379	0.982	0.191	5.145
35°	0.611	0.574	0.819	0.700	80°	1.396	0.985	0.174	5.671
36°	0.628	0.588	0.809	0.727	81°	1.414	0.988	0.156	6.314
37°	0.646	0.602	0.799	0.754	82°	1.431	0.990	0.139	7.115
38°	0.663	0.616	0.788	0.781	83°	1.449	0.993	0.122	8.144
39°	0.681	0.629	0.777	0.810	84°	1.466	0.995	0.105	9.514
40°	0.698	0.643	0.766	0.839	85°	1.484	0.996	0.087	11.430
41°	0.716	0.656	0.755	0.869	86°	1.501	0.998	0.070	14.301
42°	0.733	0.669	0.743	0.900	87°	1.518	0.999	0.052	19.081
43°	0.750	0.682	0.731	0.933	88°	1.536	0.999	0.035	28.636
44°	0.768	0.695	0.719	0.966	89°	1.553	1.000	0.017	57.290
45°	0.785	0.707	0.707	1.000	90°	1.571	1.000	0.000	undef.

Table 2   Common Logarithms

N	0	1	2	3	4	5	6	7	8	9
1.0	0000	0043	0086	0128	0170	0212	0253	0294	0334	0374
1.1	0414	0453	0492	0531	0569	0607	0645	0682	0719	0755
1.2	0792	0828	0864	0899	0934	0969	1004	1038	1072	1106
1.3	1139	1173	1206	1239	1271	1303	1335	1367	1399	1430
1.4	1461	1492	1523	1553	1584	1614	1644	1673	1703	1732
1.5	1761	1790	1818	1847	1875	1903	1931	1959	1987	2014
1.6	2041	2068	2095	2122	2148	2175	2201	2227	2253	2279
1.7	2304	2330	2355	2380	2405	2430	2455	2480	2504	2529
1.8	2553	2577	2601	2625	2648	2672	2695	2718	2742	2765
1.9	2788	2810	2833	2856	2878	2900	2923	2945	2967	2989
2.0	3010	3032	3054	3075	3096	3118	3139	3160	3181	3201
2.1	3222	3243	3263	3284	3304	3324	3345	3365	3385	3404
2.2	3424	3444	3464	3483	3502	3522	3541	3560	3579	3598
2.3	3617	3636	3655	3674	3692	3711	3729	3747	3766	3784
2.4	3802	3820	3838	3856	3874	3892	3909	3927	3945	3962
2.5	3979	3997	4014	4031	4048	4065	4082	4099	4116	4133
2.6	4150	4166	4183	4200	4216	4232	4249	4265	4281	4298
2.7	4314	4330	4346	4362	4378	4393	4409	4425	4440	4456
2.8	4472	4487	4502	4518	4533	4548	4564	4579	4594	4609
2.9	4624	4639	4654	4669	4683	4698	4713	4728	4742	4757
3.0	4771	4786	4800	4814	4829	4843	4857	4871	4886	4900
3.1	4914	4928	4942	4955	4969	4983	4997	5011	5024	5038
3.2	5051	5065	5079	5092	5105	5119	5132	5145	5159	5172
3.3	5185	5198	5211	5224	5237	5250	5263	5276	5289	5302
3.4	5315	5328	5340	5353	5366	5378	5391	5403	5416	5428
3.5	5441	5453	5465	5478	5490	5502	5514	5527	5539	5551
3.6	5563	5575	5587	5599	5611	5623	5635	5647	5658	5670
3.7	5682	5694	5705	5717	5729	5740	5752	5763	5775	5786
3.8	5798	5809	5821	5832	5843	5855	5866	5877	5888	5899
3.9	5911	5922	5933	5944	5955	5966	5977	5988	5999	6010
4.0	6021	6031	6042	6053	6064	6075	6085	6096	6107	6117
4.1	6128	6138	6149	6160	6170	6180	6191	6201	6212	6222
4.2	6232	6243	6253	6263	6274	6284	6294	6304	6314	6325
4.3	6335	6345	6355	6365	6375	6385	6395	6405	6415	6425
4.4	6435	6444	6454	6464	6474	6484	6493	6503	6513	6522
4.5	6532	6542	6551	6561	6571	6580	6590	6599	6609	6618
4.6	6628	6637	6646	6656	6665	6675	6684	6693	6702	6712
4.7	6721	6730	6739	6749	6758	6767	6776	6785	6794	6803
4.8	6812	6821	6830	6839	6848	6857	6866	6875	6884	6893
4.9	6902	6911	6920	6928	6937	6946	6955	6964	6972	6981
5.0	6990	6998	7007	7016	7024	7033	7042	7050	7059	7067
5.1	7076	7084	7093	7101	7110	7118	7126	7135	7143	7152
5.2	7160	7168	7177	7185	7193	7202	7210	7218	7226	7235
5.3	7243	7251	7259	7267	7275	7284	7292	7300	7308	7316
5.4	7324	7332	7340	7348	7356	7364	7372	7380	7388	7396
N	0	1	2	3	4	5	6	7	8	9

Table 2 (*Continued*)   Common Logarithms

N	0	1	2	3	4	5	6	7	8	9
5.5	7404	7412	7419	7427	7435	7443	7451	7459	7466	7474
5.6	7482	7490	7497	7505	7513	7520	7528	7536	7543	7551
5.7	7559	7566	7574	7582	7589	7597	7604	7612	7619	7627
5.8	7634	7642	7649	7657	7664	7672	7679	7686	7694	7701
5.9	7709	7716	7723	7731	7738	7745	7752	7760	7767	7774
6.0	7782	7789	7796	7803	7810	7818	7825	7832	7839	7846
6.1	7853	7860	7868	7875	7882	7889	7896	7903	7910	7917
6.2	7924	7931	7938	7945	7952	7959	7966	7973	7980	7987
6.3	7993	8000	8007	8014	8021	8028	8035	8041	8048	8055
6.4	8062	8069	8075	8082	8089	8096	8102	8109	8116	8112
6.5	8129	8136	8142	8149	8156	8162	8169	8176	8182	8189
6.6	8195	8202	8209	8215	8222	8228	8235	8241	8248	8254
6.7	8261	8267	8274	8280	8287	8293	8299	8306	8312	8319
6.8	8325	8331	8338	8344	8351	8357	8363	8370	8376	8382
6.9	8388	8395	8401	8407	8414	8420	8426	8432	8439	8445
7.0	8451	8457	8463	8470	8476	8482	8488	8494	8500	8506
7.1	8513	8519	8525	8531	8537	8543	8549	8555	8561	8567
7.2	8573	8579	8585	8591	8597	8603	8609	8615	8621	8627
7.3	8633	8639	8645	8651	8657	8663	8669	8675	8681	8686
7.4	8692	8698	8704	8710	8716	8722	8727	8733	8739	8745
7.5	8751	8756	8762	8768	8774	8779	8785	8791	8797	8802
7.6	8808	8814	8820	8825	8831	8837	8842	8848	8854	8859
7.7	8865	8871	8876	8882	8887	8893	8899	8904	8910	8915
7.8	8921	8927	8932	8938	8943	8949	8954	8960	8965	8971
7.9	8976	8982	8987	8993	8998	9004	9009	9015	9020	9025
8.0	9031	9036	9042	9047	9053	9058	9063	9069	9074	9079
8.1	9085	9090	9096	9101	9106	9112	9117	9122	9128	9133
8.2	9138	9143	9149	9154	9159	9165	9170	9175	9180	9186
8.3	9191	9196	9201	9206	9212	9217	9222	9227	9232	9238
8.4	9243	9248	9253	9258	9263	9269	9274	9279	9284	9289
8.5	9294	9299	9304	9309	9315	9320	9325	9330	9335	9340
8.6	9345	9350	9355	9360	9365	9370	9375	9380	9385	9390
8.7	9395	9400	9405	9410	9415	9420	9425	9430	9435	9440
8.8	9445	9450	9455	9460	9465	9469	9474	9479	9484	9489
8.9	9494	9499	9504	9509	9513	9518	9523	9528	9533	9538
9.0	9542	9547	9552	9557	9562	9566	9571	9576	9581	9586
9.1	9590	9595	9600	9605	9609	9614	9619	9624	9628	9633
9.2	9638	9643	9647	9652	9657	9661	9666	9671	9675	9680
9.3	9685	9689	9694	9699	9703	9708	9713	9717	9722	9727
9.4	9731	9736	9741	9745	9750	9754	9759	9763	9768	9773
9.5	9777	9782	9786	9791	9795	9800	9805	9809	9814	9818
9.6	9823	9827	9832	9836	9841	9845	9850	9854	9859	9863
9.7	9868	9872	9877	9881	9886	9890	9894	9899	9903	9908
9.8	9912	9917	9921	9926	9930	9934	9939	9943	9948	9952
9.9	9956	9961	9965	9969	9974	9978	9983	9987	9991	9996
N	0	1	2	3	4	5	6	7	8	9

Table 3 1.00 — Four-Place Natural Logarithms — 5.49

N	0.00	0.01	0.02	0.03	0.04	0.05	0.06	0.07	0.08	0.09
1.0	0.0000	0.0100	0.0198	0.0296	0.0392	0.0488	0.0583	0.0677	0.0770	0.0862
1.1	0.0953	0.1044	0.1133	0.1222	0.1310	0.1398	0.1484	0.1570	0.1655	0.1740
1.2	0.1823	0.1906	0.1989	0.2070	0.2151	0.2231	0.2311	0.2390	0.2469	0.2546
1.3	0.2624	0.2700	0.2776	0.2852	0.2927	0.3001	0.3075	0.3148	0.3221	0.3293
1.4	0.3365	0.3436	0.3507	0.3577	0.3646	0.3716	0.3784	0.3853	0.3920	0.3988
1.5	0.4055	0.4121	0.4187	0.4253	0.4318	0.4383	0.4447	0.4511	0.4574	0.4637
1.6	0.4700	0.4762	0.4824	0.4886	0.4947	0.5008	0.5068	0.5128	0.5188	0.5247
1.7	0.5306	0.5365	0.5423	0.5481	0.5539	0.5596	0.5653	0.5710	0.5766	0.5822
1.8	0.5878	0.5933	0.5988	0.6043	0.6098	0.6152	0.6206	0.6259	0.6313	0.6366
1.9	0.6419	0.6471	0.6523	0.6575	0.6627	0.6678	0.6729	0.6780	0.6831	0.6881
2.0	0.6931	0.6981	0.7031	0.7080	0.7129	0.7178	0.7227	0.7275	0.7324	0.7372
2.1	0.7419	0.7467	0.7514	0.7561	0.7608	0.7655	0.7701	0.7747	0.7793	0.7839
2.2	0.7885	0.7930	0.7975	0.8020	0.8065	0.8109	0.8154	0.8198	0.8242	0.8286
2.3	0.8329	0.8372	0.8416	0.8459	0.8502	0.8544	0.8587	0.8629	0.8671	0.8713
2.4	0.8755	0.8796	0.8838	0.8879	0.8920	0.8961	0.9002	0.9042	0.9083	0.9123
2.5	0.9163	0.9203	0.9243	0.9282	0.9322	0.9361	0.9400	0.9439	0.9478	0.9517
2.6	0.9555	0.9594	0.9632	0.9670	0.9708	0.9746	0.9783	0.9821	0.9858	0.9895
2.7	0.9933	0.9969	1.0006	1.0043	1.0080	1.0116	1.0152	1.0188	1.0225	1.0260
2.8	1.0296	1.0332	1.0367	1.0403	1.0438	1.0473	1.0508	1.0543	1.0578	1.0613
2.9	1.0647	1.0682	1.0716	1.0750	1.0784	1.0818	1.0852	1.0886	1.0919	1.0953
3.0	1.0986	1.1019	1.1053	1.1086	1.1119	1.1151	1.1184	1.1217	1.1249	1.1282
3.1	1.1314	1.1346	1.1378	1.1410	1.1442	1.1474	1.1506	1.1537	1.1569	1.1600
3.2	1.1632	1.1663	1.1694	1.1725	1.1756	1.1787	1.1817	1.1848	1.1878	1.1909
3.3	1.1939	1.1969	1.2000	1.2030	1.2060	1.2090	1.2119	1.2149	1.2179	1.2208
3.4	1.2238	1.2267	1.2296	1.2326	1.2355	1.2384	1.2413	1.2442	1.2470	1.2499
3.5	1.2528	1.2556	1.2585	1.2613	1.2641	1.2669	1.2698	1.2726	1.2754	1.2782
3.6	1.2809	1.2837	1.2865	1.2892	1.2920	1.2947	1.2975	1.3002	1.3029	1.3056
3.7	1.3083	1.3110	1.3137	1.3164	1.3191	1.3218	1.3244	1.3271	1.3297	1.3324
3.8	1.3350	1.3376	1.3403	1.3429	1.3455	1.3481	1.3507	1.3533	1.3558	1.3584
3.9	1.3610	1.3635	1.3661	1.3686	1.3712	1.3737	1.3762	1.3788	1.3813	1.3838
4.0	1.3863	1.3888	1.3913	1.3938	1.3962	1.3987	1.4012	1.4036	1.4061	1.4085
4.1	1.4110	1.4134	1.4159	1.4183	1.4207	1.4231	1.4255	1.4279	1.4303	1.4327
4.2	1.4351	1.4375	1.4398	1.4422	1.4446	1.4469	1.4493	1.4516	1.4540	1.4563
4.3	1.4586	1.4609	1.4633	1.4656	1.4679	1.4702	1.4725	1.4748	1.4770	1.4793
4.4	1.4816	1.4839	1.4861	1.4884	1.4907	1.4929	1.4951	1.4974	1.4996	1.5019
4.5	1.5041	1.5063	1.5085	1.5107	1.5129	1.5151	1.5173	1.5195	1.5217	1.5239
4.6	1.5261	1.5282	1.5304	1.5326	1.5347	1.5369	1.5390	1.5412	1.5433	1.5454
4.7	1.5476	1.5497	1.5518	1.5539	1.5560	1.5581	1.5602	1.5623	1.5644	1.5665
4.8	1.5686	1.5707	1.5728	1.5748	1.5769	1.5790	1.5810	1.5831	1.5851	1.5872
4.9	1.5892	1.5913	1.5933	1.5953	1.6974	1.5994	1.6014	1.6034	1.6054	1.6074
5.0	1.6094	1.6114	1.6134	1.6154	1.6174	1.6194	1.6214	1.6233	1.6253	1.6273
5.1	1.6292	1.6312	1.6332	1.6351	1.6371	1.6390	1.6409	1.6429	1.6448	1.6467
5.2	1.6487	1.6506	1.6525	1.6544	1.6563	1.6582	1.6601	1.6620	1.6639	1.6658
5.3	1.6677	1.6696	1.6715	1.6734	1.6752	1.6771	1.6790	1.6808	1.6827	1.6845
5.4	1.6364	1.6882	1.6901	1.6919	1.6938	1.6956	1.6974	1.6993	1.7011	1.7029

N	0.00	0.01	0.02	0.03	0.04	0.05	0.06	0.07	0.08	0.09

log. 0.1 = 0.6974−3   log. 0.01 = 0.3948−5   log. 0.001 = 0.0922−7

Table 3 (*Continued*)   5.50 — Four-Place Natural
Logarithms — 10.09

N	0.00	0.01	0.02	0.03	0.04	0.05	0.06	0.07	0.08	0.09
5.5	1.7047	1.7066	1.7084	1.7102	1.7120	1.7138	1.7156	1.7174	1.7192	1.7210
5.6	1.7228	1.7246	1.7263	1.7281	1.7299	1.7317	1.7334	1.7352	1.7370	1.7387
5.7	1.7405	1.7422	1.7440	1.7457	1.7475	1.7492	1.7509	1.7527	1.7544	1.7561
5.8	1.7579	1.7596	1.7613	1.7630	1.7647	1.7664	1.7681	1.7699	1.7716	1.7733
5.9	1.7750	1.7766	1.7783	1.7800	1.7817	1.7834	1.7851	1.7867	1.7884	1.7901
6.0	1.7918	1.7934	1.7951	1.7967	1.7984	1.8001	1.8017	1.8034	1.8050	1.8066
6.1	1.8083	1.8099	1.8116	1.8132	1.8148	1.8165	1.8181	1.8197	1.8213	1.8229
6.2	1.8245	1.8262	1.8278	1.8294	1.8310	1.8326	1.8342	1.8358	1.8374	1.8390
6.3	1.8405	1.8421	1.8437	1.8453	1.8469	1.8485	1.8500	1.8516	1.8532	1.8547
6.4	1.8563	1.8579	1.8594	1.8610	1.8625	1.8641	1.8656	1.8672	1.8687	1.8703
6.5	1.8718	1.8733	1.8749	1.8764	1.8779	1.8795	1.8810	1.8825	1.8840	1.8856
6.6	1.8871	1.8886	1.8901	1.8916	1.8931	1.8946	1.8961	1.8976	1.8991	1.9006
6.7	1.9021	1.9036	1.9051	1.9066	1.9081	1.9095	1.9110	1.9125	1.9140	1.9155
6.8	1.9169	1.9184	1.9199	1.9213	1.9228	1.9242	1.9257	1.9272	1.9286	1.9301
6.9	1.9315	1.9330	1.9344	1.9359	1.9373	1.9387	1.9402	1.9416	1.9430	1.9445
7.0	1.9459	1.9473	1.9488	1.9502	1.9516	1.9530	1.9544	1.9559	1.9573	1.9587
7.1	1.9601	1.9615	1.9629	1.9643	1.9657	1.9671	1.9685	1.9699	1.9713	1.9727
7.2	1.9741	1.9755	1.9769	1.9782	1.9796	1.9810	1.9824	1.9838	1.9851	1.9865
7.3	1.9879	1.9892	1.9906	1.9920	1.9933	1.9947	1.9961	1.9974	1.9988	2.0001
7.4	2.0015	2.0028	2.0042	2.0055	2.0069	2.0082	2.0096	2.0109	2.0122	2.0136
7.5	2.0149	2.0162	2.0176	2.0189	2.0202	2.0215	2.0229	2.0242	2.0255	2.0268
7.6	2.0281	2.0295	2.0308	2.0321	2.0334	2.0347	2.0360	2.0373	2.0386	2.0399
7.7	2.0412	2.0425	2.0438	2.0451	2.0464	2.0477	2.0490	2.0503	2.0516	2.0528
7.8	2.0541	2.0554	2.0567	2.0580	2.0592	2.0605	2.0618	2.0631	2.0643	2.0656
7.9	2.0669	2.0681	2.0694	2.0707	2.0719	2.0732	2.0744	2.0757	2.0769	2.0782
8.0	2.0794	2.0807	2.0819	2.0832	2.0844	2.0857	2.0869	2.0882	2.0894	2.0906
8.1	2.0919	2.0931	2.0943	2.0956	2.0968	2.0980	2.0992	2.1005	2.1017	2.1029
8.2	2.1041	2.1054	2.1066	2.1078	2.1090	2.1102	2.1114	2.1126	2.1138	2.1150
8.3	2.1163	2.1175	2.1187	2.1199	2.1211	2.1223	2.1235	2.1247	2.1258	2.1270
8.4	2.1282	2.1294	2.1306	2.1318	2.1330	2.1342	2.1353	2.1365	2.1377	2.1389
8.5	2.1401	2.1412	2.1424	2.1436	2.1448	2.1459	2.1471	2.1483	2.1494	2.1506
8.6	2.1518	2.1529	2.1541	2.1552	2.1564	2.1576	2.1587	2.1599	2.1610	2.1622
8.7	2.1633	2.1645	2.1656	2.1668	2.1679	2.1691	2.1702	2.1713	2.1725	2.1736
8.8	2.1748	2.1759	2.1770	2.1782	2.1793	2.1804	2.1815	2.1827	2.1838	2.1849
8.9	2.1861	2.1872	2.1883	2.1894	2.1905	2.1917	2.1928	2.1939	2.1950	2.1961
9.0	2.1972	2.1983	2.1994	2.2006	2.2017	2.2028	2.2039	2.2050	2.2061	2.2072
9.1	2.2083	2.2094	2.2105	2.2116	2.2127	2.2138	2.2148	2.2159	2.2170	2.2181
9.2	2.2192	2.2203	2.2214	2.2225	2.2235	2.2246	2.2257	2.2268	2.2279	2.2289
9.3	2.2300	2.2311	2.2322	2.2332	2.2343	2.2354	2.2364	2.2375	2.2386	2.2396
9.4	2.2407	2.2418	2.2428	2.2439	2.2450	2.2460	2.2471	2.2481	2.2492	2.2502
9.5	2.2513	2.2523	2.2534	2.2544	2.2555	2.2565	2.2576	2.2586	2.2597	2.2607
9.6	2.2618	2.2628	2.2638	2.2649	2.2659	2.2670	2.2680	2.2690	2.2701	2.2711
9.7	2.2721	2.2732	2.2742	2.2752	2.2762	2.2773	2.2783	2.2793	2.2803	2.2814
9.8	2.2824	2.2834	2.2844	2.2854	2.2865	2.2875	2.2885	2.2895	2.2905	2.2915
9.9	2.2925	2.2935	2.2946	2.2956	2.2966	2.2976	2.2986	2.2996	2.3006	2.3016
10.0	2.3026	2.3036	2.3046	2.3056	2.3066	2.3076	2.3086	2.3096	2.3106	2.3115

N	0.00	0.01	0.02	0.03	0.04	0.05	0.06	0.07	0.08	0.09

log. 0.0001 = 0.7897–10   log. 0.00001 = 0.4871–12   log. 0.000001 = 0.1845–14

Table 4 0.00—Values of
Exponential Functions—0.89

x	e^x	e^{-x}	x	e^x	e^{-x}
0.00	1.0000	1.00 000	**0.45**	1.5683	0.63 763
0.01	1.0101	0.99 005	0.46	1.5841	0.63 128
0.02	1.0202	0.98 020	0.47	1.6000	0.62 500
0.03	1.0305	0.97 045	0.48	1.6161	0.61 878
0.04	1.0408	0.96 079	0.49	1.6323	0.61 263
0.05	1.0513	0.95 123	**0.50**	1.6487	0.60 653
0.06	1.0618	0.94 176	0.51	1.6653	0.60 050
0.07	1.0725	0.93 239	0.52	1.6820	0.59 452
0.08	1.0833	0.92 312	0.53	1.6989	0.58 860
0.09	1.0942	0.91 393	0.54	1.7160	0.58 275
0.10	1.1052	0.90 484	**0.55**	1.7333	0.57 695
0.11	1.1163	0.89 583	0.56	1.7507	0.57 121
0.12	1.1275	0.88 692	0.57	1.7683	0.56 553
0.13	1.1388	0.87 810	0.58	1.7860	0.55 990
0.14	1.1503	0.86 936	0.59	1.8040	0.55 433
0.15	1.1618	0.86 071	**0.60**	1.8221	0.54 881
0.16	1.1735	0.85 214	0.61	1.8404	0.54 335
0.17	1.1853	0.84 366	0.62	1.8589	0.53 794
0.18	1.1972	0.83 527	0.63	1.8776	0.53 259
0.19	1.2092	0.82 696	0.64	1.8965	0.52 729
0.20	1.2214	0.81 873	**0.65**	1.9155	0.52 205
0.21	1.2337	0.81 058	0.66	1.9348	0.51 685
0.22	1.2461	0.80 252	0.67	1.9542	0.51 171
0.23	1.2586	0.79 453	0.68	1.9739	0.50 662
0.24	1.2712	0.78 663	0.69	1.9937	0.50 158
0.25	1.2840	0.77 880	**0.70**	2.0138	0.49 659
0.26	1.2969	0.77 105	0.71	2.0340	0.49 164
0.27	1.3100	0.76 338	0.72	2.0544	0.48 675
0.28	1.3231	0.75 578	0.73	2.0751	0.48 191
0.29	1.3364	0.74 826	0.74	2.0959	0.47 711
0.30	1.3499	0.74 082	**0.75**	2.1170	0.47 237
0.31	1.3634	0.73 345	0.76	2.1383	0.46 767
0.32	1.3771	0.72 615	0.77	2.1598	0.46 301
0.33	1.3910	0.71 892	0.78	2.1815	0.45 841
0.34	1.4049	0.71 177	0.79	2.2034	0.45 384
0.35	1.4191	0.70 469	**0.80**	2.2255	0.44 933
0.36	1.4333	0.69 768	0.81	2.2479	0.44 486
0.37	1.4477	0.69 073	0.82	2.2705	0.44 043
0.38	1.4623	0.68 386	0.83	2.2933	0.43 605
0.39	1.4770	0.67 706	0.84	2.3164	0.43 171
0.40	1.4918	0.67 032	**0.85**	2.3396	0.42 741
0.41	1.5068	0.66 365	0.86	2.3632	0.42 316
0.42	1.5220	0.65 705	0.87	2.3869	0.41 895
0.43	1.5373	0.65 051	0.88	2.4109	0.41 478
0.44	1.5527	0.64 404	0.89	2.4351	0.41 066
x	e^x	e^{-x}	x	e^x	e^{-x}

Table 4 (*Continued*)   0.90 – Values of
Exponential Functions – 1.79

x	e^x	e^{-x}	x	e^x	e^{-x}
0.90	2.4596	0.40 657	**1.35**	3.8574	0.25 924
0.91	2.4843	0.40 252	1.36	3.8962	0.25 666
0.92	2.5093	0.39 852	1.37	3.9354	0.25 411
0.93	2.5345	0.39 455	1.38	3.9749	0.25 158
0.94	2.5600	0.39 063	1.39	4.0149	0.24 908
0.95	2.5857	0.38 674	**1.40**	4.0552	0.24 660
0.96	2.6117	0.38 289	1.41	4.0960	0.24 414
0.97	2.6379	0.37 908	1.42	4.1371	0.24 171
0.98	2.6645	0.37 531	1.43	4.1787	0.23 931
0.99	2.6912	0.37 158	1.44	4.2207	0.23 693
1.00	2.7183	0.36 788	**1.45**	4.2631	0.23 457
1.01	2.7456	0.36 422	1.46	4.3060	0.23 224
1.02	2.7732	0.36 059	1.47	4.3492	0.22 993
1.03	2.8011	0.35 70ĺ	1.48	4.3929	0.22 764
1.04	2.8292	0.35 345	1.49	4.4371	0.22 537
1.05	2.8577	0.34 994	**1.50**	4.4817	0.22 313
1.06	2.8864	0.34 646	1.51	4.5267	0.22 091
1.07	2.9154	0.34 301	1.52	4.5722	0.21 871
1.08	2.9447	0.33 960	1.53	4.6182	0.21 654
1.09	2.9743	0.33 622	1.54	4.6646	0.21 438
1.10	3.0042	0.33 287	**1.55**	4.7115	0.21 225
1.11	3.0344	0.32 956	1.56	4.7588	0.21 014
1.12	3.0649	0.32 628	1.57	4.8066	0.20 805
1.13	3.0957	0.32 303	1.58	4.8550	0.20 598
1.14	3.1268	0.31 982	1.59	4.9037	0.20 393
1.15	3.1582	0.31 664	**1.60**	4.9530	0.20 190
1.16	3.1899	0.31 349	1.61	5.0028	0.19 989
1.17	3.2220	0.31 037	1.62	5.0531	0.19 790
1.18	3.2544	0.30 728	1.63	5.1039	0.19 593
1.19	3.2871	0.30 422	1.64	5.1552	0.19 398
1.20	3.3201	0.30 119	**1.65**	5.2070	0.19 205
1.21	3.3535	0.29 820	1.66	5.2593	0.19 014
1.22	3.3872	0.29 523	1.67	5.3122	0.18 825
1.23	3.4212	0.29 229	1.68	5.3656	0.18 637
1.24	3.4556	0.28 938	1.69	5.4195	0.18 452
1.25	3.4903	0.28 650	**1.70**	5.4739	0.18 268
1.26	3.5254	0.28 365	1.71	5.5290	0.18 087
1.27	3.5609	0.28 083	1.72	5.5845	0.17 907
1.28	3.5966	0.27 804	1.73	5.6407	0.17 728
1.29	3.6328	0.27 527	1.74	5.6973	0.17 552
1.30	3.6693	0.27 253	**1.75**	5.7546	0.17 377
1.31	3.7062	0.26 982	1.76	5.8124	0.17 204
1.32	3.7434	0.26 714	1.77	5.8709	0.17 033
1.33	3.7810	0.26 448	1.78	5.9299	0.16 864
1.34	3.8190	0.26 185	1.79	5.9895	0.16 696

x	e^x	e^{-x}	x	e^x	e^{-x}

Table 4 (*Continued*)   1.80 — Values of
Exponential Functions — 2.69

x	e^x	e^{-x}	x	e^x	e^{-x}
1.80	6.0496	0.16 530	**2.25**	9.4877	0.10 540
1.81	6.1104	0.16 365	2.26	9.5831	0.10 435
1.82	6.1719	0.16 203	2.27	9.6794	0.10 331
1.83	6.2339	0.16 041	2.28	9.7767	0.10 228
1.84	6.2965	0.15 882	2.29	9.8749	0.10 127
1.85	6.3598	0.15 724	**2.30**	9.9742	0.10 026
1.86	6.4237	0.15 567	2.31	10.074	0.09 9261
1.87	6.4883	0.15 412	2.32	10.176	0.09 8274
1.88	6.5535	0.15 259	2.33	10.278	0.09 7296
1.89	6.6194	0.15 107	2.34	10.381	0.09 6328
1.90	6.6859	0.14 957	**2.35**	10.486	0.09 5369
1.91	6.7531	0.14 808	2.36	10.591	0.09 4420
1.92	6.8210	0.14 661	2.37	10.697	0.09 3481
1.93	6.8895	0.14 515	2.38	10.805	0.09 2551
1.94	6.9588	0.14 370	2.39	10.913	0.09 1630
1.95	7.0287	0.14 227	**2.40**	11.023	0.09 0718
1.96	7.0993	0.14 086	2.41	11.134	0.08 9815
1.97	7.1707	0.13 946	2.42	11.246	0.08 8922
1.98	7.2427	0.13 807	2.43	11.359	0.08 8037
1.99	7.3155	0.13 670	2.44	11.473	0.08 7161
2.00	7.3891	0.13 534	**2.45**	11.588	0.08 6294
2.01	7.4633	0.13 399	2.46	11.705	0.08 5435
2.02	7.5383	0.13 266	2.47	11.822	0.08 4585
2.03	7.6141	0.13 134	2.48	11.941	0.08 3743
2.04	7.6906	0.13 003	2.49	12.061	0.08 2910
2.05	7.7679	0.12 873	**2.50**	12.182	0.082 085
2.06	7.8460	0.12 745	2.51	12.305	0.081 268
2.07	7.9248	0.12 619	2.52	12.429	0.080 460
2.08	8.0045	0.12 493	2.53	12.554	0.079 659
2.09	8.0849	0.12 369	2.54	12.680	0.078 866
2.10	8.1662	0.12 246	**2.55**	12.807	0.078 082
2.11	8.2482	0.12 124	2.56	12.936	0.077 305
2.12	8.3311	0.12 003	2.57	13.066	0.076 536
2.13	8.4149	0.11 884	2.58	13.197	0.075 774
2.14	8.4994	0.11 765	2.59	13.330	0.075 020
2.15	8.5849	0.11 648	**2.60**	13.464	0.074 274
2.16	8.6711	0.11 533	2.61	13.599	0.073 535
2.17	8.7583	0.11 418	2.62	13.736	0.072 803
2.18	8.8463	0.11 304	2.63	13.874	0.072 078
2.19	8.9352	0.11 192	2.64	14.013	0.071 361
2.20	9.0250	0.11 080	**2.65**	14.154	0.070 651
2.21	9.1157	0.10 970	2.66	14.296	0.069 948
2.22	9.2073	0.10 861	2.67	14.440	0.069 252
2.23	9.2999	0.10 753	2.68	14.585	0.068 563
2.24	9.3933	0.10 646	2.69	14.732	0.067 881

x	e^x	e^{-x}	x	e^x	e^{-x}

Table 4 (*Continued*)   2.70 – Values of
Exponential Functions – 3.59

x	e^x	e^{-x}	x	e^x	e^{-x}
2.70	14.880	0.067 206	**3.15**	23.336	0.04 2852
2.71	15.029	0.066 537	3.16	23.571	0.04 2426
2.72	15.180	0.065 875	3.17	23.807	0.04 2004
2.73	15.333	0.065 219	3.18	24.047	0.04 1586
2.74	15.487	0.064 570	3.19	24.288	0.04 1172
2.75	15.643	0.063 928	**3.20**	24.533	0.04 0762
2.76	15.800	0.063 292	3.21	24.779	0.04 0357
2.77	15.959	0.062 662	3.22	25.028	0.03 9955
2.78	16.119	0.062 039	3.23	25.280	0.03 9557
2.79	16.281	0.061 421	3.24	25.534	0.03 9164
2.80	16.445	0.060 810	**3.25**	25.790	0.03 8774
2.81	16.610	0.060 205	3.26	26.050	0.03 8388
2.82	16.777	0.059 606	3.27	26.311	0.03 8006
2.83	16.945	0.059 013	3.28	26.576	0.03 7628
2.84	17.116	0.058 426	3.29	26.843	0.03 7254
2.85	17.288	0.057 844	**3.30**	27.113	0.03 6883
2.86	17.462	0.057 269	3.31	27.385	0.03 6516
2.87	17.637	0.056 699	3.32	27.660	0.03 6153
2.88	17.814	0.056 135	3.33	27.938	0.03 5793
2.89	17.993	0.055 576	3.34	28.219	0.03 5437
2.90	18.174	0.055 023	**3.35**	28.503	0.03 5084
2.91	18.357	0.054 476	3.36	28.789	0.03 4735
2.92	18.541	0.053 934	3.37	29.079	0.03 4390
2.93	18.728	0.053 397	3.38	29.371	0.03 4047
2.94	18.916	0.052 866	3.39	29.666	0.03 3709
2.95	19.106	0.052 340	**3.40**	29.964	0.03 3373
2.96	19.298	0.051 819	3.41	30.265	0.03 3041
2.97	19.492	0.051 303	3.42	30.569	0.03 2712
2.98	19.688	0.050 793	3.43	30.877	0.03 2387
2.99	19.886	0.050 287	3.44	31.187	0.03 2065
3.00	20.086	0.04 9787	**3.45**	31.500	0.03 1746
3.01	20.287	0.04 9292	3.46	31.817	0.03 1430
3.02	20.491	0.04 8801	3.47	32.137	0.03 1117
3.03	20.697	0.04 8316	3.48	32.460	0.03 0807
3.04	20.905	0.04 7835	3.49	32.786	0.03 0501
3.05	21.115	0.04 7359	**3.50**	33.115	0.030 197
3.06	21.328	0.04 6888	3.51	33.448	0.029 897
3.07	21.542	0.04 6421	2.52	33.784	0.029 599
3.08	21.758	0.04 5959	3.53	34.124	0.029 305
3.09	21.977	0.04 5502	3.54	34.467	0.029 013
3.10	22.198	0.04 5049	**3.55**	34.813	0.028 725
3.11	22.421	0.04 4601	3.56	35.163	0.028 439
3.12	22.646	0.04 4157	3.57	35.517	0.028 156
3.13	22.874	0.04 3718	3.58	35.874	0.027 876
3.14	23.104	0.04 3283	3.59	36.234	0.027 598

x	e^x	e^{-x}	x	e^x	e^{-x}

Table 4 (*Continued*) 3.60 — Values of
Exponential Functions — 4.00

x	e^x	e^{-x}	x	e^x	e^{-x}
3.60	36.598	0.027 324	**3.80**	44.701	0.022 371
3.61	36.966	0.027 052	3.81	45.150	0.022 148
3.62	37.338	0.026 783	3.82	45.604	0.021 928
3.63	37.713	0.026 516	3.83	46.063	0.021 710
3.64	38.092	0.026 252	3.84	46.525	0.021 494
3.65	38.475	0.025 991	**3.85**	46.993	0.021 280
3.66	38.861	0.025 733	3.86	47.465	0.021 068
3.67	39.252	0.025 476	3.87	47.942	0.020 858
3.68	39.646	0.025 223	3.88	48.424	0.020 651
3.69	40.045	0.024 972	3.89	48.911	0.020 445
3.70	40.447	0.024 724	**3.90**	49.402	0.020 242
3.71	40.854	0.024 478	3.91	49.899	0.020 041
3.72	41.264	0.024 234	3.92	50.400	0.019 841
3.73	41.679	0.023 993	3.93	50.907	0.019 644
3.74	42.098	0.023 754	3.94	51.419	0.019 448
3.75	42.521	0.023 518	**3.95**	51.935	0.019 255
3.76	42.948	0.023 284	3.96	52.457	0.019 063
3.77	43.380	0.023 052	3.97	52.985	0.018 873
3.78	43.816	0.022 823	3.98	53.517	0.018 686
3.79	44.256	0.022 596	3.99	54.055	0.018 500
3.75	42.521	0.023 518	**4.00**	54.598	0.018 316

Table 5 Additional Antidifferentiation Formulas

1. $\displaystyle \int \frac{x\,dx}{ax+b} = \frac{x}{a} - \frac{b}{a^2}\ln\,(ax+b) + C.$

2. $\displaystyle \int \frac{x\,dx}{(ax+b)^2} = \frac{b}{a^2(ax+b)} + \frac{1}{a^2}\ln\,(ax+b) + C.$

3. $\displaystyle \int x(ax+b)^n\,dx = \frac{x(ax+b)^{n+1}}{a(n+1)} - \frac{(ax+b)^{n+2}}{a^2(n+1)(n+2)} + C.$

4. $\displaystyle \int \frac{dx}{x(ax+b)} = \frac{1}{b}\ln\frac{x}{ax+b} + C.$

5. $\displaystyle \int \frac{dx}{x(ax+b)^2} = \frac{1}{b(ax+b)} + \frac{1}{b^2}\ln\frac{x}{ax+b} + C.$

6. $\displaystyle \int \frac{dx}{a^2-x^2} = \frac{1}{2a}\ln\frac{a+x}{a-x} + C.$

7. $\displaystyle \int \frac{dx}{(ax^2+b)^2} = \frac{x}{2b(ax^2+b)} + \frac{1}{2b}\int \frac{dx}{ax^2+b}\ .$

8. $\displaystyle \int \frac{dx}{x(ax^2+b)} = \frac{1}{2b}\ln\frac{x^2}{ax^2+b} + C.$

9. $\displaystyle \int x\sqrt{ax+b}\,dx = \frac{2x}{3a}(ax+b)^{3/2} - \frac{4}{15a^2}(ax+b)^{5/2} + C.$

10. $\displaystyle \int \frac{x\,dx}{\sqrt{ax+b}} = \frac{2x}{a}(ax+b)^{1/2} - \frac{4}{3a^2}(ax+b)^{3/2} + C.$

11. $\displaystyle \int \sqrt{a^2-x^2}\,dx = \frac{1}{2}x\sqrt{a^2-x^2} + \frac{1}{2}a^2\arcsin\frac{x}{a} + C.$

12. $\displaystyle \int \sqrt{x^2\pm a^2}\,dx = \frac{1}{2}x\sqrt{x^2\pm a^2} \pm \frac{1}{2}a^2\ln\,(x+\sqrt{x^2\pm a^2}) + C.$

13. $\displaystyle \int \frac{dx}{\sqrt{x^2\pm a^2}} = \ln\,(x+\sqrt{x^2\pm a^2}) + C.$

14. $\displaystyle \int \frac{dx}{x\sqrt{a^2\pm x^2}} = \frac{1}{a}\ln\frac{x}{a+\sqrt{a^2\pm x^2}} + C.$

15. $\displaystyle \int \frac{dx}{x\sqrt{x^2-a^2}} = -\frac{1}{a}\arcsin\frac{a}{x} + C.$

Table 5 (*Continued*)   Additional Antidifferentiation Formulas

---

16. $\displaystyle\int \frac{\sqrt{x^2-a^2}}{x}\,dx = \sqrt{x^2-a^2} + a\arcsin\frac{a}{x} + C.$

17. $\displaystyle\int x^2\sqrt{a^2-x^2}\,dx = -\frac{1}{4}x(a^2-x^2)^{3/2} + \frac{1}{8}a^2x\sqrt{a^2-x^2} + \frac{1}{8}a^4\arcsin\frac{x}{a} + C.$

18. $\displaystyle\int x^3\sqrt{a^2-x^2}\,dx = \frac{1}{5}(a^2-x^2)^{5/2} - \frac{1}{3}a^2(a^2-x^2)^{3/2} + C.$

19. $\displaystyle\int x^2\sqrt{x^2\pm a^2}\,dx = \frac{1}{4}x(x^2\pm a^2)^{3/2} \mp \frac{1}{8}a^2x\sqrt{x^2\pm a^2} - \frac{1}{8}a^4\ln(x+\sqrt{x^2\pm a^2}) + C.$

20. $\displaystyle\int \frac{x^2\,dx}{\sqrt{a^2-x^2}} = -\frac{1}{2}x\sqrt{a^2-x^2} + \frac{1}{2}a^2\arcsin\frac{x}{a} + C.$

21. $\displaystyle\int \frac{x^3\,dx}{\sqrt{a^2-x^2}} = -x^2\sqrt{a^2-x^2} - \frac{2}{3}(a^2-x^2)^{3/2} + C.$

22. $\displaystyle\int \frac{x^2\,dx}{\sqrt{x^2\pm a^2}} = \frac{1}{2}x\sqrt{x^2\pm a^2} \mp \frac{1}{2}a^2\ln(x+\sqrt{x^2\pm a^2}) + C.$

23. $\displaystyle\int \frac{dx}{\sqrt{2ax-x^2}} = 2\arcsin\sqrt{\frac{x}{2a}} + C.$

24. $\displaystyle\int \sin^2 x\,dx = \frac{1}{2}x - \frac{1}{4}\sin 2x + C.$

25. $\displaystyle\int \cos^2 x\,dx = \frac{1}{2}x + \frac{1}{4}\sin 2x + C.$

26. $\displaystyle\int \sin^n x\,dx = -\frac{\sin^{n-1}x\cos x}{n} + \frac{n-1}{n}\int \sin^{n-2}x\,dx.$

27. $\displaystyle\int \cos^n x\,dx = \frac{1}{n}\cos^{n-1}x\sin x + \frac{n-1}{n}\int \cos^{n-2}x\,dx.$

28. $\displaystyle\int \cos^m x\sin^n x\,dx = \frac{\cos^{m-1}x\sin^{n+1}x}{m+n} + \frac{m-1}{m+n}\int \cos^{m-2}x\sin^n x\,dx.$

29. $\displaystyle\int \cos^m x\sin^n x\,dx = -\frac{\sin^{n-1}x\cos^{m+1}x}{m+n} + \frac{n-1}{m+n}\int \cos^m x\sin^{n-2}x\,dx.$

30. $\displaystyle\int \tan^2 x\,dx = \tan x - x + C.$

Table 5 (*Continued*)   Additional Antidifferentiation Formulas

31. $\int \cot^2 x \, dx = -\cot x - x + C.$

32. $\int \tan^n x \, dx = \dfrac{\tan^{n-1} x}{n-1} - \int \tan^{n-2} x \, dx.$

33. $\int \cot^n x \, dx = -\dfrac{\cot^{n-1} x}{n-1} - \int \cot^{n-2} x \, dx.$

34. $\int \sec^3 x \, dx = \dfrac{1}{2} \sec x \tan x + \dfrac{1}{2} \ln (\sec x + \tan x) + C.$

35. $\int \csc^3 x \, dx = -\dfrac{1}{2} \csc x \cot x + \dfrac{1}{2} \ln (\csc x - \cot x) + C.$

36. $\int \sec^n x \, dx = \dfrac{\tan x \sec^{n-2} x}{n-1} + \dfrac{n-2}{n-1} \int \sec^{n-2} x \, dx.$

37. $\int \csc^n x \, dx = -\dfrac{\cot x \csc^{n-2} x}{n-1} + \dfrac{n-2}{n-1} \int \csc^{n-2} x \, dx.$

38. $\int x \sin x \, dx = \sin x - x \cos x + C.$

39. $\int x \cos x \, dx = \cos x + x \sin x + C.$

40. $\int x^n \sin x \, dx = -x^n \cos x + n \int x^{n-1} \cos x \, dx.$

41. $\int x^n \cos x \, dx = x^n \sin x - n \int x^{n-1} \sin x \, dx.$

42. $\int x \sin^n x \, dx = \dfrac{\sin^{n-1} x (\sin x - nx \cos x)}{n^2} + \dfrac{n-1}{n} \int x \sin^{n-2} x \, dx.$

43. $\int x \cos^n x \, dx = \dfrac{\cos^{n-1} x (\cos x + nx \sin x)}{n^2} + \dfrac{n-1}{n} \int x \cos^{n-2} x \, dx.$

44. $\int \sin mx \sin nx \, dx = \dfrac{\sin (m-n)x}{2(m-n)} - \dfrac{\sin (m+n)x}{2(m+n)} + C.$

45. $\int \sin mx \cos nx \, dx = -\dfrac{\cos (m-n)x}{2(m-n)} - \dfrac{\cos (m+n)x}{2(m+n)} + C.$

46. $\int \cos mx \cos nx \, dx = \dfrac{\sin (m-n)x}{2(m-n)} + \dfrac{\sin (m+n)x}{2(m+n)} + C.$

47. $\int x e^{ax} \, dx = \dfrac{e^{ax}}{a^2} (ax - 1) + C.$

Table 5 (*Continued*)   Additional Antidifferentiation Formulas

48. $\int x^2 e^{ax} \, dx = \dfrac{e^{ax}}{a^3} (a^2 x^2 - 2ax + 2) + C.$

49. $\int x^n e^{ax} \, dx = \dfrac{x^n e^{ax}}{a} - \dfrac{n}{a} \int x^{n-1} e^{ax} \, dx.$

50. $\int e^{ax} \sin mx \, dx = \dfrac{e^{ax} (a \sin mx - m \cos mx)}{m^2 + a^2} + C.$

51. $\int e^{ax} \cos mx \, dx = \dfrac{e^{ax} (m \sin mx + a \cos mx)}{m^2 + a^2} + C.$

52. $\int \ln x \, dx = x \ln x - x + C.$

53. $\int x^n \ln x \, dx = x^{n+1} \left[\dfrac{\ln x}{n+1} - \dfrac{1}{(n+1)^2} \right] + C.$

Answers to Odd Numbered Problems

Sections 1.1–1.3

1, 3, 5.

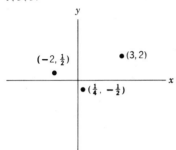

7. I and IV
9. III
11. zero
13. $d = \sqrt{2}$
15. $d = \frac{5}{4}$
17. $d = 8.2$
19. $d = 4.8$

Sections 1.4–1.5

1. -6 and 6
3. 6 and -6
5. $22a$ and $-2a$
7. 8 and 0
9. 19 and 19
11. $A = f(r)$
13. $P = F(S)$
15. $I = \phi(V, R)$

Section 1.6

1. $m = \frac{1}{2}$
   $\alpha = 27°$

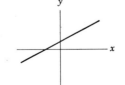

3. $m = \frac{4}{3}$
   $\alpha = 53°$

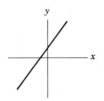

5. $m = -\frac{9}{8}$
   $\alpha = 132°$

6.

7. $m = 2$
   $\alpha = 63°$

9.

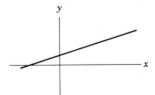

11.

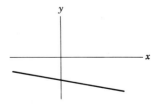

433

13. $m = 1$

15. $m = -0.577$

17. $m = 5$

19. $a = 1.5 \, \text{ft/sec}^2$

Sections 1.7–1.9

1. $2y - x = 8$

3. $7x + y = 33$

5. $x + 5y = 16$

7. $2y - 3x = 1$

9. $5y - 2x = 10$

11. $y - 0.577x = -2.577$

13. $y + 2.75x = -8.25$

15. $m = \frac{2}{3}; b = -\frac{5}{3}$

17. $m = \frac{1}{2}; b = -2$

19. $m = -\frac{5}{2}; b = -\frac{3}{2}$

21. $3y - 2x = 4$

23. $y - 3x = -3$

25. $I = \dfrac{V}{12}$

27. $50R - T = 5000; 101.8$

29.

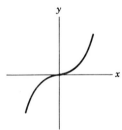

Sections 1.10–1.11

1.

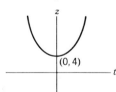

3.

5.

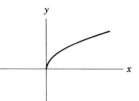

7.

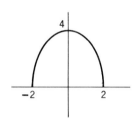

9.

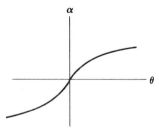

11.

13.

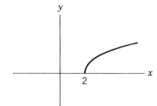

15.

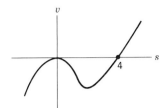

25.

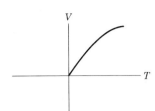

17.

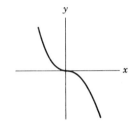

Sections 1.12–1.15

1.

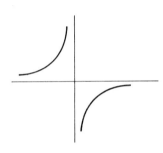

19.

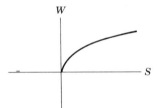

21.

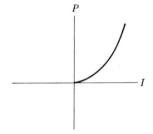

3.

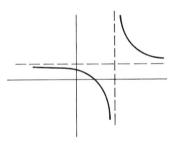

23.

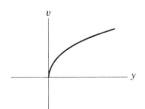

5.

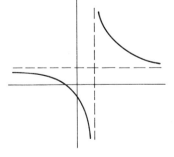

7.

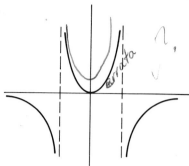

15.

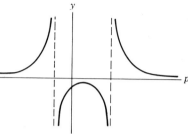

17.

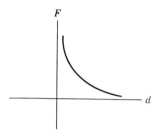

9.

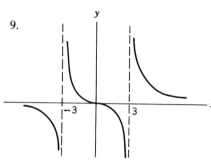

19.

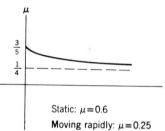

Static: $\mu = 0.6$

Moving rapidly: $\mu = 0.25$

11.

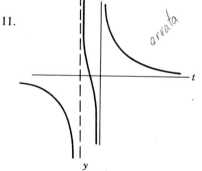

Section 2.1

1. $\frac{3}{2}$ 9. $\frac{5}{4}$

3. 2   11. $\frac{1}{2}$

5. $\frac{1}{3}$ 13. -1

7. 4

Section 2.2

1. continuous

13.

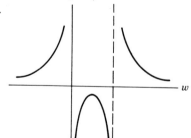

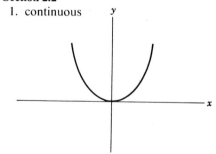

3. discontinuous at $x = 2$

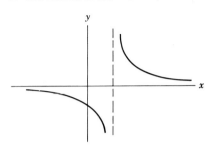

5. discontinuous at $x = 0$

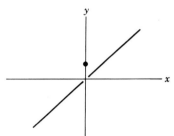

7. continuous

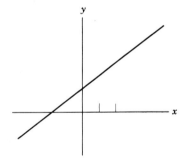

Section 2.3

1. 2

3. 2.1; 1.9

5. 3.005; 2.995

7. -6

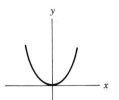

9. 2

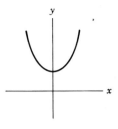

11. -2

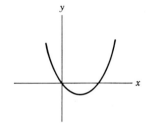

13. 0

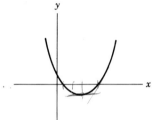

15. 12

17. $-\frac{1}{4}$

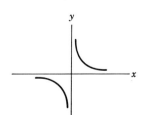

19. $6x + y = -3$
21. $v - 2x = -1$

Section 2.4
 1. 12.5 ft/sec
 3. 9.5 ft/sec
    12.5 ft/sec
 5. 9 ft/sec
 7. 6 ft/sec
 9. $-\frac{1}{4}$ ft/sec
11. 5 amp
13. 6 amp

Section 2.5
 1. $\frac{2}{3}x$
 3. $2x + 3$
 5. $6x + 5$
 7. $-3/x^2$
 9. $3x^2$
11. $3x^2 - 2x$
13. $-\frac{2}{3}; 0; \frac{2}{3}$
15. $3.2; 4$

Sections 3.1–3.3
 1. 3
 3. $3 + 8\phi$
 5. $-3/x^4$
 7. $\frac{3}{2}t^{1/2}$
 9. $\frac{\sqrt{3}}{2}t^{-1/2}$
11. $\dfrac{5at^{3/2}}{2} - \dfrac{3bt^{-2/5}}{5}$
13. $-\frac{4}{3}s^{-4/3} - \frac{5}{2}s^{-3/2}$
15. $4x^3 - 6x^2 + 10x - 7$
17. $\dfrac{-3}{2\sqrt{2t^5}}$
19. 1
21. $2x - 4$
23. $\frac{3}{2}t^{-1/2} - \frac{4}{3}t^{-2/3}$
25. $-\frac{3}{2}t^{-3/2} + \frac{3}{2}t^{1/2}$
27. $60x + 4y = 47$
29. $2x + y + 6 = 0$
31. $x = 0$ and $x = -6$
33. no real values

Section 3.4
 1. $3x^2 - 6x; 6x - 6; 6$
 3. $-2t^{-3} + 6t; 6t^{-4} + 6; -24t^{-5}$
 5. $\frac{3}{5}k^{-8/5}; -\frac{24}{25}k^{-13/5}; \frac{312}{125}k^{-18/5}$
 7. $1 - t^{-2}; 2t^{-3}; -6t^{-4}$
 9. $\frac{5}{2}x^{3/2} + \frac{3}{2}x^{-3/2};$
    $\frac{15}{4}x^{1/2} - \frac{9}{4}x^{-5/2};$
    $\frac{15}{8}x^{-1/2} + \frac{45}{8}x^{-7/2}$
11. 2
13. 0

Section 3.5
 1. $32t - 4$
 3. 32 ft/sec²
 5. 15 cm/sec²
 7. $-6t^{-3}$ ft-lb/sec
 9. 6500 ft
11. 0.35 cm/sec

Section 3.6
 1. 3.12 amp
 3. $-180t^2$
 5. -2 V
 7. 12,800 W
 9. 0.4 amp
11. $21t$ V

Section 3.7a
 1. $x > 2$
 3. $-2 < x < 2$
 5. $x < 0$ or $x > 3$
 7. $x < -3$ or $x > 2$

 9.

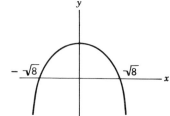

11.

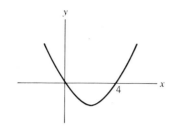

19.

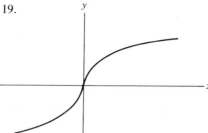

13.

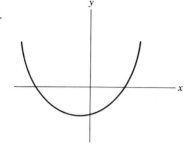

21.

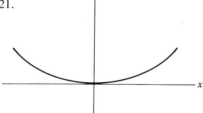

15.

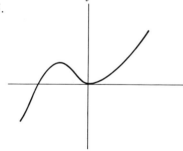

Section 3.7b
1. $x > 0$, concave up
   $x < 0$, concave down
3. concave up everywhere
5. $x < 0$ concave up
   $x > 0$ concave down
7. $x > 3$ concave up
   $x < 3$ concave down

17.

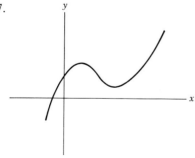

9.

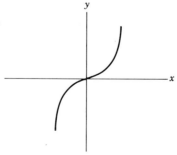

11.

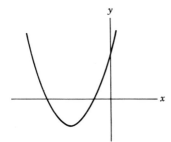

19.

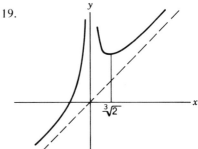

13.

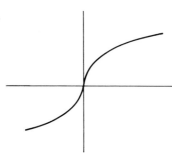

15.

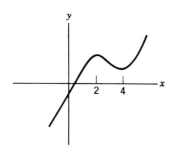

17.

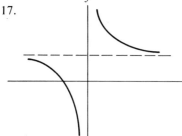

Section 3.8
1. at $x = 3$, minimum of $y = -4$
3. at $t = 0$, maximum of $i = 2$
   at $t = 2$, minimum of $i = -2$
5. no maxima or minima
7. at $x = 1$, minimum of $p = -2$
   at $x = -1$, maximum of $p = 2$
9. at $x = -4$, minimum of $y = -78$
11. when $r = 0$, test fails. Use other
   test to show neither max nor min.
   at $r = 1$, minimum of $m = 0$
13. at $t = -1$ test fails. Use other
   test to show neither max nor min.
15. at $x = \sqrt{10}$, $y = 2\sqrt{10}$ is a minimum.
   at $x = -\sqrt{10}$; $y = -2\sqrt{10}$ is a maximum.

Section 3.9
1. at $t = 2$, $v = 4$ ft/sec
3. at $t = \frac{1}{4}$, $v = \frac{7}{4}$ in./sec is a max.
5. 15, 15
7. 4, 2
9. A 2×2 square
11. 62,500 ft²
13. $b = 5.76$ in.; $h = 8.12$ in.
15. $r = 6.8$ cm and $h = 13.6$ cm

Sections 4.1–4.2
1. $8x\, dx$
3. $-\frac{1}{2}x^{-3/2}\, dx$
5. $\frac{\sqrt{3}}{2}t^{-1/2}\, dt$
7. $\frac{8}{3}b^{-1/3}\, db$
9. $(-3p^{-4} + 2p^{-3} - 4)\, dp$
11. $(3 - 2x)\, dx$
13. $\left(\frac{5}{3}s^{2/3} + 2s^{-1/3}\right)\, ds$
15. $-\frac{5}{3}t^{-6}\, dt$

Sections 4.3–4.4
1. 30
3. 0.0083
5. -6.8
7. 0.512 amp
9. 1.14×10^{-3} C
11. 312
13. 187.5 cm^3
15. 7860 ft^3
17. 0.0318 in. increase
19. -0.0318 in. decrease in both radii.

Sections 5.1–5.2
1. $3x + C$
3. $3\dfrac{z^2}{2} + C$
5. $\dfrac{x^4}{4} + \dfrac{x^3}{3} + C$
7. $\dfrac{3x^8}{8} + \dfrac{3x^{10}}{5} + C$
9. $\dfrac{3p^{4/3}}{4} + C$
11. $\dfrac{2\sqrt{3}}{3} \phi^{3/2} + C$
13. $-\dfrac{x^{-4}}{4} + C$
15. $\dfrac{x^3}{3} - \dfrac{x^5}{5} + C$
17. $\dfrac{-\theta^{-2}}{2} + 2\theta^{-1} + C$
19. $\dfrac{x^3}{3} - 3x^2 + 9x + C$
21. $\dfrac{x^3}{3} + 2x - x^{-1} + C$
23. $\frac{6}{11}x^{11/3} - \frac{9}{8}x^{8/3} + \frac{3}{5}x^{5/3} + C$
25. $\dfrac{x^4}{4} - \dfrac{x^{-2}}{2} + C$
27. $\dfrac{t^{13}}{13} + C$
29. $\frac{3}{5}\phi^{5/3} + \frac{24}{19}\phi^{19/12} + \frac{2}{3}\phi^{3/2} + C$
31. $\dfrac{8p^{9/8}}{9} - \dfrac{24p^{19/24}}{19} + C$

Section 5.3
1. $5x + C$
3. $\frac{3}{2}t^2 + C$
5. $\frac{1}{3}t^3 + C$
7. $\dfrac{x^4}{4} + C$
9. $x - \frac{3}{2}x^2 + C$
11. $y = \frac{3}{2}x^2 + \frac{1}{2}$
13. $s = \frac{3}{4}r^{4/3} - 14$
15. $p = \frac{2}{3}t^{3/2} + 2t^{1/2} - \frac{25}{3}$
17. $y = \dfrac{x^3}{6} + 2x - \frac{56}{3}$
19. $f(z) = \frac{1}{4}z^4 - 2z + 4$

Section 5.4
1. $v = \frac{2}{3}t^{3/2}$
3. 11 sec
5. 2000 ft
7. $s^2 = 2t + 5$
9. $t = 1.62$ sec
11. 32 ft/sec
13. 15.6 ft-lb
15. 0.558

Section 5.5
1. 0.3 C
3. $i = -\frac{1}{2}t^{3/2} + \frac{1}{2}$
5. $v = \frac{2}{3}t^3 + 6t^2 + 18t + 6$
7. 1920 J
9. $\frac{7}{20}t^4 + \frac{7}{25}t^3 + t + 2$
11. $\frac{1}{8}t^4 + 250t^3 - \frac{1}{5}t^2 - \frac{1997}{10}t + 1$
13. -0.105 amp
15. $v = 20t^2 + 1000t + 33.3t^3$
17. 9.5 V
19. $40t^2 + \frac{58}{5}t + \frac{121}{5}$

Sections 6.1–6.2
1. $x_1 + x_2 + x_3 + x_4 + x_5$
3. $1 + 4 + 9$
5. $1 + \frac{1}{2} + \frac{1}{3} + \frac{1}{4} + \frac{1}{5} + \frac{1}{6}$
7. $1 - \frac{1}{2} + \frac{1}{4} - \frac{1}{8}$
9. $\displaystyle\sum_{n=0}^{4} x_n$
11. $\displaystyle\sum_{n=1}^{3} \frac{1}{2^n}$

Section 6.3

1. 1.75
3. (a) 2.19
   (b) 3.19
   (c) 2.66
5. (a) 0.437
   (b) 0.938
   (c) 0.656
7. (a) 6.75
   (b) 10.75
   (c) 8.64
9. (a) $\frac{25}{4}$
   (b) $\frac{17}{4}$

Sections 6.4–6.5

1. 43.75 ft
3. $v_1 = 0$; $v_2 = 6.25$ ft/sec;
   $v_3 = 25$ ft/sec; $v_4 = 56.25$ ft/sec
5. 54.6 ft $< s <$ 79.6 ft
7. 0.0175 dyn-cm $<$ W $<$ 0.0375
   dyn-cm
9. (a) 0.0219 dyn-cm
   (b) 0.0319 dyn-cm

Sections 6.6–6.7

1. $A = \frac{8}{3}$
3. $A = 2$
5. (a) $\frac{46}{3}$
   (b) $-\frac{34}{3}$
   (c) $\frac{62}{3}$
7. $\frac{632}{3}$ dyn-cm
9. 50 in-lb

Section 6.8

1. 3
3. 4; 1.15
5. (a) $v_{min} = 0$
   (b) $v_{max} = 10$ ft/sec
   (c) $v_{avg.} = 4.33$ ft/sec
7. $s = 8.67$ ft

Sections 7.1–7.2

1. $4x^2$
3. $2\sqrt{t}$
5. 1

7. 46
9. $\frac{9}{4}$
11. $-\frac{64}{7}$
13. $\frac{7}{384}$
15. $\dfrac{8\sqrt{2}}{3}(4 - \sqrt{2})$
17. 20
19. Each has value 0.

Section 7.3

1. 5
3. $\frac{17}{4}$
5. $\frac{75}{4}$
7. 12
9. $\frac{3}{2}$
11. $\frac{2}{5}$
13. $\frac{21}{2}$
15. $\frac{21}{4}$
17. $\frac{62}{5}$

Section 7.4

1. $\frac{11}{8}$
3. $\frac{4}{15}$
5. $\frac{32}{3}$
7. $\frac{1}{3}$
9. $\frac{32}{3}$
11. $\frac{125}{6}$
13. $\frac{125}{6}$
15. 9
17. 2
19. $\frac{375}{4}$

Sections 7.5–7.7

1. $(2, 2)$
3. $\left(\frac{4}{9}, \frac{8}{3}\right)$
5. $\left(\frac{68}{30}, \frac{79}{30}\right)$
7. $\left(3, \frac{24}{5}\right)$
9. $\left(\frac{9}{8}, \frac{27}{5}\right)$
11. $\left(-2, \frac{12}{5}\right)$
13. $\left(\frac{3}{2}, \frac{14}{5}\right)$
15. $\left(\frac{1}{3}, \frac{5}{3}\right)$

Section 7.8

1. 8π
3. 72π
5. $\frac{128}{7}\pi$
7. $\frac{1296}{5}\pi$
9. $\frac{200}{3}\pi$
11. $\frac{384}{7}\pi$
13. $\frac{3}{5}\pi$
15. $\frac{512}{15}\pi$
17. $\frac{576}{5}\pi$
19. $\frac{8}{3}\pi$

Section 7.9

1. 8333π ft-lb
3. $20{,}250\pi$ ft-lb
5. 4725π ft-lb
7. $\dfrac{25125\pi}{8}$ ft-lb
9. $91{,}100\pi$ ft-lb
11. 6000π ft-lb
13. $15{,}200\pi$ ft-lb
15. $546{,}000\pi$ ft-lb

Section 7.10

1. 1050 gal
   7150 gal
3. $\frac{7}{192}$ m
5. 7.5×10^{-5} amp
7. 30 psi.
9. 35.6 ft/sec
   142.7 ft
11. $\frac{38}{3}$ C total
    $\frac{16}{3}$ C net
13. 71.75 cm total
    56.25 cm
15. $\frac{81}{4}$ gal

Section 7.11

1. 27
3. 0.446
5. 10.15
7. 2.83 cm
9. 0.25 ft/sec^2
11. 22 gal/min

Sections 8.1–8.3

1. $-15(4-3x)^4$
3. $8t(t^2+4)^3$
5. $\dfrac{1}{(5x+1)^{4/5}}$
7. $\dfrac{8t-3}{2(4t^2-3t)^{1/2}}$
9. $-6(3t+1)^{-3}$
11. $\dfrac{-30}{(2+5x)^4}$
13. $\dfrac{-3x}{(x^2+1)^{3/2}}$
15. $\dfrac{3}{x^2}\left(2-\dfrac{1}{x}\right)^2$
17. $\dfrac{\pi}{4\sqrt{y}(1+\pi\sqrt{y})}$
19. $\dfrac{4x-3}{3(4+3x-2x^2)^{4/3}}$
21. $-2(2-3x)^{-5/3}$
23. $\frac{9}{2}(3t+4)^{1/2}$
25. $\dfrac{1}{4L\sqrt{Tu}}$
27. $-C(3t+2)^{-4/3}$
29. $dG = -\dfrac{E}{2}(1+\mu)^{-2}\,d\mu$
31. $s = 6(3x-2)$
33. $s = -6x(10-x^2)^2$

Sections 8.4–8.5

1. $12x + 7$

3. $\dfrac{12}{(3 - 2x)^2}$

5. $t^2(t+1)(5t+3)$

7. $\dfrac{-24s}{(s^2 - 4)^2}$

9. $(3x+2)^2(5x+1)^3(105x+49)$

11. $\dfrac{12z(4z+2)^2}{(3z+1)^3}$

13. $-2\dfrac{(1-x)(4-x)}{(4-x^2)^2}$

15. $\dfrac{-1}{5m^{4/5}(m-1)^{6/5}}$

17. $\dfrac{-4t}{3(t^2-1)^{4/3}(t^2+1)^{2/3}}$

19. $\dfrac{-6(5x+4)}{(2x-1)^4(3x+5)^3}$

21. $12x(x^3-1)^2(11x^3-2)$

23. $\dfrac{5600}{3\sqrt[3]{12}} = 817\ \text{W}$

25. $\dfrac{m_0^2 c^3}{(p^2 + m_0^2 c^2)^{3/2}}$

27. $\dfrac{3T^2(2T-3)}{2(T^2-T)^{3/2}}$

Section 8.6

1. $-\dfrac{x}{y}$

3. $-\dfrac{4x^3}{3y^2}$

5. $-\dfrac{1+2xy^2}{1+2x^2y}$

7. $-\dfrac{y^{1/3}}{x^{1/3}}$

9. $-\dfrac{3x^2+y}{x+3y^2}$

11. $-\dfrac{y^2}{x^2}$

13. $\dfrac{1}{2(1+y)}$

15. $-\dfrac{16}{y^3}$

17. $\dfrac{(y-1)(2x+6y+6)}{(6y+x)^3}$

19. $7x + 8y = 38$

Section 8.7

1. 100 mph
3. $1.0\ \text{cm}^2/\text{min}$
5. $0.75\ \text{dyn/sec}$
7. 600 psi/sec
9. 1.1 ohms/min
11. 11.8 ft/sec
13. 6 units/sec

Section 8.8

1. $\frac{1}{8}(2x+1)^4 + C$

3. $-\frac{1}{20}(1-5x)^4 + C$

5. $\frac{1}{6}(3+4p)^{3/2} + C$

7. $\frac{1}{10}(x^2+1)^5 + C$

9. $-\frac{1}{12}(2+4x)^{-3} + C$

11. $\frac{8}{3}(r^3+1)^{1/2} + C$

13. $\frac{2}{3}(4+\sqrt{x})^3 + C$

15. $\dfrac{x^5}{5} + \dfrac{8x^3}{3} + 16x + C$

17. 6
19. $\frac{1}{2}$
21. 6
23. 0.01 lb-sec
25. 1 hp
27. 60 cm
29. 6.375 C

Section 9.1

1.

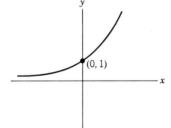

3.

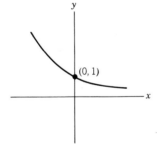

5.

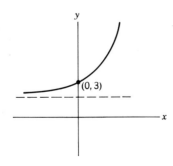

7.

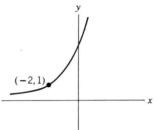

9.

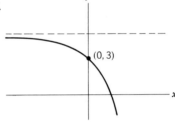

11.

13.

15.

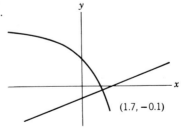

Section 9.2

1.

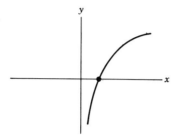

3.

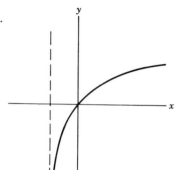

5.

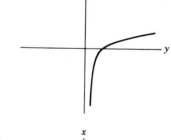

7.

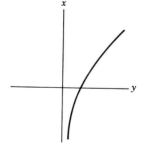

9.

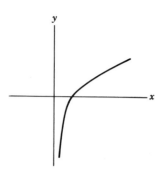

11. 15

13. 1

15. 4

17. $x = 10^y$

19. $x^2 = e^y$

21. $x = \log y$

23. $x = ye^{-2} - 2$

Section 9.3

1. $\dfrac{\log e}{x}$

3. $\dfrac{2}{2t-1}$

5. $\dfrac{2}{p}$

7. $\dfrac{3(1-2x)}{x-x^2}$

9. $\dfrac{12-5x}{x(4-x)}$

11. $\dfrac{3t^2}{2(t^3-1)}$

13. $\dfrac{-4r(r^2+2)}{r^4-1}$

15. $\dfrac{12\ln^3(3T+2)}{3T+2}$

17. $\dfrac{2R\ln(R^2-2)-2R^3/(R^2-2)}{\ln^2(R^2-2)}$

19. $\dfrac{3V^2}{3V+2}+2V\ln(3V+2)$

21. $\dfrac{-100}{r}$

23. $\dfrac{-42}{p}$

25. $\dfrac{3t(2-t^3)}{(t^3+1)^2}$ ft/sec²

27. $\dfrac{-3}{1+3T}$ psi/deg

29. at $t = 1$ sec

Section 9.4

1. $2e^{2x}$

3. $\dfrac{2(2x-1)}{(2x-3)^{2/3}(2x+3)^{1/3}}$

5. $2t(1+t)e^{2t}$
7. $-15e^{-3t}$

9. $\dfrac{-2x}{(x^2+1)^{1/2}(x^2-1)^{3/2}}$

11. $(L^2+1)^{-1/2}\left[\dfrac{8L(L^2+1)}{4L^2-5}\right.$

$\left. + L\ln(4L^2-5)\right](4L^2-5)^{(L^2+1)^{1/2}}$

13. $\dfrac{24x^2+36x+13}{(4x+3)^2}$

Section 9.5

1. $2e^{2t}$

3. $8xe^{x^2}$

5. $-12e^{-3x}$

7. $\dfrac{1}{2\sqrt{x}}e^{\sqrt{x}}$

9. $T^2e^T(T+3)$

11. $\dfrac{e^v(v-2)}{v^3}$

13. $\dfrac{-2e^r}{(e^r-1)^2}$

15. 1

17. $e^t\left(\dfrac{1}{t}+\ln t\right)$

19. e^{-x}

21. $2e^{t^2}(2t^2+1)$

23. $2te^{t^2}(2t^2+3)$

25. $\dfrac{V}{L}e^{-Rt/L}$

27. $e^x(x+1)$

29. $-kp_0e^{-kz}$

31. $3e^{-2} = 0.405$ lb

Section 9.6

1. $2\ln x + C$
3. $-\frac{1}{2}\ln(1-2y)+C$
5. $\frac{1}{2}\ln(t^2+4)+C$
7. $\ln(e^x+1)+C$

9. $-\dfrac{1}{2(x^2-3)}+C$

11. $\frac{1}{3}\ln^3 x + C$
13. $\ln(\ln x)+C$
15. $\ln(e^x+e^{-x})+C$

17. $\dfrac{\ln(2+3\ln I)}{3}+C$

19. $-\dfrac{1}{1+\ln z}+C$

21. $v = -u\ln(m_0-qt)-gt+u\ln m_0$

23. $\dfrac{\ln 2}{2} = 0.347$ lb-sec

25. $10+\ln 6 = 11.79$ cm³
27. $\frac{5}{6}\ln 13 = 2.14$ cm.

29. $\dfrac{\ln x}{2}$

Section 9.7

1. $2e^x + C$

3. $-\dfrac{1}{2^x \ln 2} + C$

5. $\frac{1}{2}e^{2x} + \frac{1}{5}e^{5x} + C$

7. $\frac{1}{2}e^{s^2} + C$

9. $\frac{1}{3}e^{x^3} + \frac{1}{4}x^4 + C$

11. $2e^{(y+1)^{1/2}} + C$

13. $\frac{2}{3}(1 - e^{-x})^{3/2} + C$

15. $\frac{1}{2}e^{2x} + 2x - \frac{1}{2}e^{-2x} + C$

17. $3e^{8/3} + C$

19. $-e^{-x} + \frac{1}{2}e^{2x} + C$

21. $-Ve^{-t/RC} + K$

23. $2e^4 - 2 = 107.2$ mi

25. $\dfrac{e^4 - 1}{2} = 26.8$ mi

27. $e^3 + 99 = 119.1$

29. 20.6 min

31. $\dfrac{1 - e^{-\pi s}}{s}$

Sections 10.1–10.2

1. $\dfrac{\pi}{4}$ 3. $\dfrac{\pi}{2}$ 5. $\dfrac{2\pi}{3}$ 7. $15°$ 9. $150°$ 11. $225°$

13.
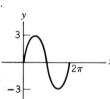

15.
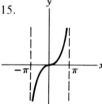

17.
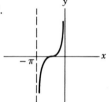

19.
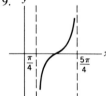

21.
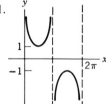

23.
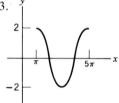

Sections 10.3–10.4

1. $\dfrac{\pi}{2}\cos\dfrac{\pi}{2}t$ 3. $-6\sin 3t$ 5. $-\frac{1}{2}\csc^2(t+1)$ 7. $3T^2\cos T^3$

9. $-e^t\sin e^t$ 11. $-3\sin t\cos^2 t$ 13. $\cos 2x/\sqrt{\sin 2x}$

15. $e^t(\cos t + \sin t)$ 17. $(-\ln x)(\csc^2 2x) + \dfrac{1}{2x}\cot 2x$ 19. $5\cot 5s$

21. $-2\sin 2t e^{\cos 2t}$ 23. $\dfrac{1 - \sin^3 x}{\sin^2 x}\cos(\cos x - \cot x)$

25. $-4x^2\sin x^2 + 2\cos x^2$ 29. $-W\sin\theta$ 31. $-L\omega I_m\cos(\omega t - \phi)$

33. $-N\csc^2 E$ 35. $-(\sigma_x - \sigma_y)\cos 2\phi - 2T\sin 2\phi$

Section 10.5

1. $\frac{\pi}{4}$ 3. $\frac{\pi}{3}$ 5. $\frac{\pi}{3}$

Section 10.6

1. $3/\sqrt{1-9t^2}$ 3. $2/(4+y^2)$ 5. $1/(2\sqrt{t-t^2})$ 7. $3/(9+x^2)$

9. $e^t/(1+e^{2t})$ 11. $1/(2\sqrt{1-x})+\dfrac{\arcsin\sqrt{x}}{2\sqrt{x}}$

13. $e^{-t}[2/(1+4t^2)-\arctan 2t]$ 15. $1/[4\sqrt{\phi-\phi^2}\,(\arccos\sqrt{\phi}\,)^{3/2}]$

17. $\arctan x$ 19. 1.101 21. $-X/(R^2+X^2)$ 23. $-T/(\omega^2+T^2)$

25. -0.025 rad/sec

Section 10.7

1. $-\frac{1}{3}\cos 3x+C$ 3. $\frac{1}{2}\ln\sec 2t+C$ 5. $\ln\sin\omega t+C$ 7. $3\sec x^2+C$

9. $\cos e^{-x}+C$ 11. $2\ln(\csc\frac{1}{2}x-\cot\frac{1}{2}x)+C$ 13. $-(1/\omega)\csc\omega t+C$

15. $\frac{1}{6}\sin^6\theta+C$ 17. $\frac{1}{3}\sec^3\phi+C$ 19. $\frac{1}{2}e^{\sin 2\phi}+C$

21. $-1/(1+\tan x)+C$ 23. $\omega t+2\ln(\sec\omega t+\tan\omega t)+\tan\omega t+C$

25. 8.98 27. 289 ft-lb   29. πh 31. 153.5

Section 10.8

1. $\dfrac{\theta}{2}+\dfrac{1}{4}\sin 2\theta+C$ 3. $\dfrac{\phi}{2}-\dfrac{1}{12}\sin 6\phi+C$ 5. $\frac{1}{10}\sin^5 2t-\frac{1}{14}\sin^7 2t+C$

7. $\frac{1}{4}\sec^4\theta+C$ 9. $\sin\omega t-\frac{1}{3}\sin^3\omega t+C$ 11. $\dfrac{t}{2}-\dfrac{1}{4}\cos 2t+\ln\sin t+C$

13. $\ln\sin x-\frac{1}{2}\sin^2 x+C$ 15. $\theta-3\cot\theta-\frac{1}{3}\cot^3\theta+C$ 17. $1/9$

19. 0.071 J

Section 10.9

1. $\frac{1}{2}\arctan y/2+C$ 3. $\frac{1}{3}\arcsin\sqrt{3}v+C$ 5. $\sqrt{6}/6\arctan\dfrac{\sqrt{6}x}{2}+C$

7. $\frac{1}{2}\arctan(x+2)/2+C$ 9. $\frac{1}{6}\ln(3z^2+5)+C$ 11. $\arcsin(x-2)/2+C$

13. $\frac{1}{6}\arctan 2(y+1)/3+C$ 15. $\frac{1}{2}\ln(x^2+2x+3)+\dfrac{1}{\sqrt{2}}\arctan\dfrac{x+1}{\sqrt{2}}+C$

17. $\dfrac{\sqrt{3}}{15}\arctan(\sqrt{3}/3)y+C$ 19. $\frac{1}{2}\ln(x^2+1)+2\arctan x+C$

21. $v=\frac{1}{3}\arctan 3t+100$ 23. 0.197 amp   25. $x=\dfrac{1}{k}\arcsin ky/c+C$

Sections 11.1–11.2

1. $(x+2)^{3/2}[(6x-8)/15]+C$ 3. $\frac{2}{15}(t-3)^{3/2}(3t+16)+C$

5. $-\frac{2}{5}\sqrt{1-s}(8+4s+3s^2)+C$ 7. $\dfrac{1}{\sqrt{17}}\arctan\dfrac{2\sqrt{y-1}}{\sqrt{17}}+C$

9. $\ln(\sqrt{x}+1)^2+C$ 11. $\frac{3}{28}(x-1)^{4/3}(4x+3)+C$ 13. $\frac{6}{5}(2t+1)^{1/4}(t-2)+C$

15. $\frac{3}{160}(2x+1)^{5/3}(10x-3)+C$ 17. 80.2 J   19. $v=2\sqrt{t}-2\arctan\sqrt{t}$

Section 11.3

1. $\ln \sqrt[3]{(x+5)^2(x-1)} + C$ 3. $\ln [x/(x+1)] + C$ 5. $\frac{1}{2}\ln z + 2\ln (z+5)$
$-\frac{1}{2}\ln (z+2) + C$ 7. $\ln \sqrt{(x+1)(x+5)} + C$ 9. $-6\ln y + 7\ln (y-1)$
$+2/(y-1) + C$ 11. $\ln [s/(s+1)] + 1/(s+1) + C$
13. $\ln [(w-1)^2/(w+1)(3-2w)^{3/2}] + C$ 15. $2\ln (x-1) - 1/(x-1) + C$
17. $\ln t + \frac{\sqrt{2}}{2}\arctan \frac{\sqrt{2}}{2}t + C$ 19. $\frac{1}{2}\ln (2x+3) - \frac{1}{\sqrt{3}}\arctan \frac{x+1}{\sqrt{3}} + C$
21. $\ln (x-1) - 1/(x-1) + \arctan x + C$ 23. $\ln 3 = 1.0986$ 25. $1 + \ln 2$

Section 11.4

1. $-x\cos x + \sin x + C$ 3. $-e^{-t}(1+t) + C$ 5. $(2-\theta^2)\cos \theta + 2\theta \sin \theta + C$
7. $\frac{1}{4}z^4 \ln z - \frac{1}{16}z^4 + C$ 9. $y\arctan y - \frac{1}{2}\ln (y^2+1) + C$
11. $x\arccos 3x - \frac{1}{3}\sqrt{1-9x^2} + C$ 13. $-\frac{1}{18}(3t^2-2)^{-3} + C$
15. $\frac{2}{3}x(2x-5)^{3/4} - \frac{4}{21}(2x-5)^{7/4} + C$ 17. $\frac{1}{2}e^{\phi^2}(\phi^2-1) + C$
19. $\frac{3}{10}e^{-x}(\sin 3x - \frac{1}{3}\cos 3x) + C$ 21. $-1 + \ln 4$ 23. 0.657
25. $y = \frac{2}{3}x^{3/2}(\ln x - \frac{2}{3}) + \frac{4}{9}$ 27. 0.438 29. $2.51\,J$

Section 11.5

1. $\frac{1}{2}$ 3. 4 5. ∞ 7. ∞ 9. ∞ 11. ∞ 13. $\pi/2$
15. $-\frac{3}{2} + \frac{3}{2}\sqrt[3]{4}$ 17. 0

Sections 12.1–12.2

1. focus $(-2,0)$
  dir. $x = 2$
  r.c. $(-2,4), (-2,-4)$
5. focus $(-4,0)$
  dir. $x = 4$
  r.c. $(-4,8), (4,8)$
9. focus $(-\frac{1}{2},0)$
  dir. $x = \frac{1}{2}$
  r.c. $(-\frac{1}{2},1), (-\frac{1}{2},-1)$
3. focus $(0,\frac{3}{2})$
  dir. $y = -\frac{3}{2}$
  r.c. $(-3,\frac{3}{2}), (3,\frac{3}{2})$
7. focus $(\frac{3}{4},0)$
  dir. $x = -\frac{3}{4}$
  r.c. $(\frac{3}{4},\frac{3}{2}), (\frac{3}{4},-\frac{3}{2})$
11. $x^2 = 8y$
13. $y^2 = 6x$ 15. $x^2 = -4y$ 17. $x^2 = y$ 19. $(0,0), (-2,1)$
21. $(1.4,4.2)$ 23. $x^2 = 1000y$

Sections 12.3–12.4

1. circle   3. ellipse   5. ellipse   7. ellipse
  $r = 5$ $a = 2$ $a = 3$ $a = \sqrt{12}$
      $b = 1$ $b = \sqrt{3}$ $b = \sqrt{8}$
      foci $(0, \pm \sqrt{3})$ foci $(0, \pm \sqrt{6})$ foci $(\pm 2, 0)$
9. ellipse   11. $9x^2 + 16y^2 = 144$ 13. $100x^2 + 9y^2 = 225$
  $a = 1$
  $b = \frac{2}{3}$
  foci $(0, \pm \frac{1}{3}\sqrt{5})$
15. $9x^2 + 25y^2 = 225$ 17. $15x^2 + 16y^2 = 240$ 19. $4x^2 + 21y^2 = 25$
21. $(1.3, -1.9), (-1.3, 1.9)$ 23. $(1.4, 0.7), (-1.4, -0.7)$ 25. $9x^2 + 100y^2 = 225$

Section 12.5

1. $a = 4$
   $b = 4$
   foci $(\pm \sqrt{32}, 0)$

3. $a = 3$
   $b = 2$
   foci $(\pm \sqrt{13}. 0)$

5. $a = 5$
   $b = 2$
   foci $(0, \pm \sqrt{29})$

7. $a = \frac{5}{2}$
   $b = \frac{5}{4}$
   foci $(\pm \frac{5}{4}\sqrt{5}, 0)$

9. $a = 1$
   $b = 1$
   foci $(\pm \sqrt{2}, 0)$

11. $9x^2 - 16y^2 = 144$

13. $4y^2 - x^2 = 4$

15. $13x^2 - 36y^2 = 117$

17. $4y^2 - x^2 = 64$

19. $y^2 - 10x^2 = 9$

Section 12.6

1. $8x = y^2 - 2y + 25$

3. $12y = -13 + 2x - x^2$

5. $4x = y^2 - 4y + 8$

7. $7x^2 + 16y^2 - 28x - 32y - 68 = 0$

9. $16x^2 + y^2 - 16x + 8y + 4 = 0$

11. $3x^2 + 4y^2 + 24x - 24y + 72 = 0$

13. $64y^2 - 49x^2 - 98x - 256y = 577$

15. $7x^2 - 9y^2 - 28x + 18y = 44$

17. $x^2 - 4y^2 + 4x - 16y - 16 = 0$

19. $(x - h)^2 + y^2 = r^2$

21. $y^2 = 4a(x - h)$

23. $(x - r)^2 + y^2 = r^2$

Section 12.7

1. $(x + 2)^2 + (y + 3)^2 = 3^2$

3. $\dfrac{(x + 1)^2}{4} + \dfrac{(y + 1)^2}{9} = 1$

5. $\dfrac{(x - 2)^2}{25} - \dfrac{y^2}{25} = 1$

7. $(x - 3)^2 = 3(y + 3)$

9. $(y - \frac{1}{2})^2 = -\frac{1}{2}(x - \frac{3}{2})$

11. $\dfrac{(x + 1)^2}{1} - \dfrac{(y - 1)^2}{4} = 1$

13. $\dfrac{(x - 1)^2}{4} + \dfrac{(y + 2)^2}{9} = 1$

15. $\dfrac{(y - \frac{1}{2})^2}{\frac{1}{4}} - \dfrac{x^2}{\frac{1}{2}} = 1$

Sections 13.1–13.2

1. $y = y'x$

3. $y = y'(x + 1)$

5. $(y')^2(x^2 - r^2) = -x^2$

7. $x - yy' = 0$

9. $ay'^2 = y$

11. $1 = 2ay''$

13. $y' - xy'' = 0$

15. $(y - k)^2(y'^2 + 1) = k^2$

Section 13.3   (No answers)

Sections 13.4–13.5

1. $xy = C$

3. $x^2 + y^2 = C^2$

5. $1 + s^2t^2 = cs^2$

7. $\sqrt{x^2 - 2} = c(1 - y)$

9. $\arctan x + \frac{1}{2}y^2 = C$

11. $-e^{-x}(x + 1) + \ln y = C$

13. $e^y - e^{-x} = C$

15. $\dfrac{(y + 1)^{1/3}(x^2 - 1)^{1/2}}{(y + 4)^{1/3}} = C$

17. $(x - 4)(y + 2) = -4$

19. $\sqrt{9 + x^2} + \ln y = 5$

21. $\arctan x + \arctan y = \frac{7}{12}\pi$

23. $I = 25e^{-0.223t}$

25. $M = 200e^{-0.25t}$

27. $P = 15e^{-0.000041h}$, 13.3 psi

29. $v = 150(1 - e^{-t/10})$

Section 13.6

1. $y = 2 + Ce^{-2x}$
3. $y = -\frac{3}{4} + Ce^{4x}$
5. $6xy = 2x^3 + 3x^2 + C$
7. $q = -\frac{1}{4}(2t+1) + Ce^{2t}$
9. $s(t^2+1) = 3t + C$
11. $y = -2 + C\sqrt{x^2+1}$
13. $v \sec r = r + C$
15. $y = -e^{-x} + Ce^x$
17. $s = \frac{t}{2} + \frac{c}{t}$
19. $y = xe^x - ex$
21. $y \sin x = -\sin x + \sqrt{3}$
23. $v = \dfrac{e^t}{k+m} + Ce^{-kt/m}$
25. $q = VC(1 - e^{-t/RC})$

Section 13.7

1. $y_0 = -\frac{1}{2}x_0 + C$
3. $\frac{2}{3}y_0^{3/2} = -a^{1/2}x_0 + C$
5. $y_0 x_0 = C$
7. $\frac{1}{2}y_0^2 = -x_0 + C$
9. $y_0^2 + x_0^2 = C$
11. $y_0^2 = Cx_0$

Section 13.8

1. $y = C_1 + C_2 e^{-x}$
3. $y = C_1 + C_2 e^{-2x}$
5. $y = C_1 e^{-x} + C_2 e^{-4x}$
7. $s = C_1 \cos 3t + C_2 \sin 3t$
9. $v = C_1 e^{3t} + C_2 t e^{3t}$
11. $x = e^t(C_1 \cos t + C_2 \sin t)$
13. $y = C_1 e^{-(1/2)x} + C_2 x e^{-(1/2)x}$
15. $s = C_1 e^{-(1/3)t} + C_2 t e^{-(1/3)t}$
17. $q = C_1 e^{\sqrt{3}t} + C_2 e^{-\sqrt{3}t}$
19. $y = C_1 \cos \frac{1}{3}x + C_2 \sin \frac{1}{3}x$
21. $y = C_1 e^{[(-1+\sqrt{5})/2]x} + C_2 e^{[(-1-\sqrt{5})/2]x}$
23. $v = e^{2s}(C_1 \cos \sqrt{3}s + C_2 \sin \sqrt{3}s)$
25. $y = \frac{1}{2}e^{2x} - \frac{1}{2}e^{-2x}$
27. $y = xe^{4x}$
29. $s = 2 \sin 3t$
31. $y = e^{3x}(2\cos x - 5\sin x)$
33. $x = 2 \cos 0.1t$

Sections 14.1–14.2

1. 13   3. $a^2 \ln a^4 b$

Sections 14.3–14.4

1. $z_x = 3x^2 y^3 + 2x$
   $z_y = 3x^3 y^2$
3. $f_x = 2x \cos t$
   $f_t = -x^2 \sin t$
5. $G_x = ye^{xy}$
   $G_y = xe^{xy}$
7. $F_u = 6v$
   $F_v = 6u$
9. $z_x = \cos x \cos y$
   $z_y = -\sin x \sin y$
11. $A_r = \arctan x$
    $A_x = \dfrac{r}{1+x^2}$
13. $H_r = 2s \sec^2 rs \tan rs$
    $H_s = 2r \sec^2 rs \tan rs$
15. $Z_X = X(X^2 + R^2)^{-1/2}$
    $Z_R = R(X^2 + R^2)^{-1/2}$
17. $P_\omega = 3t \sin^2 \omega t \cos \omega t$
    $P_t = 3\omega \sin^2 \omega t \cos \omega t$
19. $v_t = e^t(\omega \cos \omega t + \sin \omega t)$
    $v_\omega = te^t \cos \omega t$
21. $D_A = \dfrac{4}{P}, D_P = -\dfrac{4A}{P^2}$
23. $W_v = wv/g$
25. 2.5×10^{-2} mhos

Section 14.5

1. $z_{xx} = 20x^3 y^2 - \frac{1}{4}x^{-3/2}$
   $z_{yy} = 2x^5$
   $z_{xy} = z_{yx} = 10yx^4$
3. $f_{xx} = 2 \tan y$
   $f_{yy} = 2x^2 \sec^2 y \tan y$
   $f_{xy} = f_{yx} = 2x \sec^2 y$
5. $Z_{RR} = \dfrac{X^2}{(R^2 + X^2)^{3/2}}$
   $Z_{XX} = \dfrac{R^2}{(R^2 + X^2)^{3/2}}$
   $Z_{XR} = Z_{RX} = \dfrac{RX}{(R^2 + X^2)^{3/2}}$

7. $h_{ss} = -t^2 \sin st + 2$
   $h_{tt} = -s^2 \sin st$
   $h_{st} = h_{ts} = -st \sin st + \cos st$
9. $F_{zz} = 2y \sec^2 zy (zy \tan zy + 1)$
   $F_{yy} = 2z^3 \sec^2 zy \tan zy$
   $F_{zy} = F_{yz} = 2z \sec^2 zy (zy \tan zy + 1)$
11. $z_{xxx} = 60x^2y^3 - 24x$
    $z_{yyy} = 6x^5$
    $z_{xyy} = z_{yxy} = z_{yyx} = 30x^4y$
    $z_{xxy} = z_{xyx} = z_{yxx} = 60x^3y^2$
13. $\dfrac{V}{CR^2}\left(1 - \dfrac{t}{RC}\right)e^{-t/RC}$

Sections 14.6–14.7

1. 0.25 in.^{2}
3. $dV = \frac{2}{3}\pi rh\, dr + \frac{1}{3}\pi r^2\, dh$
5. $dP = \rho v\, dv + \frac{1}{2}v^2\, d\rho$
7. $dv = 32 \sin\alpha\, dt + 32t \cos\alpha\, d\alpha$
9. 80 fps, 0.29 ft/sec
11. $\dfrac{dV}{dt} = 2\pi rh\dfrac{dr}{dt} + \pi r^2\dfrac{dh}{dt}$
13. $-11,600$ in.^{3}/sec
15. a. 22 deg b. 750 psi/sec

Section 14.8

1. 9
3. 0
5. $\frac{63}{2}$
7. $\frac{16}{3}$
9. 2
11. $\frac{33}{10}$
13. $\frac{317}{60}$
15. $\dfrac{\pi}{4}$

Section 14.9

1. $\frac{9}{2}$
3. $\dfrac{-25 + 32\sqrt{2}}{24}$
5. $\frac{9}{2}$
7. $\frac{16}{3}$
9. $\frac{1}{10}$

Section 14.10

1. 6
3. $\frac{256}{3}$
5. $\frac{136}{15}$
7. $\frac{9}{2}$
9. $\frac{1}{15}$
11. $\frac{1}{12}$
13. 8 lb

Section 14.11

1. $\frac{4}{3}$
3. 51
5. $\frac{5}{42}$

Sections 15.1–15.3

11. $r(2 \cos\theta + 3 \sin\theta) = 6$
13. $r = 4 \cos\theta$
15. $r^2(\cos^2\theta + 4 \sin^2\theta) = 4$
17. $r \cos^2\theta = 4 \sin\theta$
19. $x^2 + y^2 = 25$
21. $x^2 + y^2 = 10y$
23. $x^2 + y^2 = (x^2 + y^2)^{1/2} + 2y$
25. $y^2 = 25 - 10x$

Section 15.4

1.

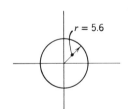

$r = 5.6$

3.

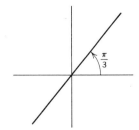

$\dfrac{\pi}{3}$

5.

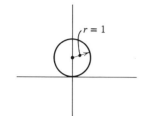

15.

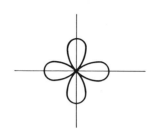

7.

17.

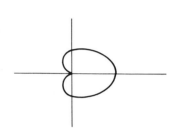

19. $3x^2 + 4y^2 - 2x - 1 = 0$

Section 15.5

1. $v_r = v_\theta = e^\pi = 23.2$ ft/sec

3. $a_r = 0$

   $a_\theta = 2e^\pi = 46.4$ ft/sec$^2 = a$

5. $v_r = 4$ ft/sec

7. $\dfrac{d\theta}{dt} = \dfrac{30}{\sqrt{3}} = 17.3$ rad/sec

   $v_\theta = \dfrac{15}{\sqrt{3}} = 8.65$ in./sec

   $v = \sqrt{300} = 17.3$ in./sec

9.

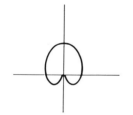

11.

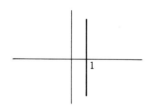

Section 15.6

1. $\dfrac{25\pi}{3}$

3. $\dfrac{31\pi^3}{128}$

5. 4

7. $\dfrac{49\pi}{4}$

9. $\dfrac{9\pi}{2}$

13.

11. $\dfrac{27\pi}{2}$

13. $\dfrac{3a^2\pi}{2}$

Section 16.1

1. $1, \frac{1}{3}, \frac{1}{5}, \frac{1}{7}$
3. $2, 4, 4, \frac{8}{3}$
5. $a_n = n/(n-1)!$
7. $a_n = 1/(2n-1)!$
9. $a_n = n/(2n+1)$
11. $a_n = x^{2n}/2^n$
13. $a_n = x^{n-1}/(n-1)!$

Section 16.2

1. converges to 0
3. converges to 1
5. diverges
7. converges to 0
9. diverges
11. converges to $\frac{1}{2}$

Sections 16.3–16.4

1. converges to $\frac{3}{2}$
3. converges to 6
5. converges
7. diverges
9. converges
11. diverges
13. diverges

Sections 16.5–16.6

1. converges
3. converges
5. converges
7. converges
9. diverges
11. converges
13. converges
15. diverges

Section 16.7

1. $-1 < x < 1$
3. all x

5. $-1 \leqslant x < 1$
7. all x
9. all x
11. $-1 \leqslant x \leqslant 1$
13. $-\frac{1}{3} < x < \frac{1}{3}$
15. all x
17. $-e < x < e$

Sections 17.1–17.2

1. $1 - \dfrac{x^2}{2!} + \dfrac{x^4}{4!} - \dots$, converges for all x

3. $x - \dfrac{x^2}{2} + \dfrac{x^3}{3} - \dots,$

   converges for $-1 < x \leqslant 1$

5. $3x - \dfrac{9x^3}{2} + \dfrac{81x^5}{40} - \dots$

7. $1 - 2x + 3x^2 - \dots$

9. $x + \dfrac{x^3}{6} + \dfrac{3x^5}{40} + \dots$

11. $1 - x + \dfrac{x^2}{2!} - \dfrac{x^3}{3!} + \dots$

13. $1 + \dfrac{x^2}{2} + \dfrac{5x^4}{24} + \dots$

15. $-\dfrac{x^2}{2} - \dfrac{x^4}{12} - \dfrac{x^6}{45} - \dots$

Section 17.3

1. $2x - \dfrac{4x^3}{3} + \dfrac{4x^5}{15} - \dots$

3. $1 - \dfrac{x^2}{8} + \dfrac{x^4}{384} - \dots$

5. $3x + \dfrac{(3x)^3}{6} + \dfrac{3(3x)^5}{40} + \dots$

7. $1 - \dfrac{(x-\pi/6)^2}{2!} + \dfrac{(x-\pi/6)^4}{4!} - \dots$

9. $1 - x + x^2 - x^3 + \dots$

11. $x - \dfrac{x^3}{3!} + \dfrac{x^5}{5!} - \dots$

Section 17.4

1. 0.9950   3. 0.08729   5. 1.22   7. −0.1053   9. 0.3103

11. −0.1865   13. $\left(x-\dfrac{x^3}{3!}+\dfrac{x^5}{5!}\right)-\dfrac{1}{3!}\left(x-\dfrac{x^3}{3!}+\dfrac{x^5}{5!}\right)^3+\dfrac{1}{5!}\left(x-\dfrac{x^3}{3!}+\dfrac{x^5}{5!}\right)^5$

15. $2\pi t-\dfrac{(2\pi t)^3}{3!}+\dfrac{(2\pi t)^5}{5!}-\cdots$

Section 17.5

1. $\dfrac{1}{2}-\dfrac{\sqrt{3}(x-\pi/3)}{2}-\dfrac{(x-\pi/3)^2}{4}+\cdots$ 5. $e\left[1+\dfrac{(x-1)}{1!}+\dfrac{(x-1)^2}{2!}+\cdots\right]$

3. $\dfrac{\sqrt{2}}{2}\left[1+(x-\pi/4)-\dfrac{(x-\pi/4)^2}{2!}-\cdots\right]$ 7. 0.5754   9. 0.8979   11. 2.749

Sections 18.1–18.3

1. $2\pi-4\sin x-2\sin 2x-\dfrac{4}{3}\sin 3x-\cdots$

3. $\dfrac{4\pi^2}{3}+4\left(\cos x+\dfrac{1}{4}\cos 2x+\dfrac{1}{9}\cos 3x+\cdots\right)$
$-4\pi\left(\sin x+\dfrac{1}{2}\sin 2x+\dfrac{1}{3}\sin 3x+\cdots\right)$

5. $\dfrac{2}{\pi}+\dfrac{4}{\pi}\left(-\dfrac{1}{3}\cos x-\dfrac{1}{15}\cos 2x-\dfrac{1}{35}\cos 3x-\cdots\right)$

7. $\dfrac{e^{2\pi}-1}{2\pi}+\dfrac{1}{\pi}\sum_{n=1}^{\infty}\left[\dfrac{e^{2\pi}-1}{1+n^2}\cos nt+\dfrac{n(1-e^{2\pi})}{1+n^2}\sin nt\right]$

Section 18.4

1. $\dfrac{1}{2}+\dfrac{2}{\pi}\left[\sin x+\dfrac{1}{3}\sin 3x+\dfrac{1}{5}\sin 5x+\cdots\right]$

3. $\dfrac{3\pi}{4}-\dfrac{2}{\pi}\left[\cos x+\dfrac{1}{9}\cos 3x+\dfrac{1}{25}\cos 5x+\cdots\right]$
$-\left[\sin x+\dfrac{1}{2}\sin 2x+\dfrac{1}{3}\sin 3x+\cdots\right]$

5. $\dfrac{3}{4}+\dfrac{1}{\pi}\left[\cos x-\dfrac{1}{3}\cos 3x+\dfrac{1}{5}\cos 5x-\cdots\right]+\dfrac{1}{\pi}\left[3\sin x+\sin 2x+\sin 3x+\cdots\right]$

7. $\dfrac{\pi^2}{48} + \left[\left(\dfrac{\pi}{4} - \dfrac{2}{\pi}\right)\cos x - \dfrac{1}{4}\cos 2x - \left(\dfrac{\pi}{12} - \dfrac{2}{27\pi}\right)\cos 3x + \cdots\right]$

$\qquad + \left[\left(1 - \dfrac{2}{\pi}\right)\sin x + \left(\dfrac{\pi}{8} - \dfrac{1}{2\pi}\right)\sin 2x + \left(-\dfrac{1}{9} - \dfrac{2}{27\pi}\right)\sin 3x + \cdots\right]$

9. $\dfrac{3}{2} - \dfrac{2}{\pi}\left[\sin x + \dfrac{1}{3}\sin 3x + \dfrac{1}{5}\sin 5x + \cdots\right]$

Index

DERIVATIVE FORMULAS

1. $\dfrac{d}{dx}(c) = 0$ (3.3)

2. $\dfrac{d}{dx}(x) = 1$ (3.3)

3. $\dfrac{d}{dx}[af(x)] = a\dfrac{d}{dx}[f(x)]$ (3.2)

4. $\dfrac{d}{dx}[f(x) \pm g(x)] = \dfrac{d}{dx}[f(x)] \pm \dfrac{d}{dx}[g(x)]$ (3.2)

5. $\dfrac{d}{dx}[f(x)g(x)] = f(x)g'(x) + g(x)f'(x)$ (8.4)

6. $\dfrac{d}{dx}\left[\dfrac{f(x)}{g(x)}\right] = \dfrac{g(x)f'(x) - f(x)g'(x)}{[g(x)]^2}$ (8.5)

7. $\dfrac{d}{dx}[x^n] = nx^{n-1}$ (3.3)

8. $\dfrac{d}{dx}[u^n] = nu^{n-1}\dfrac{du}{dx}$ (8.3)

9. $\dfrac{d}{dx}[\log_a u] = \dfrac{1}{u}\dfrac{du}{dx}\log_a e$ (9.3)

10. $\dfrac{d}{dx}[\ln u] = \dfrac{1}{u}\dfrac{du}{dx}$ (9.3)

11. $\dfrac{d}{dx}[a^u] = a^u\dfrac{du}{dx}\ln a$ (9.5)

12. $\dfrac{d}{dx}[e^u] = e^u\dfrac{du}{dx}$ (9.5)

13. $\dfrac{d}{dx}[\sin u] = \cos u\dfrac{du}{dx}$ (10.4)

14. $\dfrac{d}{dx}[\cos u] = -\sin u\dfrac{du}{dx}$ (10.4)

15. $\dfrac{d}{dx}[\tan u] = \sec^2 u\dfrac{du}{dx}$ (10.4)

16. $\dfrac{d}{dx}[\cot u] = -\csc^2 u\dfrac{du}{dx}$ (10.4)

17. $\dfrac{d}{dx}[\sec u] = \sec u\tan u\dfrac{du}{dx}$ (10.4)

18. $\dfrac{d}{dx}[\csc u] = -\csc u\cot u\dfrac{du}{dx}$ (10.4)

19. $\dfrac{d}{dx}[\arcsin u] = \dfrac{1}{\sqrt{1-u^2}}\dfrac{du}{dx}$ (10.6)

20. $\dfrac{d}{dx}[\arccos u] = -\dfrac{1}{\sqrt{1-u^2}}\dfrac{du}{dx}$ (10.6)

21. $\dfrac{d}{dx}[\arctan u] = \dfrac{1}{1+u^2}\dfrac{du}{dx}$ (10.6)